Studies in Computational Intelligence

Volume 644

Series editor

Janusz Kacprzyk, Polish Academy of Sciences, Warsaw, Poland
e-mail: kacprzyk@ibspan.waw.pl

About this Series

The series "Studies in Computational Intelligence" (SCI) publishes new developments and advances in the various areas of computational intelligence—quickly and with a high quality. The intent is to cover the theory, applications, and design methods of computational intelligence, as embedded in the fields of engineering, computer science, physics and life sciences, as well as the methodologies behind them. The series contains monographs, lecture notes and edited volumes in computational intelligence spanning the areas of neural networks, connectionist systems, genetic algorithms, evolutionary computation, artificial intelligence, cellular automata, self-organizing systems, soft computing, fuzzy systems, and hybrid intelligent systems. Of particular value to both the contributors and the readership are the short publication timeframe and the worldwide distribution, which enable both wide and rapid dissemination of research output.

More information about this series at http://www.springer.com/series/7092

Hocine Cherifi · Bruno Gonçalves
Ronaldo Menezes · Roberta Sinatra
Editors

Complex Networks VII

Proceedings of the 7th Workshop on Complex Networks CompleNet 2016

Editors
Hocine Cherifi
LE2I, UFR Sciences et Techniques
Université de Bourgogne
Dijon
France

Ronaldo Menezes
BioComplex Laboratory, Computer Sciences
Florida Institute of Technology
Melbourne, FL
USA

Bruno Gonçalves
Center for Data Science
New York University
New York, NY
USA

Roberta Sinatra
Physics Department
Northeastern University
Boston, MA
USA

ISSN 1860-949X ISSN 1860-9503 (electronic)
Studies in Computational Intelligence
ISBN 978-3-319-30568-4 ISBN 978-3-319-30569-1 (eBook)
DOI 10.1007/978-3-319-30569-1

Library of Congress Control Number: 2016932873

Printed on acid-free paper

This Springer imprint is published by Springer Nature
The registered company is Springer International Publishing AG Switzerland

Preface

The International Workshop on Complex Networks-CompleNet (www.complenet.org) was initially proposed in 2008, and the first workshop took place in 2009 in Catania. The initiative was the result of efforts from researchers from the (i) *BioComplex Laboratory in the Department of Computer Sciences at Florida Institute of Technology, USA*, and the (ii) *Dipartimento di Ingegneria Informatica e delle Telecomunicazioni, Università di Catania, Italy*. CompleNet aims at bringing together researchers and practitioners working on complex networks or related areas. In the past two decades we have indeed witnessed an exponential increase in the number of publications in this field. From Biology to Computer Science, from Economics to Social Systems, Complex Networks are becoming pervasive in many fields of science. It is this interdisciplinary nature of complex networks that CompleNet aims to address. CompleNet 2016 was the seventh event in the series and was hosted by the *Université de Bourgogne, France*, from March 23 to 25, 2016.

This book includes the peer-reviewed list of works presented at CompleNet 2016. We received an unprecedented number of 121 submissions from 20 countries. Each submission was reviewed by at least 3 members of the Program Committee. Acceptance was judged based on the relevance to the symposium themes, clarity of presentation, originality and accuracy of results and proposed solutions. After the review process, 22 papers and 6 short papers were selected to be included in this book.

The 28 contributions in this book address many topics related to complex networks and have been organized into seven major groups: (1) Theory of Complex Networks, (2) Multilayer networks, (3) Controllability of networks, (4) Algorithms for networks, (5) Community detection, (6) Dynamics and spreading phenomena on networks, (7) Applications of Networks.

We would like to thank the Program Committee members for their work in promoting the event and refereeing submissions. We are grateful to our speakers: Alain Barrat (Aix-Marseille University, France), Ernesto Estrada (University of Strathclyde, Scotland), Renaud Lambiotte (University of Namur, Belgium),

Giovanna Miritello (Zed Worldwide, Spain), Nicola Perra (University of Greenwich, England), Marco Quaggiotto (ISI Foundation, Italy), José Javier Ramasco (IFISC, CSIC-UIB, Spain), Balazs Vedres (Central European University, Hungary), Suzy Moat (University of Warwick, England); their presentation is one of the reasons CompleNet 2016 was such a success.

Special thanks also go to Chantal Cherifi, Eric Leclercq, Sylvain Rampacek, Marinnete Savonnet, Olivier Togni, from the Université de Lyon and Université de Bourgogne, for their help in organizing CompleNet 2016.

Dijon, France
March 2016

Hocine Cherifi
Bruno Gonçalves
Ronaldo Menezes
Roberta Sinatra

Contents

Part VII Applications of Networks

Part I
Theory of Complex Networks

Spanning Edge Betweenness in Practice

Andreia Sofia Teixeira, Francisco C. Santos and Alexandre P. Francisco

Abstract In this paper we present a study about spanning edge betweenness, an edge-based metric for complex network analysis that is defined as the probability of an edge being part of a minimum spanning tree. This probability reflects how redundant an edge is in what concerns the connectivity of a given network and, hence, its value gives information about the network topology. We apply this metric to distinct empirical networks and random graph models, showing that spanning edge betweenness allows us to identify those edges that are more relevant for connectivity and how removing them leads to disruption in network structure.

Keywords Spanning edge betweenness · Network analysis · Edge centrality measures

1 Introduction

Networks are the simplest representation of interactions and relations between entities. Nevertheless, a network can express very complex processes and behaviours. In this context, understanding structure and dynamics of a network is crucial to extract valuable information. The analysis of complex networks, such as social networks, biological networks, financial networks, electrical networks or even the world wide web, have gathered efforts from mathematicians, physicists, social and computer scientists to build several statistical measures and tools to evaluate the importance of each node and/or each link. Some are very well-known [1–3]: degree centrality indicates the fraction of connections that a given node has over the entire network; node/edge betweenness states how important a node/edge is through the number of

A.S. Teixeira (✉) · F.C. Santos · A.P. Francisco
INESC-ID/Instituto Superior Técnico, Universidade de Lisboa, Lisbon, Portugal
e-mail: sofia.teixeira@tecnico.ulisboa.pt

F.C. Santos
e-mail: franciscocsantos@tecnico.ulisboa.pt

A.P. Francisco
e-mail: aplf@tecnico.ulisboa.pt

H. Cherifi et al. (eds.), *Complex Networks VII*, Studies in Computational Intelligence 644, DOI 10.1007/978-3-319-30569-1_1

shortest paths between two nodes passing through it; and clustering coefficient is a key measure for social network analysis that for a given node expresses how many of its neighbours are neighbours of each other, evaluating the fraction of possible triangles that the node is a member of. All of these measures can give us information about centrality and connectivity of a network, but they are mostly focused on nodes. On the other hand, although we can evaluate the centrality of an edge by using betweenness centrality, there are many networks whose study can gain new insights if new measures are used for evaluating edge centrality that do not depended on shortest paths, as edge betweenness does. When we address phylogeny, telecommunication/electric networks, among other networks, we are often interested in studying measures that go beyond shortest path properties. If we want to know how resilient a network is, i.e., which links are fundamental to keep the network connected and which are redundant, none of the metrics described before provides that information. In algorithms for inferring phylogenies, we aim to validate the trees that are generated to represent evolution patterns and to identify bridges that connect different groups, in telecommunication/electric networks we are interested to know which links are so important that could cause a breakdown if turned off, or which of them are redundant. Recently, Morone [4] presented a work in which one of the goals is to find the minimal set of nodes that, if removed, would break down the network, but once again, the work is focused on the importance of the nodes and not on the importance of the links.

These problems can be conveniently studied by relying on minimum spanning trees. Recently, a new network measure was proposed for evaluating the importance of edges taking into account information provided by minimum spanning trees—*spanning edge betweenness* [5]. This new metric, which corresponds to the fraction of minimum spanning trees that contains an edge, has the potential to not only help on the evaluation and validation of phylogeny algorithms, for which it was originally proposed, but also to evaluate how redundant an edge is in a given network. Because of its probabilistic property, spanning edge betweenness provides direct information about an edge preventing the relativity inherent to the other measures. Contrary to what is evaluated in edge betweenness, we are not interested in knowing in how many shortest paths the edge is present, but how important the edge is to maintain the network connected. Given an edge, its spanning edge betweenness value can reflect whether the edge removal can cause a disruption in a network or if there are some alternative ways to keep the network connected, reflecting how resilient the network can be and how redundant an edge is in the network. More recently spanning edge betweenness has been object of further studies. An initial study on the importance of the metric in phylogenetic trees was reported in [6], an improvement in what concerns the efficient computation of spanning edge betweenness was presented in [7], and Qi et al. [8] introduced the concept of spanning tree centrality, that applies the same principles of spanning edge betweenness although applied to the nodes in a weighted network.

In this paper we study the applicability of spanning edge betweenness for evaluating edge redundancy on real and synthetic networks. For this aim, we compare it with previously introduced measures and evaluate how turning on/off the links with highest spanning edge betweenness can affect networks topologies and how can we identify potential bridges that are crucial to ensure networks integrity and connectivity. We use real and artificial networks and for each one we remove all the edges with three criteria: random selection, decreasing order of spanning edge betweenness values, and decreasing order of edge betweenness values. We show that removing edges with high spanning edge betweenness leads to a fast disruption in the networks, rapidly increasing the number of components of the networks.

2 Edge-Based Measures on Minimum Spanning Trees

Minimum spanning trees have been used for decades for network design, cluster analysis, among others. Given a network, a minimum spanning tree represents the set of edges with minimum weight that connect all of the nodes. Let $G = (V, E)$ be a connected, undirected and weighted graph, with weight function $w : E \rightarrow I\!R$, where V is the set of vertices and $E \subset V \times V$ is the set of edges. A minimum spanning tree $T = (V, E')$ is a subgraph of G that is a tree and contains all the vertices of G, i.e., that spans over all vertices in V, with $|E'| = |V| - 1$, and such that $\sum_{e \in E'} w(e)$ is minimum among all spanning trees. For generality, we can assume an unweighed graph as a graph with all edges' weights equal to 1. If the network is a tree, then there is only one minimum spanning tree, otherwise the network can have multiple minimum spanning trees.

When constructing certain networks—such as electrical, computer, transportation, and telecommunication networks—the major concern is to choose the cheaper path for laying the connections. On the other hand, if we already have a network, how can we know which are the links whose presence is imperative to connect all the nodes and which provide a more flexible choice? On other perspective: given a computer network, which connections should we choose to assure its resilience preventing a massive disruption? Which connections/edges are critical? The study of spanning edge betweenness on a network allows us to give some answers for these questions.

The first known edge-based centrality, edge betweenness, was initially proposed by mathematician Anthonisse and later formalized and published by Freeman in 1977 [9]. It was developed in the context of communication networks. For a given edge e it measures how central the edge is, i.e., how many geodesic paths transverse that edge. In 2002, Girvan and Newman [10] applied this metric to the study of finding and evaluating community structures in networks, but little has been done in what concerns exploring new edge importance measures in a network. In 2012, Meo et al. [11] developed a k-path centrality, initially developed for nodes, which is based on random walks and is defined as the sum of the frequency with which a message traverses an edge e from a given source to all k-edges-distance possible destinations. These two centrality measures play a central role in reporting knowledge about data flow in a network but few about the structure/topology of the network.

In fact, when analysing minimum spanning trees, shortest-paths or random walks approaches yield insufficient information to infer how much resilient a network is or how redundant are some connections, depending on the subject in study. Recently, Teixeira et al. [5] introduced an edge-based centrality measure that relies on minimum spanning trees to evaluate how important is an edge in the structure of a network. Here, we extend the evaluation made on phylogenetic trees [6], providing information about the metric behaviour in real well-known networks, including social, technological and electric networks. The fact that it tells directly the probability of an edge being in a minimum spanning tree, thus reflecting how important it is for the network structure, ensures a high confidence in the analysis of network resilience and edge redundancy.

2.1 Spanning Edge Betweenness

For a given edge e, the *spanning edge betweenness* is defined as:

$$\delta_G(e) = \frac{\tau_G(e)}{\tau_G},$$

where τ_G is the number of different minimum spanning trees for G and $\tau_G(e)$ is the number of different minimum spanning trees for G where e occurs.

There are many applications for this new measure, as exemplified by Teixeira et al. [6] in the context of inferring phylogenies. As we said before, a network can have many minimum spanning trees. Spanning edge betweenness comes to help in the confidence evaluation of the tree generated. Because this metric takes values between [0,1] we can infer: (1) if spanning edge betweenness is 1 than the edge has to be on the network to keep it connected; (2) if it is 0, which only can occur in weighted networks, than the edge is completely redundant; (3) being the value between 0 and 1 it means that there are other edges that can keep the network connected, i.e., there is a different minimum spanning tree for the network, thus expressing the redundancy of an edge. As we will see, the proportion of these values can provide information about the network topology.

3 Methods and Results

To evaluate the significance of the spanning edge betweenness we chose eight different networks, with different sizes and from different contexts. Four are real well-known networks (Karate, Power Grid, Political Blogs and NetScience),[1] and four are random networks: two generated from Barabási-Albert model [12] and two networks with community structure.[2] The properties of these networks are in Tables 1, 2

[1] http://www-personal.umich.edu/~mejn/netdata/.

[2] https://sites.google.com/site/santofortunato/inthepress2.

Table 1 Detail for real networks

Network	# nodes	# edges
Karate	34	78
PowerGrid	4941	6594
Polblogs	1490	2742
NetScience	1589	1252

Table 2 Barabási-Albert model parameters for generating random networks

# nodes	# edges	Average degree
1000	2975	4
1000	4939	4

Table 3 Model paraemters for generating random networks with community strcuture

# nodes	# edges	Min/Max degree	Min/Max community size
1000	2222	4/8	20/40
1000	3985	8/16	20/40

and 3. In practice, we computed five measures: node degree centrality, node betweenness, edge betweenness, cluster coefficient and spanning edge betweenness. Than we correlated spanning edge betweenness with the other metrics. Spanning edge betweenness and edge betweenness were directly correlated; for the other node-based metrics—node betweenness, degree centrality and cluster coefficient—we correlated with the minimum/maximum/average metrics between the source and destination nodes of each edge.

The first conclusion is that spanning edge betweenness has no correlation with the other measures. When we correlated it with the other measures mentioned, none of them showed meaningful correlation values. This reinforces the idea that this measure provides novel information that was not given before. In Fig. 1, we show that spanning edge betweenness has a different expression than edge betweenness. While spanning edge betweenness took values between 0 and 1, expressing directly the importance of an edge, edge betweenness took all of its values below 0.3. Comparing directly both measures, it is possible to see that the values of edge betweenness do not allow to infer clear information about network structure. Edge betweenness is about how much information flow passes through an edge in shortest paths, while spanning edge betweenness is about the significance of and edge, potentially identifying edges that can break the network and reflecting if the network has a strong or weak redundancy. We can also see that PowerGrid has a very different behaviour from other three chosen networks. This is because the topology of the network is like a tree, or a star, with only ten redundant edges, being one example that if a link is disconnected, most probably the network will break. The other networks illustrate the redundancy that

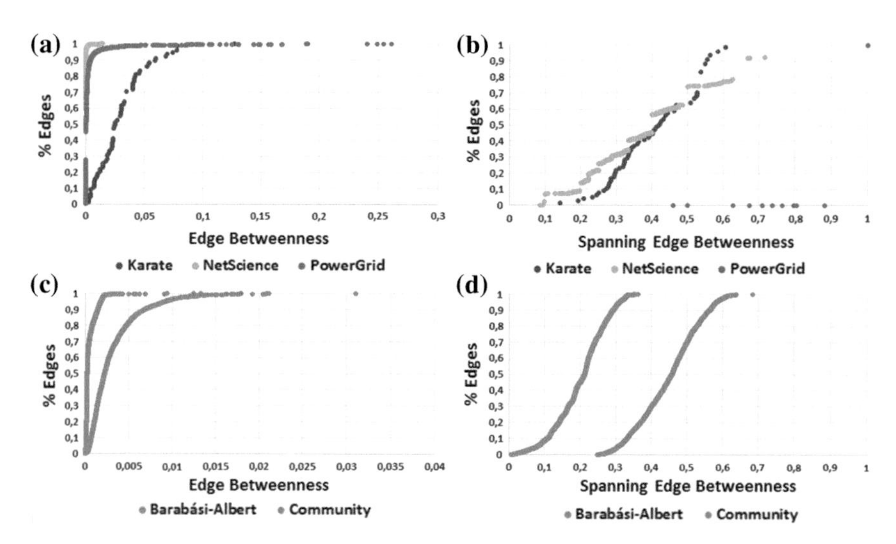

Fig. 1 Edge betweenness versus spanning edge betweenness. In *panels a* and *c* we show the values of edge betweenness for three empirical networks and two random generated networks. In *panels b* and *d* we show the values of spanning edge betweenness for the same networks. While spanning edge betweenness shows a wide range of values, expressing edge significance in network structure, edge betweenness is limited to a very small set of values not being possible to infer directly information about network structure

is expected from that kind of networks. As friends are friends from each other, as one cites another, there can be much alternatives to maintain the network connected and reachable between all nodes.

To reinforce the idea that spanning edge betweenness provides information about the redundancy and the connectivity of a network, we also present an evaluation on how removing edges from a network affects network structure. For all networks mentioned before, after we calculated the values of each measure, we sorted them by decreasing order and then, one by one, we removed each edge from the networks, registering the number of connected components after each removal. The result was as expected, removing by decreasing order of spanning edge betweenness speeds up the disruption of the networks when comparing with decreasing order of edge betweenness. In Fig. 2, we show four examples—two from real networks and two from generated networks—but for all networks the results were similar on what concerns the number of connected components growth. For the same proportion of edges removed, removing edges with decreasing order of spanning edge betweenness breaks the network structure into more components than with decreasing order of edge betweenness.

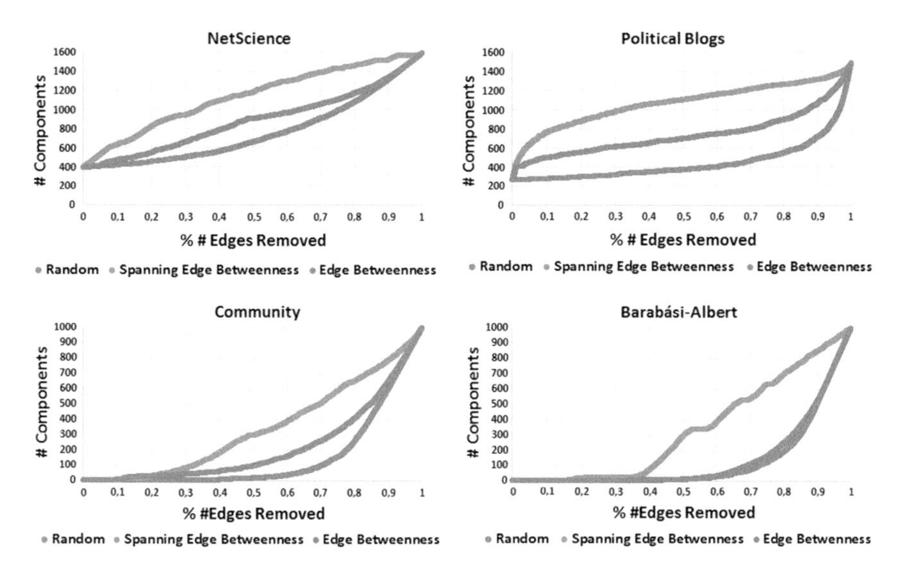

Fig. 2 Analysis of removing edges: randomly, in decreasing order of spanning edge betweenness and edge betweenness values in NetScience, PoliticalBlogs, Barabási-Albert and Community networks. It is possible to see that for all networks, empirical and random generated networks, removing edges in decreasing order of spanning edge betweenness leads to a earlier break down into more components of each network when comparing with the other two methods

4 Final Remarks

Centrality measures are important in a large number of graph applications, from search and ranking to social and biological network analysis. Most of these measures are calculated upon the nodes/vertices, but sometimes our interest is to study the importance of links/edges on a network. Spanning edge betweenness is a useful measure that can be applied both in weighted and unweighed graphs, allowing different types of evaluations—from confidence in phylogenetic trees to the identification of edges that are critical to keep the network connected, passing through the ones that express redundancy and alternative network configurations. In this paper we compared it with another measures, namely with traditional edge betweenness, and on several real and synthetic networks, concluding that spanning edge betweenness performs better at identifying the relevance of edges for maintaining networks connectivity. Since spanning edge betweenness gives direct information about the importance of a link, on further research we plan to investigate other application fields as epidemic spreading, identifying which links are critical in the spreading process, following some of the ideas introduced in [13].

Acknowledgments This work was partly supported by national funds through FCT—Fundação para a Ciência e Tecnologia, under projects Incentivo/EEI/LA0021/2014, EXCL/EEI-ESS/0257/2012, UID/CEC/50021/2013, PTDC/EEI-SII/5081/2014, and PTDC/MA T/STA/3358/2014.

References

1. Borgatti, S.P., Everet, M.G.: A Graph-theoretic perspective on centrality. Soc. Netw. **28**(4), 466–484 (2006)
2. Costa, L.A., Rodrigues, F.A., Travieso, G., Villas, P.R.: Boas, characterization of complex networks: a survey of measurements. Adv. Phys. **56**, 167–242 (2007)
3. Albert, R., Barabási, A.: Statistical mechanics of complex networks. Rev. Mod. Phys. **74**, 47 (2002)
4. Morone, F., Makse, H.A.: Influence maximization in complex networks through optimal percolation. Nature **524**, 65–68 (2015)
5. Teixeira, A.S., Monteiro, P.T., Carriço, J.A., Ramirez, M., Francisco, A.P.: Spanning edge betweenness. In: Eleventh Workshop on Mining and Learning with Graphs (2013)
6. Teixeira, A.S., Monteiro, P.T., Carriço, J.A., Ramirez, M., Francisco, A.P.: not seeing the forest for the trees: size of the minimum spanning trees (msts) forest and branch significance in mst-based phylogenetic analysis. PLOS one **10**(3), e0119315 (2015)
7. Mavroforakis, C., Garcia-Lebron, R., Koutis, I., Terzi, E.: Spanning edge centrality: large-scale computation and applications. In: Proceedings of the 24th International Conference on World Wide Web (WWW '15), PP. 732–742 (2015)
8. Qi, X., Fuller, E., Luo, R., Zhang, C.: A novel centrality method for weighted networks based on the Kirchhoff polynomial. Pattern Recogn. Lett. **58**, 51–60 (2015)
9. Freeman, L.C.: A set of measures of centrality based on betweenness. Sociometry **40**, 35–41 (1977)
10. Girvan, M., Newman, M.E.: Community structure in social and biological networks. Proc. Natl. Acad. Sci. USA **99**, 7821–7826 (2002)
11. De Meo, P., Ferrara, E., Fiumara, G., Ricciardello, A.: A novel measure of edge centrality in social networks. Know. -Based Syst. **30**, 136–150 (2012)
12. Barabási, A., Albert, R.: Emergence of scaling in random networks. Science **286**(5439), 509–512 (1999)
13. Grady, D., Thiemann, C., Brockmann, D.: Robust classification of salient links in complex networks. Nat. Commun. **3**, 864 (2012)

Predictive Partitioning for Efficient BFS Traversal in Social Networks

Damien Fay

Abstract In this paper we show how graph structure can be used to significantly reduce the computational bottleneck of the Breadth First Search algorithm (the foundation of many graph traversal techniques) for social networks. In particular, we address parallel implementations where the bottleneck is the number of messages between processors emitted at the peak iteration. First, we derive an expression for the expected degree distribution of vertices in the frontier of the algorithm which is shown to be highly skewed. Subsequently, we derive an expression for the expected message along an edge in a particular iteration. This skew suggests a weighted, iteration based, partition would be advantageous. Empirical simulations show that such partitions can reduce the message overhead in the order of 20 % *for graphs with common social network structural properties*. These results have implications for graph processing in multiprocessor and distributed computing environments.

Keywords BFS · Graph structure · Social network properties

1 Introduction

Breadth First Search (BFS) is a fundamental graph algorithm which is applied constantly to huge social network graphs in distributed and parallel systems consuming large amounts of energy and resources. BFS is central to several more complicated graph algorithms such as identifying connected components, testing for bipartiteness, belief propagation, finding community structures in social networks and computing the max flow-min cut for a graph [1]. As such it has drawn much attention from the parallel processing community as a benchmark algorithm with several competing variants focused on efficient implementation [1–14]. However, despite its importance *known structural properties* of social networks have not been leveraged to improve the algorithms efficiency. The aim of this paper is to *prove the concept* that a simple adjustment of the partitioning vector based on common graph structure can

D. Fay (✉)
Department of Computing, Bournemouth University, Poole, UK
e-mail: dfay@bournemouth.ac.uk

H. Cherifi et al. (eds.), *Complex Networks VII*, Studies in Computational Intelligence 644, DOI 10.1007/978-3-319-30569-1_2

greatly improve the efficiency at the algorithms bottleneck. We also show (theoretically and empirically) that it is the graph properties that result in this improvement, and *absence* of these properties can lead to little or no improvement. The graph theoretic analysis of the BFS frontier in this paper is novel and should be of interest to researchers in the parallel and distributed graph traversal communities.

The setting here envisages that BFS is performed repeatedly on an unweighted, undirected graph from random root vertices. In addition, we assume basic statistics about the graph can be collected after each run or alternatively offline. It is assumed the graph is traversed in parallel by several processors thus requiring a-priori a partition of the graph vertices across each processor. In this setting the basic computation step of BFS is dominated by the communications costs (messages) between processors after each iteration (as noted in [6] amongst others). The messages emitted after the peak iteration further dominate the communication costs amounting to $\sim$70 % of the total (Sect. 4), thus this is the *bottleneck* of the whole algorithm.

With the exception of a few papers (Sect. 2) most approaches ignore information about the structure of the graph focusing instead on CPU-GPU architecture specifics. We show that the incident degree distribution per iteration is highly skewed away from a power law distribution. Thus the number of edges crossing a partition is a biased estimate of the messages between partitions at the peak iteration. Further we propose a new weighted graph construction which reflects the expected number of messages per edge. Finally, we show empirically that using a weighted partitioning algorithm that the subsequent reduction in messages emitted across partitions can be reduced in some individual cases by $\sim$50 %, for some graphs on average by $\sim$20 %.

The paper is laid out as follows. Section 2 discusses related work, Sect. 3 gives the background behind the BFS algorithm, partitioning and develops the theory showing that the degree distributions are highly skewed. Section 4 presents empirical results and finally in Sect. 5 we mainly focus on future work and discussing the consequences of the findings.

2 Related Work

Implementing BFS in parallel is a well established approach which generally consists of three stages: graph pre-ordering, graph partitioning and parallel architecture specific implementation. This research is most pertinent to graph partitioning however there are several aspects of architecture specifics of interest.

Graph partitioning seeks to reduce the number of messages sent between partitions during processing which can be achieved in several ways. The most obvious mechanism is to use a 1-D partition; each vertex and associated edges are sent to an individual processor [1, 4, 9, 15]. An excellent overview of 1-D graph partitioning methods can be found in [13] with techniques designed specifically for scale-free networks exist such as [16]. Although [16] considers partitioning for social network graphs they do not do so in the context of BFS, indeed the two approaches are complimentary as here we provide a weighted social network graph for partitioning.

Shang and Kitsuregawa [4] consider partitioning edges across processors (as opposed to vertices). The edges may be uniformly distributed by either the source or the target vertex. They propose that when the degree of the target vertex exceeds a pre-defined threshold the algorithm performs best by switching to a target vertex partitioning, while Hong et al. [17] note that for low degree vertices partitioning should be based on vertex but for large degree vertices the partitioning should be based on edge. In contrast a 2-D partition [2, 8, 10, 11] distributes the edges of a vertex across several processors. The 2-D approach is based on the observation that an exploration from a set of vertices is equivalent to the product of the adjacency matrix and a vector of the vertices touched. Thus they partition the adjacency matrix into two dimensions (blocks along the rows and columns) and then collect the row products in one set of messages and the unique column entries in another. Thus the messages produced are between particular processors and not *all to all* as in the 1-D case. It would appear from the literature that the 2D partitioning approach results in more efficient BFS traversals but we do not consider this approach in this research (see future work, Sect. 5).

Skewed graph structure is a central topic in many papers [1, 2, 5, 12, 17]. The non-locality of neighbours in a graph, and the fact that some vertices can have degrees several factors larger than the average, leads to load imbalances across processors and random memory access patterns. Yuan et al. [12] examines the expected distance between two pairs of nodes being explored in a BFS and show that they can predict the vertex locality. This is perhaps the closest work to this research. In contrast our approach looks at the expected use of a vertex of a given degree in a particular iteration, though the two approaches are similar in spirit. Alternative approaches include implementing BFS from multiple sources [18]. However, to the best of our knowledge this research is the first to use the non-uniform frontier distribution to improve the parallel BFS algorithm.

3 Background

Given a graph $G(V, E)$ and a source vertex s, where V, E refer to the vertex and edge sets respectively the BFS algorithm returns a route from s to every reachable vertex in G. The BFS algorithm begins with a set $V_0 = \{s\}$ and explores the graph by identifying neighbours of s, denoted as the set V_0^+, where + denotes neighbour expansion. At the next iteration all vertices connected to V_0^+ minus those already visited are $V_1 = V_0^+ \setminus \{V_0\}$. We call the set of unique vertices in the τth iteration, V_τ, the *frontier* set. In general the frontier consists of

$$V_\tau = V_{\tau-1}^+ \setminus \{\bigcup_{i=0}^{\tau-1} V_i\} \tag{1}$$

and the set of vertices already visited, $\{\bigcup_{i=0}^{\tau-1} V_i\}$, are said to be *touched*. The algorithm continues until $V_\tau = \{\emptyset\}$ and all vertices have been explored.

The BFS algorithm may be implemented on P parallel processors by partitioning V into P subsets $\mathscr{V}_1, \ldots, \mathscr{V}_P$, where $\mathscr{V}_i \bigcap \mathscr{V}_j = \{\emptyset\} \quad \forall i \neq j$, and $\bigcup_i \mathscr{V}_i = V$, such that each vertex is assigned a processor which performs the neighbour expansion of that vertex. This is the basic format of most parallel BFS (P-BFS) algorithm implementations. At the end of each iteration the processor owning each element in the next frontier must be notified that this vertex is now to be explored. We define a message $\mathscr{M}^\tau_{\mathscr{V}_i \to \mathscr{V}_j}(u, v)$ to be a notification from processor i to processor j that vertex u has identified vertex v to be a member of the next frontier set. If u and v reside in the same processor then there is no communication cost and thus the communications cost for P-BFS is here defined as the sum of all messages that cross a partition:

$$C^\tau = \sum_{u \in V_{\tau-1}, v \in V_\tau} \mathscr{M}^\tau_{\mathscr{V}_i \to \mathscr{V}_j}(u, v) \quad \forall i \neq j \tag{2}$$

The aim of a partitioning algorithm is to partition a graph into equal sets, $|\mathscr{V}_i| \approx |\mathscr{V}_j|$, such that a specific objective is achieved such as the number of edges that cross the partitions, the *edge-cut*, is minimized as:

$$\underset{\mathscr{V}_1, \ldots, \mathscr{V}_P}{\operatorname{argmin}} C = \sum_{u \in \mathscr{V}_i, v \in \mathscr{V}_j, \forall i \neq j} w_{u,v} \tag{3}$$

There are several methods for graph partitioning (a recent review of such methods may be found in [13]) and the one adopted here is the popular METIS [19] multi-level k-way algorithm. Like many partitioning algorithms METIS can operate on weighted graphs; the weights themselves are the core of our technique as now discussed.

The development here initially follows that of Kurant et al. [20] who derive expressions for the *observed* degree distribution of a graph sampled by BFS (i.e. a different problem). The configuration model [21] is a construct which allows construction of graphs with a desired degree distribution. N vertices are each assigned k *stubs* sampled uniformly from a desired degree distribution, p_k, i.e. $k \sim p_k$. The configuration model then pairs these stubs at random thus constructing edges and thus a graph with the desired degree distribution. The order in which these stubs are connected is irrelevant as the pairing is random. Thus we may assign to each stub an arbitrary time, $t \in [0, 1]$ and moving from $t = 0 \to 1$ connect the stubs as their randomly assigned time is passed. This converts a discrete graph generation process into a continuous time process and is a useful framework to derive expressions for the bias inherent in BFS sampling [20, 22]. Kurant et al. interweave the stub matching step with the exploration phase of BFS. Thus the stubs are connected only when the frontier is being explored and the unconnected stubs with the lowest time are chosen first. A vertex enters the frontier when all of its stubs have been paired and this happens with probability $(1 - t)^k$ therefore the expected fraction of vertices of degree k touched before time t is [20]:

$$f_k(t) = p_k(1 - (1 - t)^k) \quad (4)$$

where p_k is the probability a vertex has degree k (i.e. the degree distribution). The fraction of nodes of any degree visited before time t is [20]:

$$f(t) = 1 - \sum_k p_k(1 - (1 - t)^k) \quad (5)$$

Kurant investigated the bias of BFS samples but here we are concerned with the degree distribution of the frontier. In addition, we are only interested in particular times; those that correspond to the iterations. It is assumed that the number of vertices touched up to each iteration $n_\tau = \sum_{i=1}^{\tau} |V_i|$ is known.[1] Here we depart from Kurant's analysis [20]. Define times, t_τ:

$$t_\tau = f^{-1}(n_\tau / N) \quad (6)$$

where f^{-1} denotes the inverse of $f(t)$ as (5) cannot be inverted explicitly. This inverse consists of finding the minimum of a smooth function in one dimension and may be solved easily using gradient descent or any similar search algorithm. The frequency of degrees of type k in the τth frontier, n_k^τ, can be calculated iteratively by removing those seen in the previous frontiers as:

$$n_k^\tau = \frac{p_k(1 - (1 - t_\tau)^k)}{\sum_k p_k(1 - (1 - t_\tau)^k)} n_\tau - \sum_{i=1}^{\tau-1} p_k^i n_i \quad (7)$$

where $n_0 = 0$ and p_k^τ is the frontier degree distribution defined as:

$$p_k^\tau = \frac{n_k^\tau}{\sum_k n_k^\tau} \quad (8)$$

The probability that a vertex of degree k is used in the frontier is then the number of vertices of degree k in the frontier divided by the total number in the graph:

$$\pi_k^\tau = \frac{p_k^\tau n_\tau}{p_k N} \quad (9)$$

Figure 1 shows $f_k(t)$ for $p_k \propto k^{-2}$.[2] Up to iteration 3, 25 % of the degrees touched are of degree 1 which rises to ~50 % by iteration 5. That is, BFS is biased (proportionately) towards higher degree vertices initially, moving towards lower degree vertices

[1] A good estimate of the number of vertices expected in each iteration of BFS can be obtained from a single graph traversals.

[2] Here we use the YouTube friendship graph as an example: the power law exponent = -2 and $t_\tau = \{0.0006, 0.02, 0.19, 0.53, 0.81, 0.93, 0.97, 0.99, 1\}$, the results are similar for the other graphs we examined.

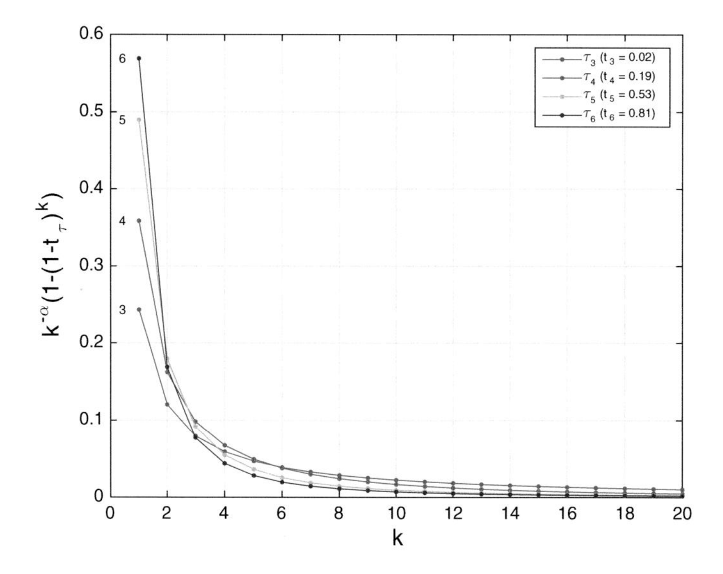

Fig. 1 Proportion of vertices of degree k seen before iteration τ ($\alpha = -2$)

at later iterations. Note that Fig. 1 shows the *accumulated* proportion as the algorithm progresses, however, it is the difference in these proportions that are touched at each iteration and this has a very different shape (Fig. 2).

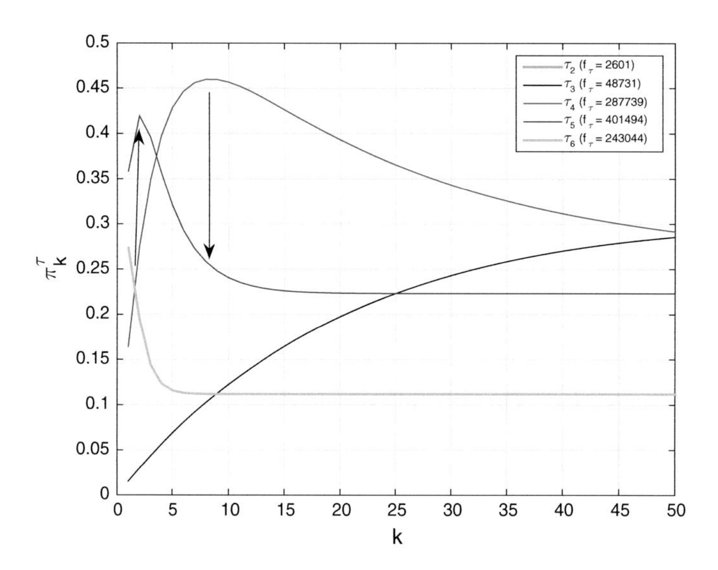

Fig. 2 Probability a vertex of degree k will be used in iteration τ (theoretical)

Figure 2 shows π_k^τ for the YouTube friendship graph (Sect. 4). The distribution of nodes used in iterations 2 and 3 is biased towards high degree nodes. In iteration 4 the bias centres on vertices of degree 10 with 40 % being touched but only 15 % of degree 1 nodes are touched. In iteration 5 the bias switches, ∼40 % of degree 1 vertices are touched but only ∼25 % of degree 10 vertices are touched. There is a similar switch between iteration 5 and 6. The interesting thing about this behaviour is that the degree distribution of vertices used is highly skewed and during the main iterations (4, 5, 6) those used in one iteration tend not to be used in the next and visa versa (as illustrated with arrows in Fig. 2). Thus at a specific iteration we have a prior probability over the vertices that will be used *and* a different prior over the vertices they are connected to in the next frontier, and these distributions are different from the initial power-law distribution, i.e. $\pi_k^\tau \neq \pi_k^{\tau+1} \not\propto p_k$. The transition from $\pi_k^\tau \to \pi_k^{\tau+1}$ involves connecting vertices with degree distribution π_k^τ to those with $\pi_k^{\tau+1}$. It would be tempting to assume that the probability of a node of degree k connects to a node of degree k' is just the product of π_k^τ and $\pi_k^{\tau+1}$, however the two events are not independent. Real-world graphs are generally assortative and as has been shown graph generators that take into account the correlation structure in the joint degree distribution $p_{k,k'}$ produce far better approximations to real-world graphs [21] and have very different properties from those that assume independence [23]. Here we assume that the joint degree distribution, $p_{k,k'}$, [21] gives a good approximation of the expected edges between the vertices in iteration τ and $\tau + 1$, therefore we may define the probability of transitioning from a vertex with degree k to an edge with degree k' in iteration τ, $p_{k,k'}^\tau$ as:

$$p_{k,k'}^\tau = \pi_k^\tau p_{k,k'} \pi_k'^{\tau+1} \tag{10}$$

The probability of using a particular edge, $\{u, v\}$, in iteration τ is equal to the probability of passing from $u \to v$, or from $v \to u$ but not both, $u \leftrightarrow v$, as this would imply u and v have already been touched in iteration τ, therefore:

$$w_{k,k'}^\tau = p_{k,k'}^\tau + p_{k',k}^\tau - p_{k,k'}^\tau p_{k',k}^\tau \tag{11}$$

where $w_{k,k'}^\tau$ can be used to weight each edge in G where the weights represents the expected message along that edge in iteration τ. The total number of expected messages given a particular partition is then:

$$E[C^\tau] = \sum_{u \in V_\tau, v \in V_{\tau+1}} w_{k_u,k_v}^\tau \mathcal{I}_{\mathcal{V}_i \to \mathcal{V}_j}(u, v) \tag{12}$$

where $\mathcal{I}_{\mathcal{V}_i \to \mathcal{V}_j}(u, v)$ is an indicator variable s.t. $u \to v$ crosses a partition. To implement this approach requires estimates of; p_k, $p_{k,k'}$, n_τ. Given these a weighted version of G, $W(V, E)$, may be constructed, and partitioned using a weighted partitioning algorithm (here we use the popular METIS algorithm).

4 Results

The simulations presented below consist of randomly choosing a source node, performing a BFS using the competing algorithms (described below), and recording the number of messages generated. Code and examples may be found on the project webpage.[3] The simulations are based on 500 randomly chosen root vertices. There are four competing algorithms which represent different levels of knowledge:

1. The original graph with no weighting is used as a baseline,
2. Using the results from 1. We calculate the actual messages counts and use these to give an empirical weighted matrix, W_{emp}. Note that in essence we are using the answer to derive the partition which is unrealistic. The aim here is to give an upper bound on the algorithms performance,
3. Using the $p^{\tau}_{k,k'}$ from all 500 iterations we combine and smooth these estimates to produce a single weighted graph called, W_{smooth}. The aim here is to give an estimate of performance without the approximation error inherent in Eq. 9, and
4. Using Eqs. (9,11,12) we form a single weighted graph, W_{avg}.

The joint degree distribution, $p_{k,k'}$, can present problems of storage and estimation especially when the maximum degree is high. However, as the graphs studied have a power law distribution, the number of vertices with a high degree falls rapidly. In this paper we calculate $p_{k,k'}$ where nodes with $k \geq 300$ are counted in a single bin. Therefore, $p_{k,k'}$ is formed of a, 300×300 grid. We choose the number of partitions to be 100 as this reflects the order of processors in a GPU (the number of processors varies greatly depending on the machine; the NVIDIA GeForce GTX280, for example, has 30 [9] while the NVidia Kepler architecture has 4,096 GPU's [10]).

The datasets used in this study are taken from the Konect graph repository.[4] We are specifically interested in social network graphs and so the RMAT graphs used in studies such as [1, 7] are not included though we do include a synthetically generated ER graph with a single large component. We also did not consider graphs with $N > 2$M for computational reasons. These graphs are listed in Table 1.

Figure 3 shows the empirical distribution of π^{τ}_{k} (based on a sample of 500 random root nodes) for the YouTube Graph versus the theoretical (Fig. 2). As can be seen for low degrees the approximation is excellent but deviates at higher degrees, especially during iteration 4. This occurs because high degree nodes in real networks cluster together in the network core (breaking the uniform assumption in the configuration model). That said, most nodes in power-law network are of low degree where the approximation is excellent and as will be seen the results are not effected adversely.

Figure 4 shows the average number of messages per iteration using the 4 algorithms above applied to the YouTube graph. As can be seen the three weighted graph versions perform better than the unweighted graph. The average number of messages (over all iterations) transmitted using W_{avg} is the lowest at 681 K while those for the unweighted graph are 790 K. The results differ on closer inspection however.

[3] https://sites.google.com/site/structuralgraphproperties/home.

[4] http://konect.uni-koblenz.de.

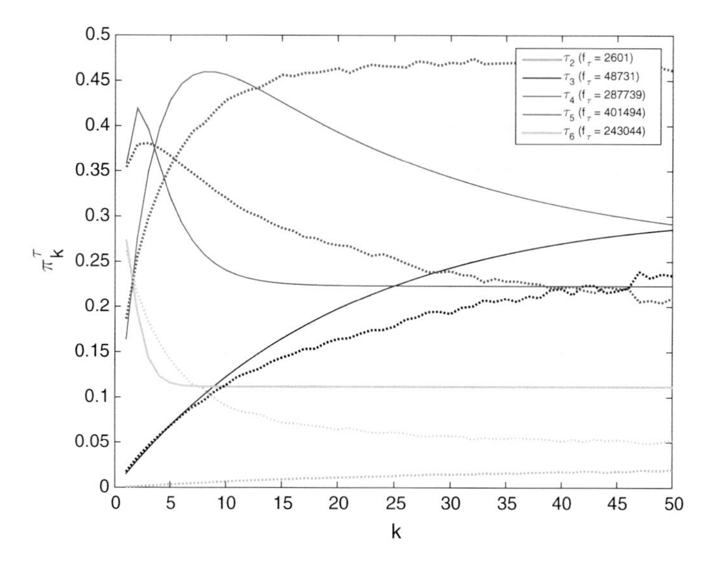

Fig. 3 π_k^τ theoretical (*solid*) versus empirical (*dashed*) (YouTube Graph)

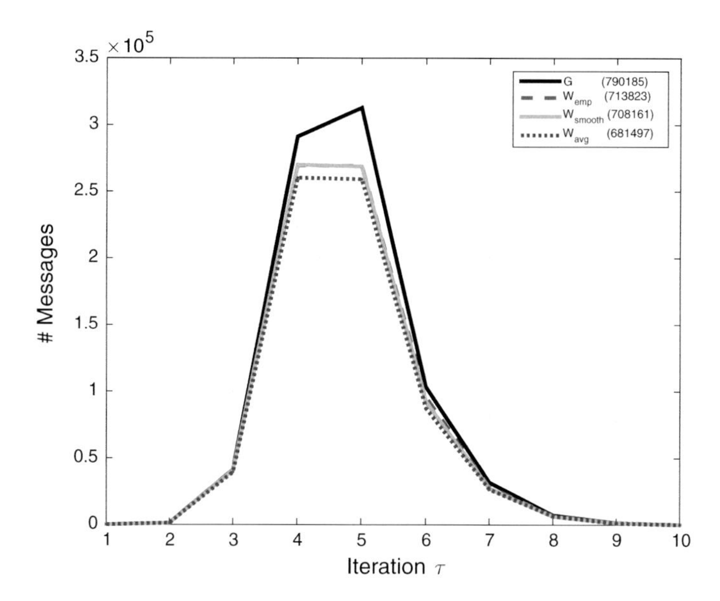

Fig. 4 Average number of messages per iteration (YouTube; totals in brackets)

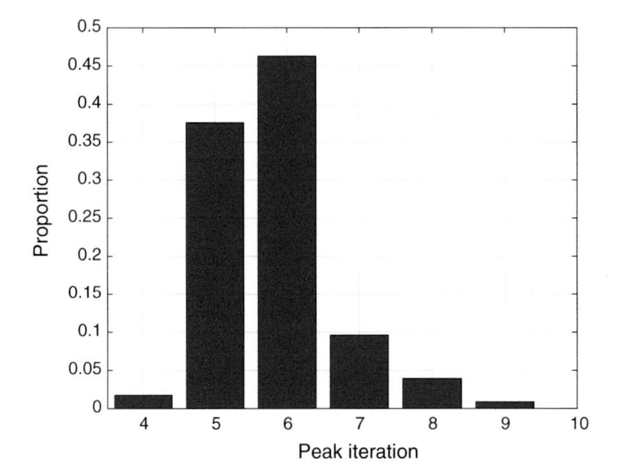

Fig. 5 Histogram of iteration at which the number of vertices in the frontier reached a peak

Figure 5 shows the histogram of the iteration at which the peak iteration occurred in each BFS run. For most source vertices the iteration at which the number of vertices in the frontier reaches a peak is 5 or 6.

Figure 6 shows the distribution of messages for a particular root node and as can be seen here the peak occurs at iteration 4 and the number of messages in the peak far

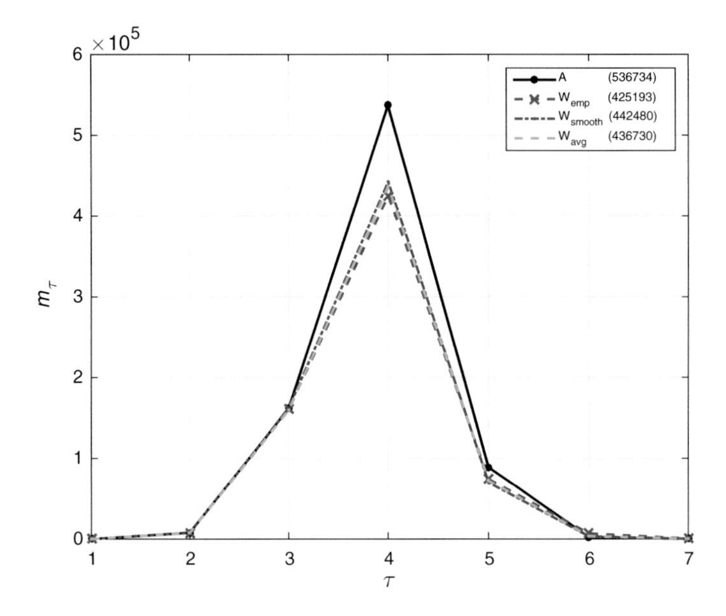

Fig. 6 Example showing number of messages per iteration for YouTube graph (root u=157, 298)

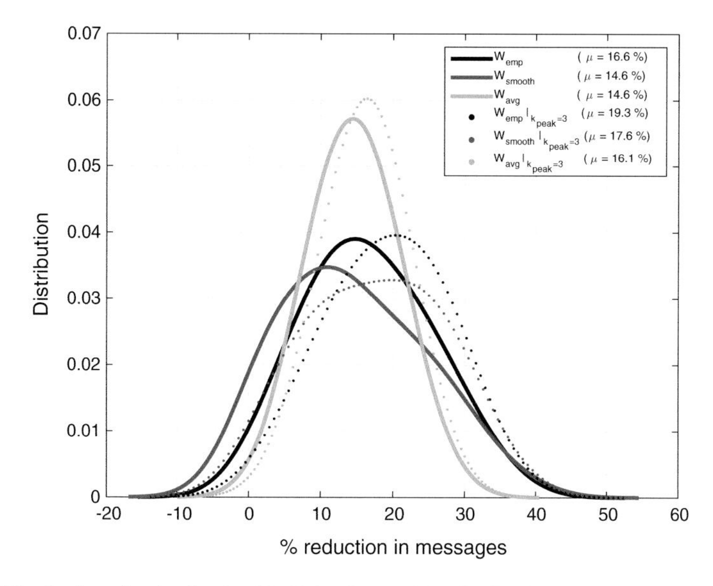

Fig. 7 Distribution of reduction for YouTube dataset (the distribution using those with peak ≥ 6 is shown using the *dotted line*, 500 samples)

exceeds those in the other iterations. Next we turn our attention to how the algorithm *performs* relative to the baseline. Figure 7 shows the *percentage improvement in messages* over the baseline algorithm. The savings are in the order of 15 % for this graph which is quite significant. In this particular case the three algorithms perform reasonably similarly but note that W_{avg} leads to the lowest improvement in messages at the peak but interestingly the highest improvement in the overall number of messages (Fig. 4).

Figure 8 shows the improvements observed with the Epinions graph. Here there is a distinct bi-modal distribution, with one distribution centred around 4 % and another centred ~35 %. For this graph about half the iterations peak at $\tau = 3$ and the remainder at $\tau = 4$. If we look at the improvement for those that peak at $\tau = 3$ alone then a clearer picture emerges. For these vertices the improvement is very small (the 4 % mode in the distribution). One possibility is that vertices which reach the peak at $\tau = 3$ lie in the core of the graph and have less hops to the periphery; thus the BFS algorithm has less time to achieve the random mixing assumed in Eq. 11 (Kurant similarly notes that the starting vertex can significantly effect their estimates [20]).

Next we examine a graph with no structure, an Erdos Renyi (ER) graph, where the joint degree distribution is uniform and the degree distribution is concentrated around the mean. As there is no structure in the graph we expect the algorithm to fail and this is exactly what is seen in Fig. 9.[5] The % (dis)improvement is a distinctive Gaussian distribution centred on zero. Moving onto a collection of graphs, Table 1 summarizes

[5]Alternatively one could insert a concentrated degree distribution for p_k in (4) and see that $\pi_k^{tau} = p_k$.

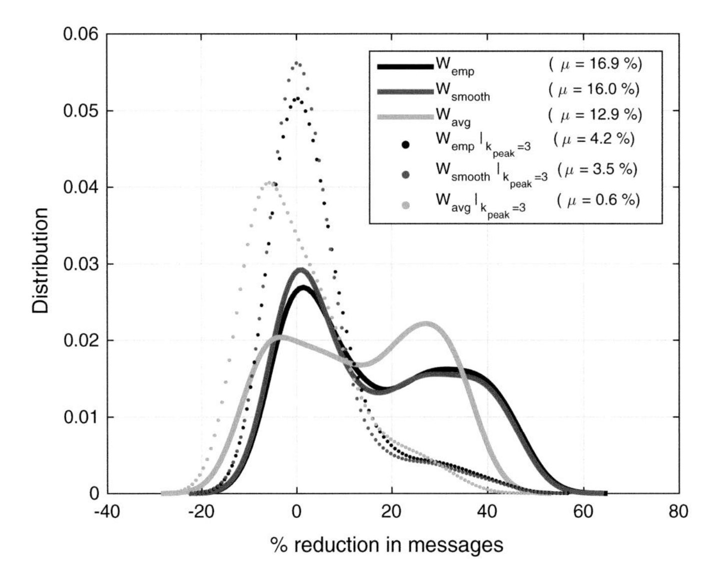

Fig. 8 Distribution of reduction for epinions dataset (the distribution using those with peak ≤ 3 is shown using the *dotted line*)

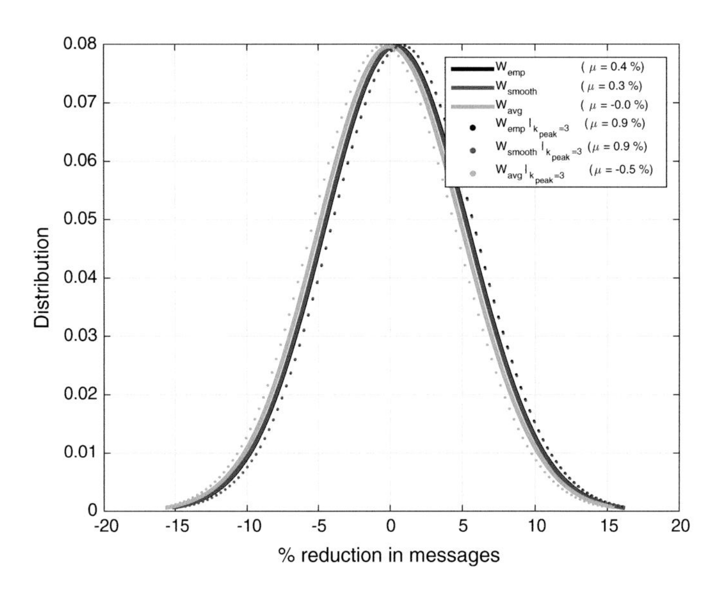

Fig. 9 Distribution of reduction for ER dataset (the distribution using those with peak ≥ 3 is shown using the *dotted line*)

Table 1 Summary of results for a collection of graphs (http://konect.uni-koblenz.de) values for ρ in brackets exclude core nodes, results averaged over 500 simulations

Name	Type	$\|V\|$	$\|E\|$	r	ρ_{emp} %	ρ_{smooth} %	ρ_{avg} %
YouTube friendship	Social	1,134,890	2,987,624	−0.03	16.57 (12.59)	14.61 (10.18)	14.59 (12.37)
Epinions	Social	75,879	508,837	−0.04	16.9 (21.4)	15.9 (20.3)	12.8 (17.2)
Gowalla	Social	196,591	950,3279	−0.02	11.79 (10.38)	9.52 (8.02)	7.2 (6.62)
DBLP	Coauthorship	1,314,050	18,986,618	0.10	8.75 (8.76)	8.11 (7.99)	6.74 (6.56)
Wikipedia En	Hyperlink	1,853,493	39,953,145	−0.05	7.75 (15.15)	6.69 (13.62)	4.80 (13.96)
Catster friendship	Social	149,700	5,449,275	−0.16	4.62 (3.09)	3.51 (2.36)	3.86 (2.61)
Google	Hyperlink	875,713	5,105,039	−0.05	−3.14(−3.75)	−0.83(−0.34)	−0.52(−0.89)
ER graph	Synthetic	100,000	1,151,281	0.00	0.41 (0.23)	0.34 (0.14)	−0.01 (0.16)

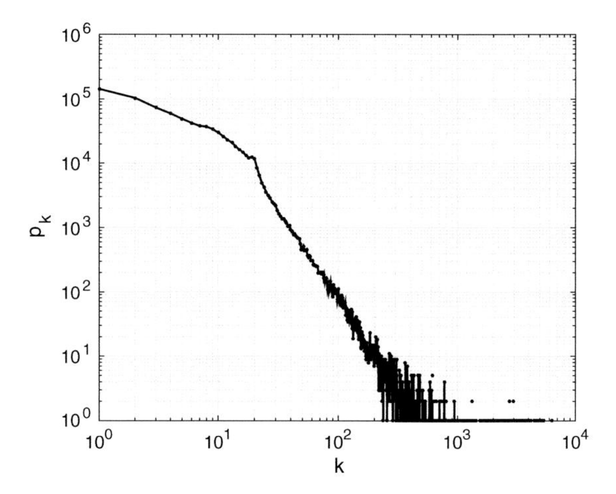

Fig. 10 Degree distribution for Google hyperlink graph (*loglog scale*)

our results. These results are quite mixed; for some graphs the reduction in messages can be very significant and in the order of ∼15 % while for others it can be quite low. For the epinions and YouTube graphs the improvement is 12.80 and 14.59 % on average which is not far from the upper bound of 16.90 and 16.57 %. For the Catster, Wikipedia, and DBLP graph the results are reasonable and in the region of 5 % (3.8, 4.8 and 6.7 %). The Google graph does not show any improvement as the degree distribution for this graph is not power law (Fig. 10). While the distinctive power law tail exists the distribution for low degree nodes is more uniformly distributed breaking the underlying assumption required for the algorithm to work.

For the Epinions graph the result for the non-core vertices increases to 17.20 % but for the YouTube graph it actually decreases to 12.37 %. For the DBLP graph, there is no difference. For Wikipedia the difference is quite significant with non-core vertices reporting a reduction in messages up from 4.80 to 13.96 %. The main conclusion here is that the position of a vertex in the graph certainly has an effect on the performance but it is unclear what the effect will actually be.

5 Conclusion

Social networks with power law characteristics are an important and common class of real-world graphs. This paper has clearly demonstrated the principle that their structure can be leveraged to improve the efficiency of BFS; in some cases significantly by up to 20 %. The computational overhead is minimal; the quantities required for the algorithm to work $\{p_k, p_{k,k'}, n_\tau\}$ can be easily estimated from an initial burn in period (several BFS runs). Future work will look at extending this approach to weighted, directed graphs, we also note that as vertices and edges are added to a real-world graph its degree distribution does not change rapidly and so there is scope for application in dynamic and streaming graph analysis. The skew present in π_k^τ is such that (the standard) unweighted edge partition is not optimal for any iteration. This is why in Fig. 4 we see that the total number of messages (not just at the peak) can also be significantly reduced.

As the techniques mentioned in the Related work (Sect. 2) are not graph structure dependent, it would be interesting to examine if the highly skewed BFS frontier statistics can be usefully incorporated. Future work will investigate a GPU implementation, collecting a taxonomy of graphs for which the technique gives significant improvement and integration with 2-D approaches for improved performance. Finally, further work is required to determine why the algorithm works better for some start vertices, if those vertices can be identified in advance, and in a computationally efficient manner. It is also possible that Eq. 4 could be made conditional on known information about the root vertex.

References

1. Merrill, D., Garland, M., Grimshaw, A.: Scalable GPU graph traversal. In: 17th ACM SIGPLAN Symposium on Principles and Practice of Parallel Programming, vol. 47, no. 8, pp. 117–128, Feb 2012
2. Buluç, A., Madduri, K.: Graph partitioning for scalable distributed graph computations. Contemp. Math. **588**, 83 (2013)
3. Krzywdzinski, K., Kowalski, D.: On the complexity of distributed BFS in ad hoc networks with spontaneous wake-up. Discrete Math. Theor. Comput. Sci. **15**(3), 101–118 (2013)
4. Shang, H., Kitsuregawa, M.: Efficient breadth-first search on large graphs with skewed degree distributions. In: Joint 2013 EDBT/ICDT Conferences, EDBT'13 Proceedings, Genoa, Italy, March 18–22, 2013, pp. 311–322 (2013)
5. Chen, R., Shi, J., Chen, Y., Chen, H.: PowerLyra: differentiated graph computation and partitioning on skewed graphs. In: Proceedings of the Tenth European Conference on Computer Systems, EuroSys'15, pp. 1:1–1:15. ACM, New York, NY, USA (2015)
6. Fu, Z., Dasari, H., Berzins, M., Thompson, B.: Parallel breadth first search on GPU clusters. SCI Institute, University of Utah, SCI Technical report UUSCI-2014-002 (2014)
7. Wang, Y., Owens, J.D.: Large-scale graph processing algorithms on the GPU. UC Davis, Technical report (2013)
8. Buluç, A., Beamer, S., Madduri, K., Asanović, K., Patterson, D.: Distributed-memory breadth-first search on massive graphs. In: Bader, D. (ed.) Parallel Graph Algorithms. CRC Press, Taylor-Francis (2016). https://www.crcpress.com/Parallel-Graph-Algorithms/Bader/9781466573260
9. Luo, L., Wong, M., Hwu, W.-M.: An effective GPU implementation of breadth-first search. Proceedings of the 47th Design Automation Conference. DAC'10, pp. 52–55. ACM, New York, NY, USA (2010)
10. Bisson, M., Bernaschi, M., Mastrostefano, E.: Parallel distributed breadth first search on the Kepler architecture (2014). arXiv:1408.1605
11. Buluç, A., Madduri, K.: Parallel breadth-first search on distributed memory systems. In: Proceedings of 2011 International Conference for High Performance Computing, Networking, Storage and Analysis, SC'11, pp. 65:1–65:12. ACM, New York, NY, USA (2011)
12. Yuan, L., Ding, C., Tefankovic, D., Zhang, Y.: Modeling the locality in graph traversals. In: ICPP. IEEE Computer Society, pp. 138–147 (2012)
13. Buluç, A., Meyerhenke, H., Safro, I., Sanders, P., Schulz, C.: Recent advances in graph partitioning (2013). arXiv:1311.3144

14. Akiba, T., Iwata, Y., Yoshida, Y.: Fast exact shortest-path distance queries on large networks by pruned landmark labeling. In: Proceedings of the 2013 ACM SIGMOD International Conference on Management of Data, SIGMOD'13, pp. 349–360. ACM, New York, NY, USA (2013). http://doi.acm.org/10.1145/2463676.2465315
15. Idwan, S., Etaiwi, W.: Computing breadth first search in large graph using hmetis partitioning. Eur. J. Sci. Res. **29**(2), 215–221 (2009)
16. Abou-Rjeili, A., Karypis, G.: Multilevel algorithms for partitioning power-law graphs. Proceedings of the 20th International Conference on Parallel and Distributed Processing. IPDPS'06, pp. 124–124. IEEE Computer Society, Washington, DC, USA (2006)
17. Hong, S., Kim, S.K., Oguntebi, T., Olukotun, K.: Accelerating CUDA graph algorithms at maximum warp. SIGPLAN Not. **46**(8), 267–276 (2011)
18. Then, M., Kaufmann, M., Chirigati, F., Hoang-Vu, T.-A., Pham, K., Kemper, A., Neumann, T., Vo, H.T.: The more the merrier: Efficient multi-source graph traversal. In: Proceedings of VLDB Endow., vol. 8, no. 4, pp. 449–460 (2014). http://dx.doi.org/10.14778/2735496.2735507
19. Karypis, G., Kumar, V.: Multilevel k-way partitioning scheme for irregular graphs. J. Parallel Distrib. Comput. **48**, 96–129 (1998)
20. Kurant, M., Markopoulou, A., Thiran, P.: On the bias of BFS (breadth first search). In: 22nd International Teletraffic Congress (ITC), pp. 1–8 (2010)
21. Mahadevan, P., Hubble, C., Krioukov, D., Huffaker, B., Vahdat, A.: Orbis: rescaling degree correlations to generate annotated Internet topologies. SIGCOMM Comput. Commun. Rev. **37**(4), 325–336 (2007)
22. Achlioptas, D., Clauset, A., Kempe, D., Moore, C.: On the bias of traceroute sampling: or, power-law degree distributions in regular graphs. In: STOC'05: Proceedings of the 37th Annual ACM Symposium on Theory of Computing, Baltimore, MD, May 2005, pp. 694–703
23. Fay, D., Haddadi, H., Uhlig, S., Kilmartin, L., Moore, A.W., Kunegis, J., Iliofotou, M.: Discriminating graphs through spectral projections. Comput. Netw. **55**(15), 3458–3468 (2011)

Part II
Multilayer Networks

Temporal Multi-layer Network Construction from Major News Events

Borut Sluban, Miha Grčar and Igor Mozetič

Abstract Good news should answer the following questions: *'Who?'*, *'Where?'*, *'When?'*, *'What?'*, and possibly *'Why?'*. We present an approach which extracts interesting events from thousands of daily news. We construct a time-varying, three-layer network where the nodes are entities of interest in the news. The temporal aspect of the network answers the *'When?'* question. The layers are: (1) the co-occurrence of entities which answers the *'Who?'* or *'Where?'*, (2) the summary layer which answers the *'What?'*, and (3) the sentiment layer which labels the links as 'good' or 'bad' news. We demonstrate the news network evolution over a period of four years in an interactive web portal.

Keywords Multi-layer networks · Temporal networks · Sentiment · Summarization

1 Introduction

News inform people about interesting events around the world. We monitor a large number of news web sites around the globe and analyze the structure and the contents of the news. The paper addresses the following question: how to characterize and extract the 'unusual', highly publicized events? We apply a set of network analysis, text mining, sentiment analysis and visualization methods to extract and highlight major news.

The theory of complex networks characterizes systems in the form of entities (nodes) connected by some interactions (links) [1, 3, 11]. A special case of networks extracted from the data are co-occurrence networks, used in diverse fields, such as linguistics [7], bioinformatics [5], ecology [10], scientometry [21], and socio-technological networks [4]. Co-occurrence networks are defined as networks in which nodes represent some entities (for example persons, companies, countries, etc.), and links represent an observation that these entities exist together in some data collection

B. Sluban (✉) · M. Grčar · I. Mozetič
Jožef Stefan Institute, Jamova 39, 1000 Ljubljana, Slovenia
e-mail: borut.sluban@ijs.si

H. Cherifi et al. (eds.), *Complex Networks VII*, Studies in Computational Intelligence 644, DOI 10.1007/978-3-319-30569-1_3

(for example database, news article, etc.). For textual sources, it is important to extract the links between the entities that represent real relations, and are not created by chance.

In our previous work, we have developed a method to estimate the significance of co-occurrences, and a benchmark model against which their robustness is evaluated [14]. The method was applied to analyze the contents of financial news in comparison to empirical networks, constructed from other data sources, like geographical proximity, trade volumes, and correlations between financial indicators [19].

The above co-occurrence detection method models well the persistent, 'everyday' contents of news. However, one is often interested in unusual events, reported by multiple media sources in high news volume. In this paper, we report on a method which detects days with news peaks, and construct a time-varying multilayer network. In particular, at the daily time resolution, we construct a three-layer network with the co-occurrence layer, the summary layer, and the sentiment layer. The co-occurrence layer consists of all the links at peak days, when the news volume is significantly higher than in the past. The summary layer consists of top news for the peak days, where the top news are summarized by the most distinguished titles. The sentiment layer might have a longer time span. We aggregate the sentiment of the top news over all the peak days within a time period. Finally, we are concerned with the presentation of such a temporal multi-layer network. The network evolution over time, with drill-down inspection of details, is demonstrated in a public, interactive web portal at http://newsstream.ijs.si/occurrence/major-news-events-map. The portal facilitates access to over 35 million news, predominantly financial, collected from 170 English news sites, over a period of the last four years.

The paper is organized as follows. In Sect. 2 we describe the entity recognition in news, detection of days with news peaks, and identification of distinguishing topics which summarize major events at peak days. We also describe a lexicon-based approach to sentiment analysis in the news, and the network construction method. In Sect. 3 we give details about the financial news collected, and illustrate the detection of significant events. Some interesting topics recently reported in the news are highlighted, together with the estimated sentiment. We compare the sentiment distribution of all the news, peak news, and top news, and show that there are small, but statistically significant differences. Finally, we show the network visualization implemented in our web portal. We conclude in Sect. 4 with ideas for future work.

2 Methods

We describe a multi-stage approach to construct a multi-layer network of major news events. The stages consist of entity recognition (which identify nodes of the network), event detection (which identify links between the nodes and the co-occurrence layer), content identification (the summary layer), and sentiment analysis (the sentiment layer).

2.1 Entity Recognition

News are about events related to politicians, countries, companies, etc., which we call entities. The process of identifying entities in textual documents requires three components: an ontology of entities and terms, gazetteers of the possible appearances of the entities in the text, and a semantic annotation procedure that finds and labels the entities. We describe the entity recognition approach, as implemented in our NEWSSTREAM portal (http://newsstream.ijs.si) [12].

The ontology we use for information extraction consists of three main categories: financial entities, financial terms, and geographical entities. Most of the ontology is automatically constructed from various data sources. The geographical entities (continents, countries, cities, organizations) were extracted from GeoNames (http://www.geonames.org). MSN Money (http://money.msn.com) was used to organize stock indices and link them to the companies that issue these stocks. The hierarchy of financial terms related to the financial crisis was developed in collaboration with financial experts.

Each entity in the ontology has associated a gazetteer, which is a set of rules that specify the lexicographic information about possible appearances of the entity in text. For example, 'The United States of America' can appear in text as 'USA', 'US', 'the United States', etc. The rules include capitalization, lemmatization, POS tag constraints, must-contain constraints (i.e., another gazetteer must be detected in the document or in the sentence) and followed-by constraints.

Finally, a semantic annotation procedure recognizes the entities of interest. It traverses each document and searches for entities from the ontology. The gazetteers of the entities in the ontology provide information required for the disambiguation of different appearances of the observed entities.

2.2 Event Detection—Peak Days

The next step of content analysis of news is detection of relevant events in the news. We use the daily volume of news articles as a proxy for identifying exceptional events in the news. Given a set of entities of interest $E = \{e_1, \ldots, e_l\}$, we identify all events related to all pairs of entities (e_i, e_j). We monitor the volume of news about these pairs and construct a network of exceptional events between the observed entities.

In [14] we proposed to establish a co-occurrence link between a pair of entities (e_i, e_j) when the number of observed co-occurrences is significantly greater than expected by chance. The probability of a random co-occurrence was estimated from the observed individual occurrences. In this paper, we propose an alternative approach, where we compare the daily number of observed co-occurrences to a longer time period.

We construct a time series of co-occurrence volumes $\mathbf{v}_{ij} = \{v_{ij}(t)\}$ for a pair (e_i, e_j) and a time period T, $t \in T$. At a given time point t, we consider a window

$W_h(t) = \{v_{ij}(t-h-1), \ldots, v_{ij}(t-1)\}$ of length h as a historical baseline, from which we calculate the expected volume at the time point t.

We assume, for a pair of entities, that the volume of their co-occurrences fluctuates around the average volume for a given time period, and that the fluctuations have Gaussian distribution around the average. As the value of the average changes through time, we use a sliding window W_h to compute a moving average.

Given the co-occurrence volume time series $\mathbf{v}_{ij}$, and the size h of the sliding window, we calculate the mean co-occurrence volume $\bar{v}_{ij}(t)$ in $W_h(t)$ and its standard sample deviation $\sigma_{ij}(t)$. Let $z_{ij}(t)$ denote the multiple of $\sigma_{ij}(t)$-deviations from the mean $\bar{v}_{ij}(t)$:

$$z_{ij}(t) = \frac{v_{ij}(t) - \bar{v}_{ij}(t)}{\sigma_{ij}(t)}.$$

For a given Z_0, we say that the co-occurrence volume, such that $z_{ij}(t) > Z_0$, is unexpected and represents an exceptional event between the entities e_i and e_j at day t. Such day t is named a *peak* day. The co-occurrence links between the entities at peak days constitute the *co-occurrence layer* of the constructed network.

2.3 Identification of Relevant Topics—Top News

The goal of the next stage in network construction is to attribute a shallow semantics to the links. The semantics is actually a summary of the top news at peak days, in the form of the most relevant titles.

First we select all the news related to a particular link on a particular day. For example, to attribute semantics to the link between the U.S. and China on a particular day, we consider only the news that contain both these two entities and were published on that day. All the titles of these news are merged into a single text document. One such merged document is created for each day in the past two months (excluding weekends). We apply the standard text preprocessing approach to compute the bag-of-words (BOW) vectors of these documents [9]. In this process, we employ tokenization, stop word removal, stemming, and the TF-IDF weighting scheme [17], standard in text mining. The TF (term frequency) weight, $TF_{d,k}$, denotes the number of times the word k occurs in the document d. The TF-IDF weight is a combination of the TF weight and the IDF weight, where IDF stands for inverse document frequency. IDF of the word k is computed as follows:

$$IDF_k = log\frac{|D|}{n_k},$$

where n_k is the number of documents in the collection D that contain the word k. The TF-IDF weight is then:

$$TFIDF_{d,k} = TF_{d,k} \times IDF_k.$$

The TF-IDF scheme weights a word higher if it occurs often in the same document (the TF component), and if it occurs in only a few documents from the corpus (the IDF component).

The BOW vector for the current day contains information about how important a certain word is with respect to the most relevant events on that day. Instead of showing the top-ranked words, we propagate the weights to the news titles and thus rank the titles by their relevance. The weight-propagation formula computes the average of the word-weights in a title c. The weight of the title, w_c, is computed as follows:

$$w_c = \frac{1}{|c|} \sum_{k \in c} TFIDF_{d^*,k},$$

where k denotes the words in the title c, and d^* represents the merged document for the day in question. Note that the weight w_c penalizes long titles since it is inversely proportional to the title length $|c|$. In our case, this is a desirable property because we would like to find short and to-the-point titles that best describe the most important events. The most distinguished titles at peak days represent the *summary layer* of the constructed network. This layer enriches the co-occurrence links between a pair of entities, by summarizing the news published at the peak days.

2.4 Lexicon-Based Sentiment Analysis

The final stage of the network construction attributes sentiment to the links. We construct the *sentiment layer* of the news network by detecting sentiment orientation and strength of news articles which mention pairs of entities. The sentiment attached to a link between two entities indicates whether the news were 'good' or 'bad' for a given day. However, in contrast to the co-occurrence and summary layers, which have daily time granularity, it is often convenient to aggregate the sentiment links over a longer time period, encapsulating all the top news at peak days.

A sentiment polarity is calculated by a lexicon-based approach. The sentiment polarity of a document is computed from the counts of predefined sentiment terms (positive and negative) in the document. The sentiment terms are from the Harvard-IV-4 sentiment dictionary [22]. For a document d, the sentiment polarity s is calculated by the following formula:

$$s_d = \frac{pos_d - neg_d}{pos_d + neg_d},$$

where pos and neg are the numbers of positive and negative dictionary terms found in the document d, respectively. The sentiment polarities of a set of documents can then be aggregated. An aggregate sentiment for a pair of entities (e_i, e_j) is computed from the top news documents d at peak day t, and from several peak days in a period T:

$$s_{ij}(T) = \frac{1}{N} \sum_{t \in T} n(t) \times s_{ij}(t) \ , \ s_{ij}(t) = \frac{1}{n(t)} \sum_{d} s_d \ , \ \text{where: } (e_i, e_j) \in d,$$

where N is the total number of documents selected at peak days t in the time period T.

2.5 Network Construction Parameters

The temporal news network consists of nodes and three layers of links, at daily resolution. The nodes are entities of interest, $E = \{e_1, \ldots, e_l\}$. The links are pairs of entities (e_i, e_j), with different properties attached at each layer.

The *co-occurrence layer* links entities detected during unusual events, i.e., the peak days. A link (e_i, e_j) is created when the volume of documents containing e_i and e_j significantly exceeds the average volume observed in the previous h days. Assuming the volume of entity co-occurrences in news has Gaussian distribution around the average for a given time period, the significance threshold is set to $Z_0 = 3$, and $h = 44$ (the number of weekdays in the past two months). In a document, each entity must occur at least three times, and at least one entity must occur in the document title. These constraints eliminate most of the noise due to the entity occurrence in a boilerplate. The *summary layer* links the two entities by extracting the most relevant news contents at peak days, thus providing a shallow semantics of the links. A summary link consists of the titles of the top three news articles. The *sentiment layer* presents the emotional attitude of the top news, in terms of the balance between the positive and negative words used. A sentiment link value ranges between -1 and $+1$, where -1 denotes the 'bad' news, 0 the neutral or balanced news, and $+1$ the 'good' news.

3 Results and Discussion

We have been collecting articles from 170 English financial news and blog sites, from November 2011. On average, there are about 35,000 articles per day, a total of over 35 million articles collected until September 2015. This data holds information about temporal relations between different types of entities, such as people, companies, stocks, countries, etc. In this paper, we describe how to detect major events involving different countries, and the construction of the corresponding temporal network. The network captures the major news events detected over the previous four years, and reveals the semantics of the relations between the countries in terms of the contents and sentiment.

3.1 Significant Events

The NEWSSTREAM portal provides an API to collect all the news about any pair of entities. We detect significant events by comparing the daily news volume to the volume of the past two months. If on a particular day the news volume exceeds the average volume of the past two months by more than three standard deviations, i.e., $\bar{v}_{ij} + 3 \cdot \sigma_{ij}$, this day is identified as a significant event day for the observed pair of countries (e_i, e_j).

Figure 1 shows the volume of the news involving 'China' and 'United States' from November 2011 to September 2015. The significant increases in the news volume are identified as volume peaks above the gray line.

In the period between January 2012 and September 2015, 17,702 significant events between 217 countries were detected. We analyze these events in terms of the most relevant content and the associated sentiment.

3.2 Most Relevant Contents and Its Sentiment

We focus on the news related contents published at the peak days. We identify the most relevant and distinguishing topics for each significant event day, as described in Sect. 2.3. We compared our top news results to the major news timeline of *Europe Media Monitor* (EMM, http://emm.newsexplorer.eu/NewsExplorer/timelineedition/en/timeline.html), and reached an overlap of 45 % with all EMM major news, and 60 % overlap with major news topics mentioning at least two countries in the topic title. These differences are mostly due to the following reasons. The major news events of EMM are not limited to country relations (links), therefore they include also news events mentioning only one country or none at all. Topics persisting for

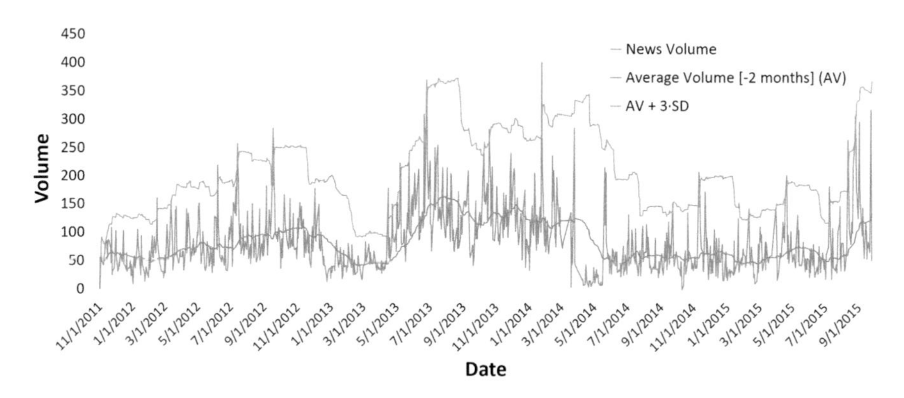

Fig. 1 Volume of news articles about 'China' and 'United States' (*blue line*). The *orange line* denotes the moving average volume over a two months window, and the *gray line* is three standard deviations above the average. Significant events occur at days peaking above the *gray line*

several days with little development are avoided by our approach as we are looking for significant new events. Some countries that are involved in certain topics may be overlooked in the evaluation process due to unresolved indirect mentioning, like 'Merkel' or 'VW' instead of Germany.

For each news article at a peak day we also compute its sentiment, as described in Sect. 2.4. Some significant event days in August and September 2015 for three country pairs are in Table 1. Each event is characterized by the top news headlines and the associated sentiment.

Table 1 illustrates three examples of breaking news about a pair of countries. The first two examples are about the events concerning French-built warships, which were not delivered to Russia, but were later sold to Egypt. The third news example highlights the 'emissions scandal' of a German automobile producer VW, which broke out in the United States.

Table 1 Content and sentiment of the most relevant news on significant event days

Link (Sentiment)	Day	News	Sentiment
FR–RU (0.265)	Aug 6 2015	France to pay Russia under $1.31 billion over warships	0.286
		France to pay Russia under 1.2 billion euros over warships	0.256
		France says several nations interested in Mistral warships	0.254
FR–EG (−0.072)	Sep 23 2015	France sells 2 disputed warships to Egypt	−0.091
		France sells warships to Egypt after Russia deal scrapped	−0.020
		France to sell warships to Egypt after Russia deal scrapped	−0.103
	Sep 21 2015	VW rocked by US emissions scandal as stock slides 17 %	0.039
		VW Rocked by U.S. emissions scandal as stock Slides 17 %	−0.036
		VW shares plunge on emissions scandal US widens probe	−0.026
DE–US (−0.015)	Sep 24 2015	Will Volkswagen scandal tarnish Made in Germany image?	0.007
		After year of stonewalling Volkswagen stunned U.S. regulators with confession	−0.042
		Insight—After year of stonewalling Volkswagen stunned U.S. regulators with ...	−0.030

Shown are significant links between France and Russia, France and Egypt, and between Germany and the United States, in August and September 2015

3.3 Sentiment Distribution

We examine the differences in the sentiment distribution over different sets of news articles. The goal is to compare the sentiment distribution of 'everyday' news with the sentiment at peak days, and with the top news at the peak days. Figure 2 shows the three sentiment distributions.

All three sentiment distributions are approximately Gaussian, and very similar. There is a minor positive sentiment bias in all news, while the peak news are slightly negative. The top news at peak days also seem relatively less positive than the all news. However, the top news contain proportionally more extremely positive and extremely negative news articles. The statistics are in Table 2.

We test the null hypothesis that a pair of news populations (all, peak, or top) has equal mean sentiment. We apply Welch's t-test [23] which is robust for skewed distributions [8]. The results are in Table 3. With t values > 10, the degrees of freedom $\gg 100$, and the p-values $\ll 0.0001$, the null hypothesis can be rejected for all pairs of news populations. We conclude, with high confidence, that the three populations of news have significantly different sentiment means, though these differences are small. Of course, with such large samples one always detects differences. Nevertheless, the results are useful to hint at meaningful differences which are exploited by introducing the neutral zone.

To distinguish 'bad' news from 'good' news, we introduce a neutral zone around the sentiment mean. The $\bar{s}$ value is the sample mean, and the population mean is in the interval $\bar{s} \pm 9SEM$ with very high confidence. We take this interval band around

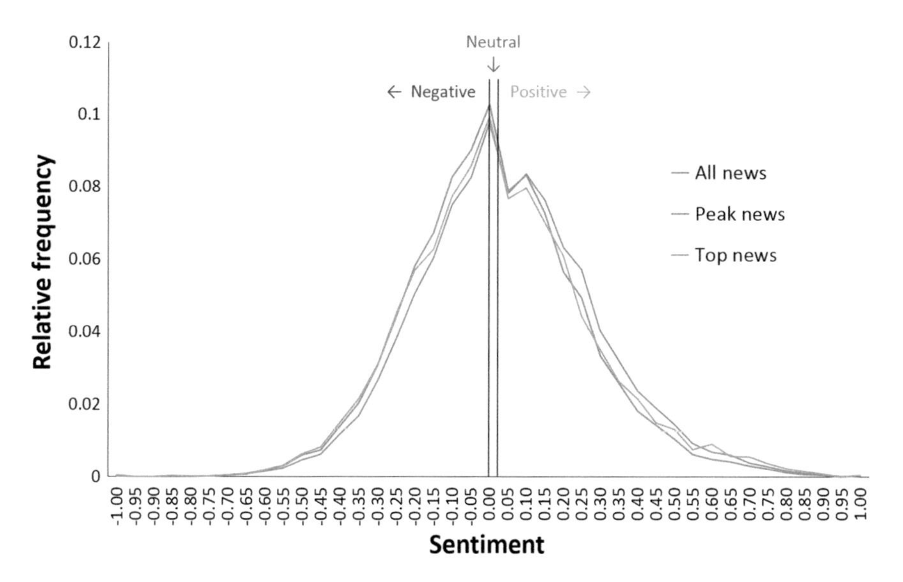

Fig. 2 Comparison of the sentiment distribution of all news articles, peak day articles, and top news, i.e., most relevant articles at peak days

Table 2 The sentiment distributions for different sets of news relating pairs of countries. $\bar{s}$ is sentiment mean, SD standard deviation, and SEM standard error of the mean

	Documents	$\bar{s}$	SD	SEM
All news	1,567,396	0.028	0.239	0.0002
Peak news	274,806	−0.003	0.231	0.0004
Top news	48,097	0.010	0.249	0.0011

Table 3 The results of t-tests for comparison of sentiment means. DF is the estimated degrees of freedom

	t	DF
All news versus peak news	65.55	385,256
Top news versus peak news	10.67	63,425
All news versus top news	15.63	50,852

$\bar{s}$ as the neutral zone. We classify the sentiment of the top news into three discrete classes: *negative* if $-1 \leq s < 0$, *neutral* if $0 \leq s \leq +0.02$, and *positive* if $+0.02 < s \leq +1$. The neutral zone is very narrow, as shown in Fig. 2, and is used just to clearly distinguish between the negative and positive sentiment of top news. This classification is used to label the links in the network visualization with different colors.

3.4 Network Visualization

A network visualization offers a unique way to understand and analyze complex systems by enabling the user to easily inspect and comprehend relations between individual units and their properties [16]. In addition to single layer network visualization [2], also multi-layer visualization is increasingly popular [6, 13].

We have implemented a spatio-temporal visualization of the country co-occurrence network, constructed from the detected major news events, their most relevant content, and the associated sentiment. The visualization facilitates the inspection of various aspects of the network: time dimension, news content, news sentiment, and geography. We have embedded the network into the world map and included functionality to explore the different aspects of the network. Figures 3 and 4 show two instances of the network in time and space. The visualization is an extension of the NEWSSTREAM portal, publicly accessible at http://newsstream.ijs.si/occurrence/major-news-events-map.

Fig. 3 Temporal country co-occurrence network of major news events during Sep 2015

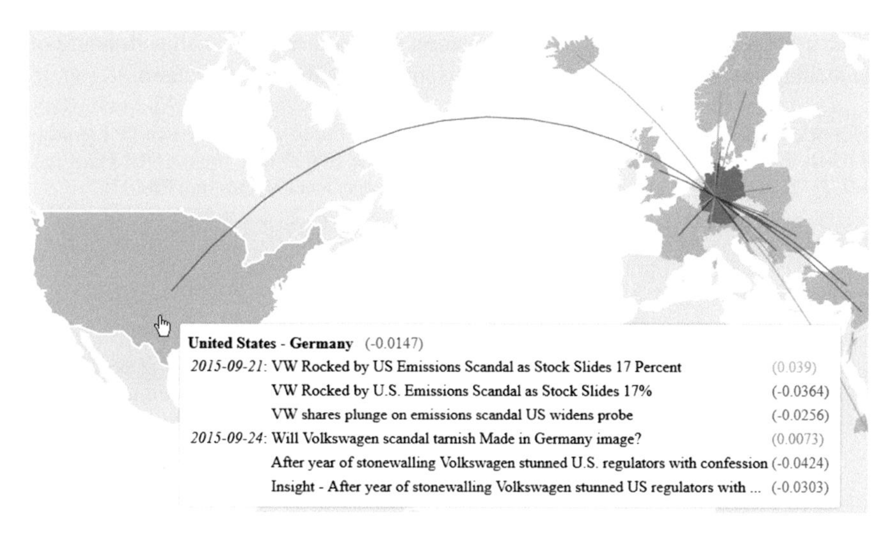

Fig. 4 The most significant news about Germany and the United States in Sep 2015

4 Conclusions

We describe a methodology for the construction of temporal multi-layer news networks. The network captures the links between entities during major news events, summarizes the relations between them, and assigns the sentiment to them. In an experimental setup, we have constructed a time-varying network of countries mentioned in the news over the past four years. We have detected 17,702 major news events involving 217 countries, in the period from January 1, 2012 until September 30, 2015. The interactive visualization of the network supports the spatio-temporal exploration of the major news events.

One of the weaknesses of this approach is a simple, lexicon-based sentiment analysis. We have already implemented much more sophisticated sentiment classification approaches, based on the SVM models, and applicable to short texts in different languages and in various domains [15, 18, 20, 24]. In the future, we plan to focus on key sentences around the entities of interest, and combine the model-based and lexicon-based approaches to sentiment classification for evaluating longer news articles.

Another direction of future research is to study the role of news in the policy making process. In the context of policy debates about complex global issues, such as climate change, financial crises, sustainable development, or migrations, there is an antagonism between the public and private interests. The news play an important role in shaping the policy debates but they might be influenced by the ownership structure of media companies and industrial corporations. We plan to analyze the news and create multi-layer networks of corporations and legal issues, or corporations and environmental issues, and study their sentiment leaning towards different issues. At the same time we will take into consideration the (in)direct ownership structure of the media companies, and analyze how this influences the reported news.

Acknowledgments This work was supported in part by the European Commission FP7 projects MULTIPLEX (no. 317532) and SIMPOL (no. 610704), the H2020 FET project DOLFINS (no. 640772), and by the Slovenian ARRS programme Knowledge Technologies (no. P2-103).

References

1. Albert, R., Barabási, A.L.: Statistical mechanics of complex networks. Rev. Mod. Phys. **74**(1), 47 (2002)
2. Bastian, M., Heymann, S., Jacomy, M.: Gephi: an open source software for exploring and manipulating networks (2009)
3. Caldarelli, G.: Scale-Free Networks: Complex Webs in Nature and Technology. Oxford University Press, Oxford (2007)
4. Cattuto, C., Schmitz, C., Baldassarri, A., Servedio, V.D., Loreto, V., Hotho, A., Grahl, M., Stumme, G.: Network properties of folksonomies. AI Commun. **20**(4), 245–262 (2007)
5. Cohen, A.M., Hersh, W.R., Dubay, C., Spackman, K.: Using co-occurrence network structure to extract synonymous gene and protein names from medline abstracts. BMC bioinformatics **6**(1), 103 (2005)
6. De Domenico, M., Porter, M.A., Arenas, A.: MuxViz: a tool for multilayer analysis and visualization of networks. J. Complex Netw. (2014)
7. Edmonds, P.: Choosing the word most typical in context using a lexical co-occurrence network. In: Proceedings of 35th Annual meeting of ACL, pp. 507–509. Association for Computational Linguistics (1997)
8. Fagerland, M.W.: t-tests, non-parametric tests, and large studies a paradox of statistical practice? BMC Med. Res. Methodol. **12**(78) (2012)
9. Feldman, R., Sanger, J.: Text Mining Handbook: Advanced Approaches in Analyzing Unstructured Data. Cambridge University Press, New York (2006)

10. Freilich, S., Kreimer, A., Meilijson, I., Gophna, U., Sharan, R., Ruppin, E.: The large-scale organization of the bacterial network of ecological co-occurrence interactions. Nucleic Acids Res. **38**(12), 3857–3868 (2010)
11. Jackson, M.O.: Social and Economic Networks. Princeton University Press, Princeton (2010)
12. Kralj Novak, P., Grčar, M., Sluban, B., Mozetič, I.: Analysis of financial news with newsStream. Technical report IJS-DP-11965, (2015). arXiv:1508.00027
13. Piškorec, M., Sluban, B., Šmuc, T.: MultiNets: web-based multilayer network visualization. In: Proceedings of European Conference on ML and KDD. LNCS, vol. 9286, pp. 298–302. Springer (2015)
14. Popović, M., Štefančić, H., Sluban, B.: Kralj Novak, P., Grčar, M., Puliga, M., Mozetič, I., Zlatić, V.: Extraction of temporal networks from term co-occurrences in online textual sources. PLoS ONE **9**(12), e99515 (2014)
15. Ranco, G., Aleksovski, A., Caldarelli, G., Grčar, M., Mozetič, I.: The effects of Twitter sentiment on stock price returns. PLoS ONE **10**(9), e138441 (2015)
16. Rossi, L., Magnani, M.: Towards effective visual analytics on multiplex and multilayer networks. Chaos, Solitons Fractals **72**, 68–76 (2015)
17. Salton, G.: Automatic Text Processing: The Transformation, Analysis, and Retrieval of Information by Computer. Addison-Wesley Longman Publishing Co., Inc, Boston (1989)
18. Sluban, B., Smailović, J., Battiston, S., Mozetič, I.: Sentiment leaning of influential communities in social networks. Comput. Soc. Netw. **2**(9), 1–21 (2015)
19. Sluban, B., Smailović, J., Mozetič, I.: Understanding financial news with multi-layer network analysis. In: Proceedings of European Conference on Complex Systems, ECCS-14. Springer (2015)
20. Smailović, J., Kranjc, J., Grčar, M., Žnidaršič, M., Mozetič, I.: Monitoring the Twitter sentiment during the Bulgarian elections. In: Proceedings of IEEE International Conference on Data Science and Advanced Analytics. IEEE (2015)
21. Su, H.N., Lee, P.C.: Mapping knowledge structure by keyword co-occurrence: a first look at journal papers in technology foresight. Scientometrics **85**(1), 65–79 (2010)
22. Tetlock, P.C., Saar-Tsechansky, M., Macskassy, S.: More than words: quantifying language to measure firms' fundamentals. J. Finan. **63**(3), 1437–1467 (2008)
23. Welch, B.L.: The generalization of "Student's" problem when several different population variances are involved. Biometrika **34**(1–2), 28–35 (1947)
24. Zollo, F.: Kralj Novak, P., Del Vicario, M., Bessi, A., Mozetič, I., Scala, A., Caldarelli, G., Quattrociocchi, W.: Emotional dynamics in the age of misinformation. PLoS ONE **10**(9), e138740 (2015)

Part III
Controllability of Networks

Sensitivity of Network Controllability to Weight-Based Edge Thresholding

Barnabé Monnot and Justin Ruths

Abstract In this study we investigate the change in network controllability, the ability to fully control the state of a network by applying external inputs, as edges of a network are removed according to their edge weight. A significant challenge to analyzing real-world networks is that surveys to capture network structure are almost always incomplete. While strong connections may be easy to detect, weak interactions, modeled by small edge weights are the most likely to be omitted. The incompleteness of network data leads to biasing calculated network statistics—including network controllability—away from the true values [16]. To get at the sensitivity of network control to these inaccuracies, we investigate the evolution of the minimum number of independent inputs needed to fully control the system as links are removed based on their weights. We find that the correlation between edge weight and the degrees of their adjacent nodes dominates the change in network controllability. In our surveyed real networks, this correlation is positive, meaning the number of controls increases quickly when weaker links are targeted first. We confirm this result with synthetic networks from both the scale-free and the Erdös-Rényi types. We also look at the evolution of the control profile, a network statistic that captures the ratio of the different functional types of controls.

1 Introduction

The topic of network controllability has yielded a variety of structural and functional insights and, thus, has been added to the pantheon of statistics available for network analysis. Like the diameter of a graph (of all the shortest paths between all pairs of nodes, the diameter is the longest), network controllability depends on paths through

B. Monnot (✉) · J. Ruths
Singapore University of Technology and Design, 8 Somapah Road,
487372 Singapore, Singapore
e-mail: monnot_barnabe@mymail.sutd.edu.sg

J. Ruths
e-mail: justinruths@sutd.edu.sg

H. Cherifi et al. (eds.), *Complex Networks VII*, Studies in Computational
Intelligence 644, DOI 10.1007/978-3-319-30569-1_4

a network and although it is a global statistic about a network, it is sensitive to small changes in structure. In this work, we investigate the sensitivity of network control to a specific type of preferential edge loss, namely edges with low weight. This type of edge loss or omission is motivated from three different perspectives: omission in sampling of real-world networks, robustness to failure or attack, and feasibility of network controllability.

In order to digitize real-world networks, researchers must survey or sample them, however, oftentimes these methods fall short of capturing a perfect reproduction of the network. In some cases explicit thresholds are imposed as in, for example, when social networks are constructed by asking participants to list their top k acquaintances. In biochemical networks, scientists must run experiments to establish links in the network of proteins. Due to the sensitivity of the experiments or the sheer number that are required, scientists are limited in the scope of these experiments to only the most significant interactions. Fundamentally, all recorded networks are an approximation of the much more complete set of interactions that exist in a real-world system. In all of these examples, the networks that are produced tend to omit the weaker, more subtle dynamical connections that exist in the systems.

There is a growing literature of studies that investigate the robustness of network controllability to node and edge failure or attack [13, 14, 18, 23]. Researchers seek control configurations (the placement of controls in the network) such that under node or edge removal, the property of network controllability is maintained as much as possible. While a variety of factors may contribute to the failure of an edge, one with low weight is likely to be more susceptible due to inherent weakness or overloading.

While network controllability has provided a valuable tool for network science, there is a valid debate as to how to interpret these results specifically for the purpose of controlling a network—i.e., driving the network to a desired state by applying external inputs. In particular, several studies have pointed out that, in the presence of self-loops in the graph, structural controllability analysis becomes trivial [6]. Other studies have characterized the input signals that would need to be used in order to control a network using the minimum number of controls [15, 24]. The general claim made by some of this work is that for real-world networks, the input signals could be unrealistic. The triviality of the analysis and the singularity of the input synthesis is again tied to the scale of the edge weights. Self-loops represent self-dynamics, which are often characterized by timescales that are much longer (therefore, their edge weights would be much smaller) than the inter-state dynamics. This type of timescale separation can be seen in, for example, physics and sociology [3]. Input synthesis requires calculation of the inverse of a matrix, which becomes more singular based on properties of structure and edge weight combined [15, 24]. Thus in the context of these studies, our work here aims to determine to what extent can low weight edges be safely omitted from the structural controllability analysis without dramatically changing the outcomes.

This work, then, requires a database of both directed and weighted networks. There are surprisingly few publicly available datasets fitting these simple requirements. This work is based on all of the networks that we could find; we then supplemented these real-world datasets with tests on synthetic networks to support our empirical observations. There is likewise very little work done on the correlations with edge weights. A notable exception is the study conducted by Li et al., which looks at a very large network dataset of Chinese phone records [8]. We expect this reality is due to the challenge of being able to accurately capture the strengths of the interactions in a large-scale network. In addition, networks are often used so that attention is paid only to structure and not to edge weights. We anticipate that our work and others like it will begin to emphasize the need for weighted network datasets.

2 Network Controllability

By studying network controllability, we implicitly assume a corresponding linear dynamical system whose states evolve according to the structure of the connectivity of the graph. A linear system $\dot{x}(t) = Ax(t) + Bu(t)$ is said to be controllable if for any initial condition $x(0)$ the state can be driven the system to any terminal condition $x(T)$ in finite time T, using the set of controls $u(t)$. We let $x(t) = [x_1(t), \ldots, x_N(t)]^T \in \mathbb{R}^N$ represent the state of the N nodes at time t, the network adjacency matrix $A \in \mathbb{R}^{N\times N}$ codes the dynamical interaction between the nodes (states) and the weight of these connections, and $u(t) = [u_1(t), \ldots, u_M(t)]^T$, $M \leq N$ is the set of external controls. Each control is attached to one or more nodes according to the input matrix $B \in \mathbb{R}^{M\times N}$.

The analytic test for controllability involves checking the rank of a $N \times NM$ matrix and so is computationally costly [4]. In addition, issues of numerical precision become important when the rank of very large matrices is computed. Structural control provides a more computationally efficient approach that scales well with system size. Since it is based on structure, this approach also avoids the numerical issues tied to taking edge weights into account. Structural control was originally introduced to study the relationship between the dynamic structure of a system and its control properties; namely to investigate the properties of a system even when physical parameters, such as stiffnesses and moments of inertia, were not known [9, 21]. More recently the opportunity to use this machinery to draw insights about large-scale real-world networks has brought renewed attention to structural control from both the control and network science communities. In [10], Liu et al. leveraged the wealth of datasets developed in network science to apply structural controllability theory on real-world complex networks. The number and location of the minimum set of independent inputs required to fully control the system can be determined by computing a maximum matching on the bipartite representation of the directed graph [5, 11]. Structural controllability states that when separate controls are applied to the unmatched nodes, the system can be driven to any state. The minimum number of inputs N_c needed to control the system is a fixed statistic based on the structure of the network, even though the maximum matching (and, therefore, the exact location of the controls) is in general not unique.

Ruths and Ruths developed a statistic that categorizes controls according to one of three functional roles in controlling a network [20]. A control may arise because a node only possesses outbound neighbors and no inbound neighbors; this source node must be controlled (because there is no matching that can include this node) and we call this a *source* control. The remaining controls are due to dilations in the network, some *internal* and some *external*. Controls arising from external dilations are added so that sink nodes can be driven independently, while controls arising due to internal dilations are those not due to source or sink nodes. In both situations, one more control will be required. A useful finding is that in spite of the non-uniqueness of the maximum matching, the number of source (N_s), external dilation (N_e) and internal dilation (N_i) controls remain constant. This leads the authors to define a control profile statistic η discriminating between source dominated, external dilation dominated or internal dilation dominated networks according to the largest fraction of the control profile,

$$\eta = (\eta_s, \eta_e, \eta_i) = \left(\frac{N_s}{N_c}, \frac{N_e}{N_c}, \frac{N_i}{N_c}\right).$$

2.1 Robustness of Network Control

This work is closely related to the topic of robustness of network controllability, which seeks to quantify the loss of network controllability as edges (and/or nodes) are removed from the network. Like robustness of network connectivity (e.g., [1]) addition of edges cannot reduce controllability but removal may (although it is possible to have "redundant" edges for which network controllability does not change if they are removed). As with the studies on connectivity, resilience to both failures (random events) and attacks (targeted events) are of interest.

In [14], the authors provide a measure of how robust network controllability is to node and edge removal, including both random failure and targeted attacks. They make the distinction between control-based robustness, which reports the increase in the number of controls required to control the network, and reachability-based robustness, which reports the decrease in the number of controllable nodes given the same fixed set of external inputs. To date, most of the work on robustness of network controllability has centered on the control-based approach (e.g., [13, 18, 23]), however, it was shown that these two metrics for robustness yield different results since the relationship between the two metrics is, in general, nonlinear [14].

For this work, we aim to understand how the omission or loss of low weight edges changes the characteristics of network controllability. As in the robustness discussion above, we could define this in terms of control-based or reachability-based robustness. We determined that in this context, we are most interested in understanding how both the control profile defined previously and the number of controls required to drive the system to any point evolve as edges are removed from the graph.

3 Edge Weights of Real Networks

While structural control does not explicitly depend on the link weights of a network, in this paper we seek to understand the relationship between edges and their importance in network controllability, based on their weight. We operationalize this hypothesis by using structural control to analyze different snapshots of the network as edges are removed according to their weight. Our results point to a strong connection related to the correlation that edge weight has with node degree. Therefore, in this section we discuss this correlation in general and specifically for the networks in our datasets.

We consider real networks arising from a variety of applications. Two networks capture the volume of air traffic between airports internationally and in the United States; one is the graph of neuronal connections of the C. Elegans worm; several correspond to food webs in different environments; and the remaining represent communication traffic in a social network including email, twitter mentions, and messaging. These networks were selected because they provided edge weights, capturing the intensity of the connectivity between nodes, such as the number of passengers transiting between two airports, or the number of emails exchanged between two addresses.

Our first observation is that in most of these networks and like many statistics (e.g., clustering, centrality) of real-world networks, the distribution of edge weights follows a power law, with a large share of lightweight edges and relatively few heavyweight edges (Fig. 1a, b) [12]. More precisely, denoting w weight of an edge, we have that the frequency of edge weight w will be follow the formula $p(w) = Cw^{-\gamma}$, where C is a constant and γ the exponent of the power law. We give in Table 1 under the column γ, the computed exponents of these power laws when they make sense. We plot two of these weight distributions, for the Twitter mention network (Fig. 1a) and the UC Irvine exchanged messages network (Fig. 1b), on a logarithmic scale to better appreciate the power law property.

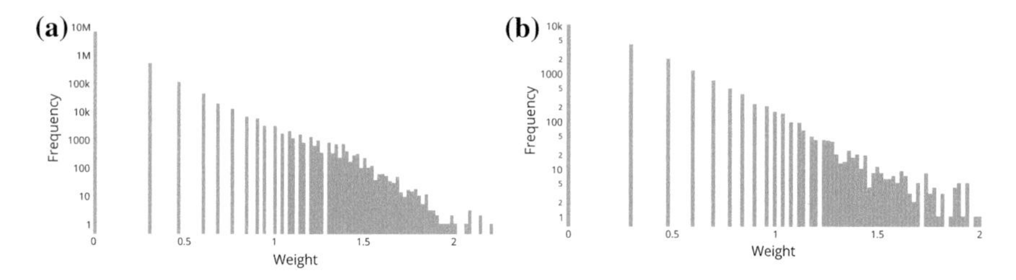

Fig. 1 Edge weight distribution of two empirical networks, demonstrating the scale-free nature of the distributions. **a** Twitter mention network. **b** UC Irvine exchanged messages dataset *one_mode_message*

Table 1 Statistics of the real-world networks used in this study

Name	Description	Nodes	Edges	γ	ρ	ρ_{in}	ρ_{out}
intl	International airport connections	2939	30501	—	0.442	0.436	0.427
us	US airport connections	1574	28236	3.518	0.585	0.530	0.529
celegans	Neural network of *C.Elegans* bacteria	297	2345	2.653	0.037	0.502	−0.133
Chesapeake	Foodwebs	39	177	1.785	0.109	−0.227	0.151
ChesLower	Foodwebs	37	178	—	0.514	0.054	0.669
ChesMiddle	Foodwebs	37	209	—	0.235	−0.219	0.470
ChesUpper	Foodwebs	37	215	2.060	0.206	−0.282	0.520
CrystalC	Foodwebs	24	125	—	0.177	−0.185	0.465
CrystalD	Foodwebs	24	100	—	0.357	−0.015	0.517
Everglades	Foodwebs	69	916	—	0.112	−0.304	0.409
Florida	Foodwebs	128	2106	—	−0.070	−0.653	0.312
Maspalomas	Foodwebs	24	82	1.803	0.071	−0.318	0.554
Michigan	Foodwebs	39	221	1.874	0.492	−0.025	0.710
Mondego	Foodwebs	46	400	—	−0.055	−0.472	0.537
Narragan	Foodwebs	35	220	1.738	−0.073	−0.359	0.169
Rhode	Foodwebs	19	53	1.764	−0.033	−0.371	0.113
StMarks	Foodwebs	54	356	1.847	0.205	−0.413	0.303
Days 1–12	Mail logs (all days)	16547	28539	2.434	0.200	0.137	0.059
Days 1–4	Mail logs	5943	9700	2.494	0.191	0.246	0.059
Days 5–8	Mail logs	9106	14545	2.636	0.218	0.212	0.060
Days 9–12	Mail logs	5776	8987	2.366	0.202	0.269	0.070
mention_network	Twitter network of mentions	5345140	7067931	5.702	0.346	0.304	0.304
one_mode_char	Messages exchanged at UC Irvine	1899	20296	2.329	0.625	0.585	0.467
one_mode_message	Messages exchanged at UC Irvine	1899	20296	3.118	0.732	0.673	0.593

3.1 Edge Weight Correlations with Node Degree

We find substantial correlations between edge weights and other network properties. These connections seem to be the main drivers of the behavior we observe regarding sensitivity of network controllability to edge weights.

In a directed graph $G = (V, E)$, where V is the set of nodes and $E \subseteq V \times V$ the set of edges, we say that a node $u \in V$ is an in-neighbor of $v \in V$ if there exists a link $e \in E$ such that $e = (u, v)$, i.e., there is a link pointing from u to v. Similarly, in this situation we call v an out-neighbor of u. The *in-degree* of a node $v \in V$ is the number of in-neighbors of v. The *out-degree* of a node $v \in V$ is the number of out-neighbors of v. The *degree* of $v \in V$ $\deg(v)$ is then the sum of the in-degree and the out-degree, thus the total number of edges entering and exiting v. We call a node *disconnected* if it has degree zero, i.e., no edges are entering or leaving the node.

One observation made in most of our real networks under analysis is the concentration of heavier links (edges with relatively large weights) around the nodes that possess higher degree. To validate this, we compute the Spearman rank correlation coefficient [7], giving us a measure of how good a monotonic function could capture the relation between node degree and total average link weights around a node. A value close to 1 means there is a positive correlation between the two quantities, close to -1 indicates a negative correlation, and close to 0 reveals no correlation.

The correlation computations are collected in Table 1 under the column ρ. The positive correlation is particularly striking in the airport networks and some of the food webs (although the smaller size of these networks means this value is less significant here). This table also reports the size of the network (nodes and edges) and the computed measures of the power law exponent approximation γ.

Let ρ_{in} denote the coefficient of correlation between the in-degrees and the average weights of incoming edges per node and ρ_{out} the coefficient of correlation between the out-degrees and the average weights of outgoing edges per node. The series of average weights is obtained by iterating over the nodes and computing for each node the average weight of their incoming/outgoing edges. We will later see that in addition to ρ, these correlation coefficients also play a role in the evolution of the control profile. Both ρ_{in} and ρ_{out} are given in Table 1.

4 Results and Discussion

To study the sensitivity of network controllability to the omission or loss of low weight edges, we employ a thresholding method. We first compute and store the control characteristics of the original network. We then iteratively remove a set of edges and compute and store the new control statistics until all the edges in the network have been removed. We sort the edges of the network in terms of ascending weight (lower weight edges first) and remove them in this order, adjusting the *step* (i.e., the number of edges removed per iteration) depending on the size of the graph.

In the case where several links possess the same weight, we randomly choose the order of edges removed.[1] At the end of the procedure, we obtain the evolution of the control characteristics as the fraction of removed edges increases from 0 to 100 %. To eliminate the randomness in this procedure (arising when several links have the same weight), we repeat this algorithm twenty times and average the results. In all cases, we find the standard deviation due to this randomness to be imperceptible[2] and so we leave it out of the remaining discussion. This whole process can be described as *weight cuts*.

To provide a meaningful comparison, we apply the same thresholding procedure, however, instead of sorting links based on their weights, we randomly select the edges to remove (equal in number to the same step size set in the weight cuts). We repeat this process for each network again twenty times and present the averaged results. We call this procedure *random cuts*.

4.1 Evolution of the Number of Controls

We first look at how the number of controls evolves as the thresholding level is increased, i.e., as we remove more and more edges from the graph. A useful benchmark is that if a single edge is removed from the network, the number of controls will increase at most by one (it is possible that the number controls remains the same). Therefore, by removing k edges, we expect the number of controls to increase by some $\kappa \leq k$.

Some networks yield a very clear distinction between the outcome of weight cuts and that of random cuts, such as the *one_mode_message* or the *us* (Fig. 2a, b). In these graphs, we plot the fraction of minimum number of controls $n_c = N_c/N$ against the percentage of edges that have been removed from the graph. We note that this fraction n_c increases much faster in the weight cuts (i.e., when lighter edges are removed first) than in the random cuts. We also give the weight of the edge that was removed from the graph, as the thresholding level grows.

4.1.1 Correlation Between Degree and Edge Weight

We argue that the clear distinction between the two curves, with the weight cuts curve being above the random cuts curve, is the case due to the node degree/average edge weight positive correlation discussed above. Indeed, for the *one_mode_message* the empirical coefficient of correlation was of 0.593 and for *us*, 0.585. If lighter edges are removed first from the graph and if these edges are mainly attached to nodes that have a few connections (a node with lower degree), then we can expect that these nodes

[1] The library used to work on the networks is accessible here: Weight Analysis.

[2] To compare two sequences of weight cuts $\{w\}_{i=1}^{n}$ and $\{\tilde{w}\}_{i=1}^{n}$, we use the l^1 distance between them, $d(w, \tilde{w}) = \sum_{i=1}^{n} |w_i - \tilde{w}_i|$.

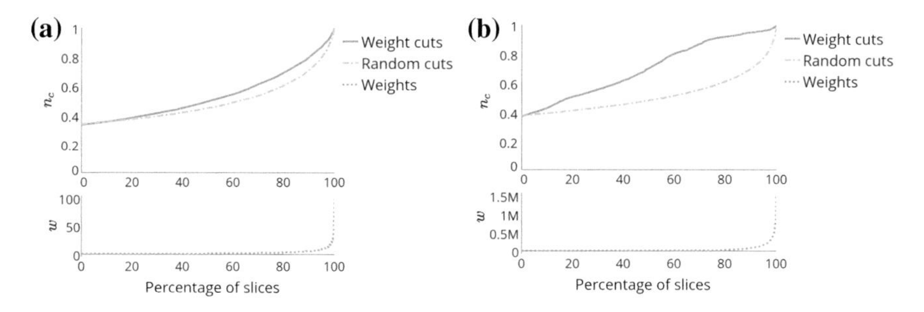

Fig. 2 The evolution of the fraction of minimum number of controls $n_c = N_c/N$ compared between the weight cut procedure and the random cut procedure. **a** *one_mode_message* email network **b** *us* airport network

will be disconnected from the network relatively fast. To control a disconnected node, a new source control must be added, thus increasing N_c by one. The fast increase is then explained by the premature appearance of many source controls.

Posfai et. al study the degree-degree correlations to explain the minimum number of required controls, namely for each edge they investigate the correlations between the in/out-degree of the source node and the in/out-degree of the target node [17]. They find that a high level of correlation, either positive or negative, between the in-degree (resp. out-degree) of the source node and the in-degree (resp. out-degree) of the target node increases the fraction n_c. We argue here that if edges with high weights are connected to these high degree nodes, as is the case with the two previous graphs, we get the same kind of correlation obtained by Posfai et. al.

To make this assertion clearer, we generate random graphs and assign weights to the links following three different methods: either as an increasing/decreasing function of the sum of the degrees of their incident nodes, or randomly. If weights follow an increasing function of the sum of the degrees, then links attached to more connected nodes will receive a higher weight. We do this in a randomized way so that the correlations are comparable with what we observe in the real datasets (i.e., not a perfect correlation). More specifically, let w_e be the weight attached to edge e, and u_e, v_e denote the two node extremities of e. The three ways of assigning weights are the following:

1. Increasing: $w_e = k \times (\deg(u_e) + \deg(v_e))$.
2. Decreasing: $w_e = k/(\deg(u_e) + \deg(v_e))$.
3. Randomly: $w_e = k$.

where k is a random variable uniformly distributed over the integers from 1 to 100.

We generate graphs from two distinct families: Barabási-Albert (BA) preferential attachment networks that exhibit scale-free degree distributions [2] or Erdös-Rényi (ER) random graphs [19]. All of these graphs were generated with 1000 nodes and approximately 2000 edges.

The results are different for the two families. Scale-free graphs exhibit a faster increase in the number of controls required for the control of the graph (Fig. 3). As before, we explain this by the appearance of many source controls attached to

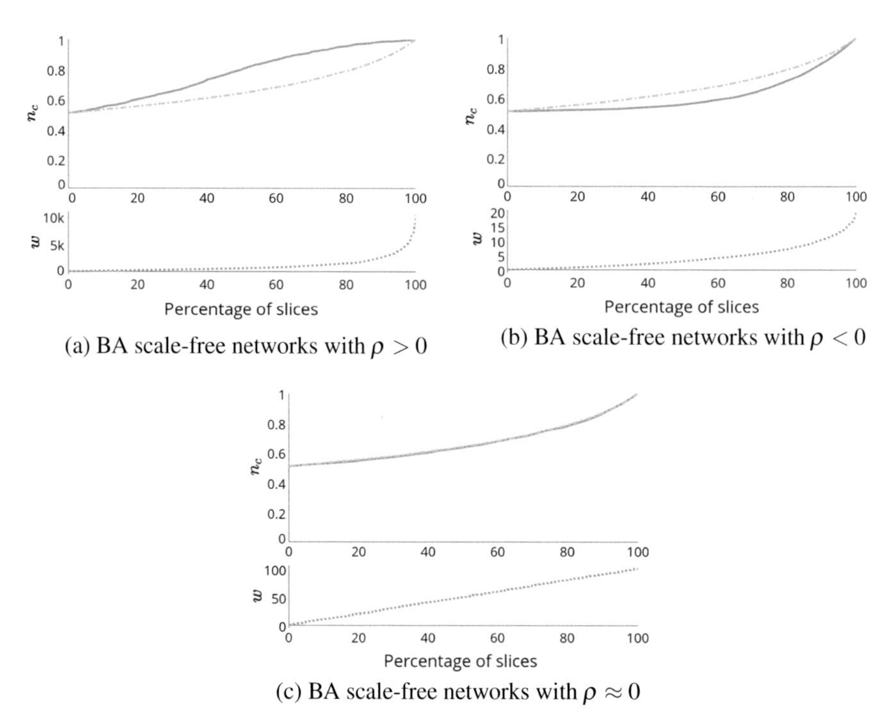

Fig. 3 Barabási-Albert scale-free networks generated with link weights **a** positively correlated to the degrees of their adjacent nodes, **b** negatively correlated, and **c** randomly chosen

the nodes with low average weighted edges (correlated with low degree) which get disconnected first. The reverse phenomenon is observed when more connected nodes receive less weight on their adjacent links. However, when weights are assigned randomly, we unsurprisingly find a curve very similar to that of the random cuts.

Erdös-Rényi graphs exhibit the same properties but at a different scale (Fig. 4). As is known from the theory of random graphs, the degree of the nodes in an Erdös-Rényi graph does not follow the kind of power law observed in scale-free graphs. Indeed their degrees concentrate around a mean value (Poisson distribution), with few outliers in the degree distribution. These tighter connections explain why the evolution of the number of controls is close to that of random cuts.

4.1.2 Correlation Between In/Out-Degree and Edge Weight

This result drives us to find a more precise explanation for this behavior, in particular investigating whether the series of node in/out-degrees and the series of average weights of incoming/outgoing edges around the nodes are correlated, and if so, how this correlation impacts the evolution of n_c.

We also generate a random graph of type Erdös-Rényi (*er*) and another one of scale-free type (*ba*), both of size 100 nodes and approximately twice as many edges.

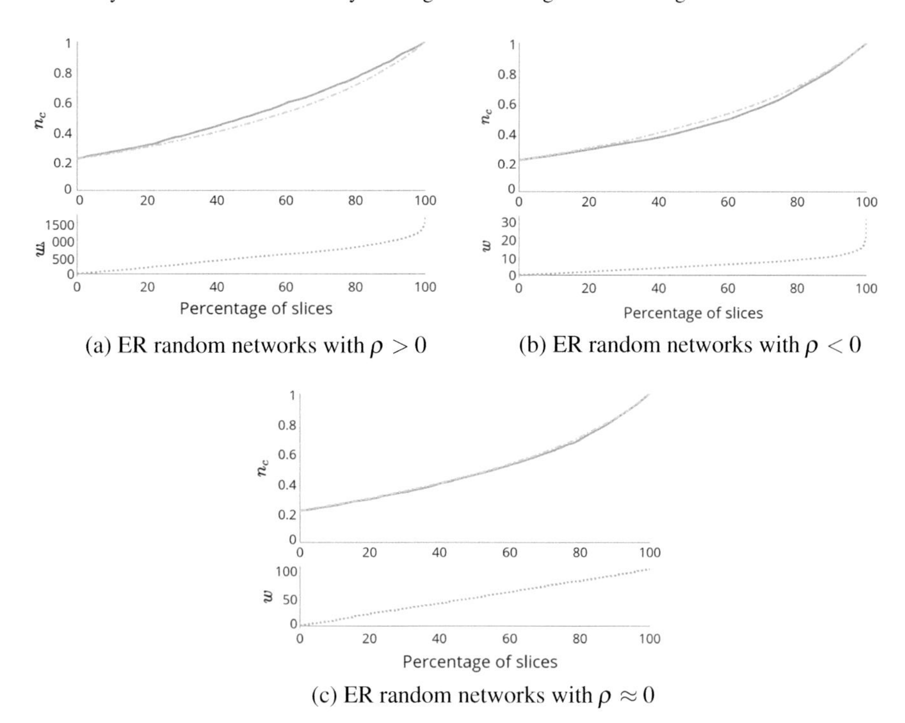

(a) ER random networks with $\rho > 0$

(b) ER random networks with $\rho < 0$

(c) ER random networks with $\rho \approx 0$

Fig. 4 Erdös-Rényi random networks generated with link weights **a** positively correlated to the degrees of their adjacent nodes, **b** negatively correlated, and **c** randomly chosen

For these graphs we apply 4 different weight assignments. For an edge $e \in E$, let $s(e),\ t(e) \in V$ denote respectively the source and the target of the edge.

1. Degree-based assignment: this is the same as in the previous section. This is done in the graphs **-inc*, **-inv*, **-ran* (where * can be replaced with the name of graph, *inc* denotes the use of the increasing function, *inv* that of the decreasing function and *ran* that of the random assignment).
2. In-degree-based assignment: Assign as a weight for e either an increasing or a decreasing function of the in-degree of $t(e)$. This is done in the graphs **-in-inc* and **-in-inv*.
3. Out-degree-based assignment: Assign as a weight for e either an increasing or a decreasing function of the out-degree of $s(e)$. This is done in the graphs **-out-inc* and **-out-inv*.
4. Reverse in-degree/out-degree assignment: Assign as a weight for e either an increasing (resp. decreasing) or a decreasing (resp. increasing) function of the in-degree (resp. out-degree) of $t(e)$ (resp. $s(e)$). This is done in the graphs **-in_out-inc* and **-in_out-inv*.

The main observation is that in-degree/average weight of incoming edges correlation seems to be a stronger driver for the shape of the weight cut curve than that of

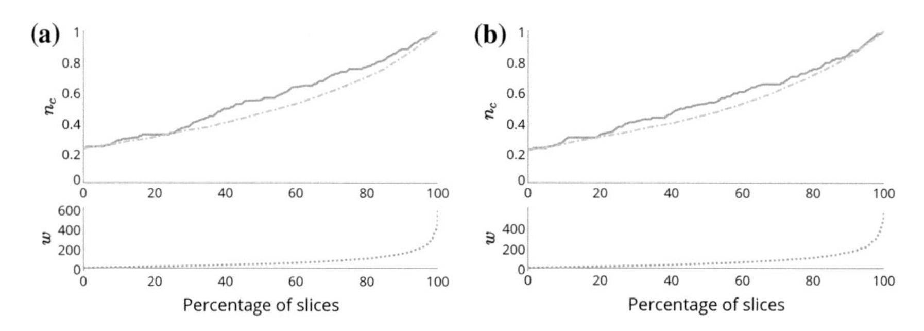

Fig. 5 Erdös-Rényi networks ($N = 100$) with weights distributed inversely on incoming and outgoing edges. **a** *er-in_out-inc*: $\rho_{in} > 0$ and $\rho_{out} < 0$ **b** *er-in_out-inv*: $\rho_{in} < 0$ and $\rho_{out} > 0$

the out-degree/average weight of outgoing edges one. However, it is sufficient that one of our two coefficients of correlation ρ_{in} or ρ_{out} be large enough for the weight cuts curve to be over the random cuts curve.

We give as an example the result of the fourth procedure on a synthesized Erdös-Rényi graph. We see that in spite of the negative correlation between either the in-degree and average weight of incoming edges or the out-degree and average weight of outgoing edges, the weight cuts curve is still above the random cuts curve (Fig. 5a, b).

In order to succinctly capture the gap between the weight cuts and the random cuts curves, we introduce the *control distance* denoted $d(w, r)$, where $w = (n^i_{c,w})$ is the series of fraction of controls determined by the weight cut procedure and $r = (n^i_{c,r})$ is the series of fraction of controls determined by the random cut procedure. The statistic

$$d(w, r) = \frac{1}{N} \sum_{i=1}^{|E|} (n^i_{c,w} - n^i_{c,r}),$$

effectively computes the average distance between the weight cut and random cut curves over the entire sequence of cuts. A positive $d(w, r)$ indicates a tendency for the series of weight cuts to increase faster than that of the random cuts. This appears to be a valid approach because in all of our tests, we observe that the weight cut curve is in general either always above, always below, or roughly equal to the random cut curve. Since we do not observe significant crossings between these two curves, this distance metric is sufficient to capture the general trend of the curves.

Armed with the control distance, we compute this statistic on our real networks with reassigned weights, using the fourth procedure of reverse in-degree/out-degree assignment. The results are given in Fig. 6. It can be seen that for all networks, the control distance is higher in the case of positive ρ_{in} and negative ρ_{out} than in the case of negative ρ_{in} and positive ρ_{out}. This leads us to believe that ρ_{in} has a higher impact than ρ_{out} on the speed with which new controls appear in the weight cuts, compared with the random cuts.

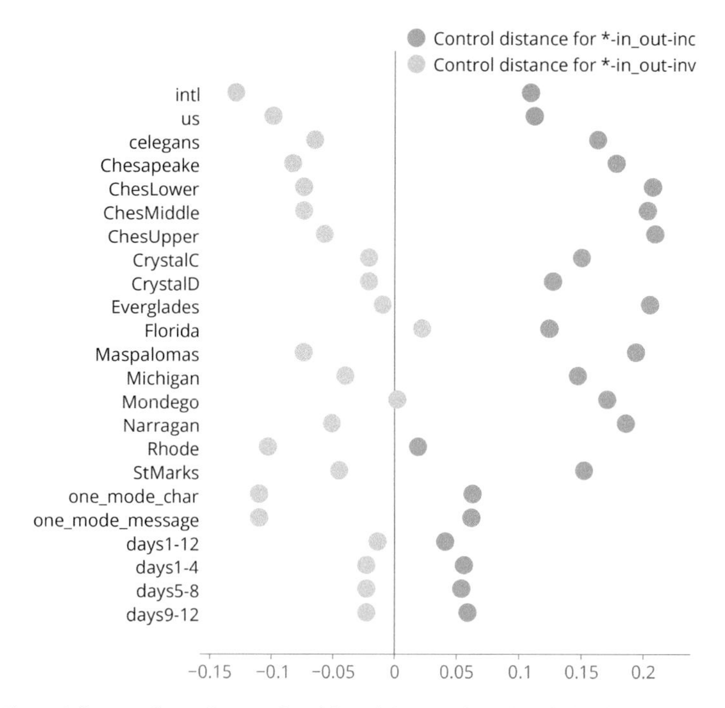

Fig. 6 Control distance for real networks with weights reassigned such that the in-degree and out-degree correlations are opposite, i.e., *inc* networks: $\rho_{in} > 0$ and $\rho_{out} < 0$; *inv* networks: $\rho_{in} < 0$ and $\rho_{out} > 0$

4.2 Evolution of the Control Profiles

We now focus our attention to a finer measure of the controllability: the control profile statistic giving at every step the ratio of source, external dilation and internal dilation controls compared to the overall minimum number of controls.

The resulting path is given in a ternary plot where the lower-left, lower-right, and upper corners correspond respectively to to graphs only possessing source controls, external-dilation controls, and internal-dilation controls. The trajectory of the path corresponds to the evolution of the control profile statistic. We have put in bold and lighter colors the first 20 % of the slices.

Again we observe a discrepancy between the evolution of this control profile in the weight cut procedure and the random cuts. Since at the end of the algorithm, all edges have been removed and hence all nodes are disconnected, both sequences of control profiles eventually converge to the lower left corner, i.e., all controls are source controls. However, the paths taken to arrive at this point vary following the method. We give below two examples of these ternary plots (Fig. 7a, b).

Using our previously constructed *celegans* graph with newly assigned weights, we notice the following. When we have high ρ_{in} (close to 1), such as in

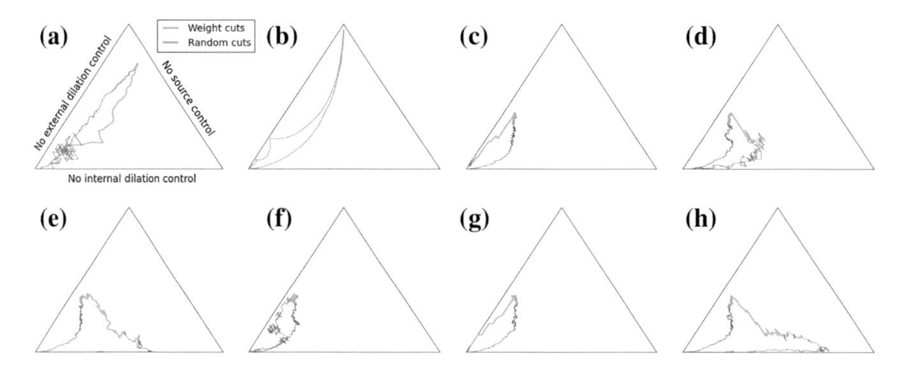

Fig. 7 The evolution of the control profile yields a trajectory that eventually leads to source domination of the controls. The transience of the evolution is determined by the correlation between weights and the in- and out-degree of the nodes. In degree tends to dominate this effect. **a** *Ches-Lower* food web network **b** *intl* airport network **c** *celegans-in-inc* ρ_{in} >0 **d** *celegans-in-inv* $\rho_{in} < 0$ **e** *celegans-out-inc* $\rho_{out} > 0$ **f** celegans-out-inv $\rho_{out} < 0$ **g** *celegans-in_out-inc* $\rho_{in} > 0$ and $\rho_{out} < 0$ **h** *celegans-in_out-inv* $\rho_{in} < 0$ and $\rho_{out} > 0$

celegans-in-inc, the control profile moves fast towards the no external dilation side of the ternary plot. However, when ρ_{in} is low (close to -1) as in *celegans-in-inv*, the profile evolves first towards the lower center of the ternary plot (Fig. 7c, d).

The situation is reversed when looking at ρ_{out}. A high ρ_{out} (*celegans-out-inc*) results in a trajectory where the internal dilation controls disappear first. With a low ρ_{out} these controls are removed after the external dilation controls (Fig. 7e,f).

The last series of graphs (Fig. 7g, h) confirm our intuition that in-degree/average weight of incoming edges correlation impacts more the growth of n_c than its ρ_{out} counterpart. Indeed these graphs show realizations close to the *celegans-in* ternary plots. This observation also holds for our other networks.

4.3 Estimating the Error Due to Missing Edges

Our original objective in this paper was to determine scale of the adverse effects of systematically omitting edges with low edge weights, as we have argued may arise during empirical surveys of networks. The results we have presented in this paper have begun to answer this question, connecting the answer to the correlation between edge weights and node degree. While the full analysis of this next extension is beyond the scope of this paper, We now highlight some intriguing observations that suggest it may be possible to quantify the discrepancy in the control statistic n_c between the true network and its approximation (with missing light-weight edges).

In order to address this question, we then applied the following procedure to two real and two synthetic networks to generate graphs that have a range of correlation ρ values.

i. Assign weights following the formulas for w_e given in Sect. 4.1.1, i.e., with the edge weight following an increasing, decreasing or random function of the adjacent nodes' degrees.
ii. Choose a fraction $\delta \in [0, 1]$ and for $\lfloor \delta \times |E| \rfloor$ iterations, select an edge e at random and assign it a random weight contained between the minimum and maximum weight found in the graph (note that an edge can be chosen multiple times).

The rationale behind this procedure is to obtain graphs with varying ρ values. For example, in the increasing function case, we start from a graph that has high weights attributed to edges connected to high degree nodes and progressively assign random weights to randomly chosen edges, thus reducing the level of correlation.

With these networks with different levels of ρ, we then compute the control distance $d(w, r)$ for these graphs. Figure 8 presents a scatter plot showing the relationship between ρ and the control distance for Erdös-Rényi, Barabási-Albert, *celegans*, and *us* (airport) networks. For each network, a very clear trend indicates that as ρ increases, the distance between the weight cuts curve and the random cuts curve also increases, in most cases in a nonlinear fashion. Interestingly, the trends are not consistent across networks, despite a definitive pattern for each network. The *us* network for example, while showing some correlation between ρ and the control distance, also presents outliers. We anticipate that this variation in trends is due to varying aspects network structure, possibly the degree distribution.

While the exact nature of these patterns is not yet clear, it does highlight several useful findings. For networks exhibiting negative or near zero correlations between edge weights and node degrees, the effect of omitting edges is much reduced compared with those with high correlations. Although the control distance is measured

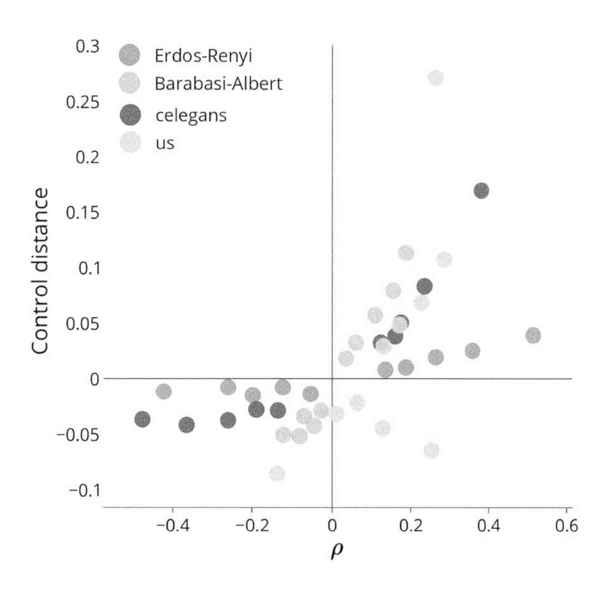

Fig. 8 Control distance for real and synthesized networks with assigned weights plotted against the rank correlation ρ. Increasing ρ leads to an increase in the control distance. While the trend for each network is clear, there is variation between networks, suggesting a dependence on network structure

relative to the random cut outcome, in several cases, the absolute increase in n_c is small for the first 10-20 % of edges cut (see, for example Figs. 2b and 3b). More exciting is the potential ability to predict the biasing induced by the omission of edges. Consider a true real-world graph G, the true nature of it is unknown, and the sampled and digitized version $\tilde{G}$, which is known. Network scientists will analyze $\tilde{G}$ in efforts to infer attributes of the original graph G, for example computing $\tilde{n}_c$, the number of controls required for $\tilde{G}$ in order to try to know n_c, the number of controls required for G. The findings here suggest that researchers could compute the correlation ρ for their empirical network $\tilde{G}$ and use that estimate to infer the effect of omitting k of the light-weight edges of G.

5 Conclusion

In this paper we have looked at how ordered removal of edges according to edge weights influence the change of two network controllability statistics: the minimum number of controls and the control profile. The level of correlation between node degrees and their associated average link weights is the driving factor in understanding how these statistics evolve under targeted edge removal. We identify that positive in-degree correlations, where on average high weight edges are found incident to nodes with high in-degree, is the dominant weight-degree correlation. We hope these results introduce another viewpoint from which to study the field of network controllability, namely that of the link weights. As the community captures more networks with edge weight information, insight in topics like this one will become increasingly more rich and valuable [22].

Our motivation for this study was to understand how the omission of light weight edges may distort network statistics, specifically those related to network control. Our findings identify that the general correlation between edge weight and node degree can provide an estimate of whether the omission will bias the number of controls higher or lower than actual value. When these correlations are high and positive, the network control statistics are relatively sensitive to edge omission, whereas networks without a strong correlation or with negative correlation are more neutral to omitting light weight edges.

An interesting by-product of this study is that assigning edge weights with certain correlations is a compelling way to on-average test characteristics of network controllability and more specifically robustness of network controllability. A variety of work we have cited in this paper has looked at the effects of certain types of edge removal/attack as well as the impact of degree correlations. Assigning weights to edges and removing edges in a given order provides a relatively systematic tool that has a soft edge—averaging out idiosyncratic network structures that might otherwise obscure the main effect.

References

1. Albert, R., Jeong, H., Barabási, A.-L.: Error and attack tolerance of complex networks. Nature **406**(6794), 378–382 (2000)
2. Barabási, A.-L., Albert, R.: Emergence of scaling in random networks. Science **286**(5439), 509–512 (1999)
3. Bloch, F., Hansen, W.W., Packard, M.: Nuclear induction. Phys. Rev. **70**(7–8), 460–474 (1946)
4. Brockett, R.: Finite Dimensional Linear Systems. Wiley (1970)
5. Commault, C., Dion, J.-M., Van der Woude, J.W.: Characterization of generic properties of linear structured systems for efficient computations. Kybernetika **38**(5), 503–520 (2002)
6. Cowan, N.J., Chastain, E.J., Vilhena, D.A., Freudenberg, J.S., Bergstrom, C.T.: Nodal dynamics, not degree distributions, determine the structural controllability of complex networks. PLoS ONE **7**(6), e38398 (2012)
7. Daniel, W.: Applied nonparametric statistics. In: The Duxbury advanced series in Statistics and Decision Sciences. PWS-Kent Publ. (1990)
8. Li, M.-X., Jiang, Z.-Q., Xie, W.-J., Micciché, S., Tumminello, M., Zhou, W.-X., Mantegna, R.N.: A comparative analysis of the statistical properties of large mobile phone calling networks. Sci. Rep. 4 (2014)
9. Lin, C.-T.: Structural controllability. IEEE Trans. Autom. Control 19(3) (1974)
10. Liu, Y.-Y., Slotine, J.-J., Barabási, A.-L.: Controllability of complex networks. Nature **473**, 167–173 (2011)
11. Murota, K.: Matrices and Matroids for Systems Analysis. Springer, Berlin (2000)
12. Newman, M.E.J.: Networks: An Introduction. Oxford University Press, Oxford (2010)
13. Nie, S., Wang, X., Zhang, H., Li, Q., Wang, B.: Robustness of controllability for networks based on edge-attack. PloS one **9**(2), e89066 (2014)
14. Parekh, D., Ruths, D., Ruths, J.: Reachability-based robustness of network controllability under node and edge attacks. In: Workshop on Complex Networks and their Applications, Marrakech, Morocco (2014)
15. Pasqualetti, F., Zampieri, S., Bullo, F.: Controllability metrics, limitations and algorithms for complex networks. IEEE Trans. Control Netw. Syst. **1**(1), 40–52 (2014)
16. Platig, J., Ott, E., Girvan, M.: Robustness of network measures to link errors. Phys. Rev. E **88**(6), 062812 (2013)
17. Posfai, M., Liu, Y.-Y., Slotine, J.-J., Barabási, A.-L.: Effect of correlations on network controllability. Sci. Rep. 3 (2013)
18. Pu, C.-L., Pei, W.-J., Michaelson, A.: Robustness analysis of network controllability. Physica A **391**(18), 4420–4425 (2012)
19. Renyi, A., Erdos, P.: On random graphs. Publicationes Mathematicae **6**(290–297), 5 (1959)
20. Ruths, J., Ruths, D.: Control profiles of complex networks. Science (New York, N.Y.) **343**, 1373–1376 (2014)
21. Shields, R.W., Pearson, J.B.: Structural controllability of multi-input linear systems. IEEE Trans. Autom. Control **AC-21**(2), 203–212 (1976)
22. Toivonen, R., Kumpula, J.M., Saramki, J., Onnela, J.-P., Kertsz, J., Kaski, K.: The role of edge weights in social networks: modelling structure and dynamics (2007)
23. Wang, B., Gao, L., Gao, Y., Deng, Y.: Maintain the structural controllability under malicious attacks on directed networks. EPL (Europhys. Lett.) **101**(5), 58003 (2013)
24. Yan, G., Ren, J., Lai, Y.-C., Lai, C.-H., Li, B.: Controlling complex networks: how much energy is needed? Phys. Rev. Lett. **108**(21), 218703 (2012)

Part IV
Algorithms for Networks

Where Is My Next Friend? Recommending Enjoyable Profiles in Location Based Services

Riccardo Guidotti and Michele Berlingerio

Abstract How many of your friends, with whom you enjoy spending some time, live close by? How many people are at your reach, with whom you could have a nice conversation? We introduce a measure of *enjoyability* that may be the basis for a new class of location-based services aimed at maximizing the likelihood that two persons, or a group of people, would enjoy spending time together. Our enjoyability takes into account both topic similarity between two users and the users' tendency to connect to people with similar or dissimilar interest. We computed the enjoyability on two datasets of geo-located tweets, and we reasoned on the applicability of the obtained results for producing friend recommendations. We aim at suggesting couples of users which are not friends yet, but which are frequently co-located and maximize our enjoyability measure. By taking into account the spatial dimension, we show how 50 % of users may find at least one *enjoyable* person within 10 km of their two most visited locations. Our results are encouraging, and open the way for a new class of recommender systems based on enjoyability.

1 Introduction

Recommending people with similar properties, or similar interest, has been at the basis of many applications, including commercial services like Twitter, Facebook, or Amazon. The underlying assumption common to most of them is that one would like, or be interested in, something or someone similar for properties or interests. This concept is known as *homophily* in social contexts [4, 10]. However, in many scenarios, social diversity fosters new relationships, opens new horizons for collaboration, or, more in general, *enriches* our cultural experience by bringing new topics or new people in our lives.

R. Guidotti (✉)
KDDLab, University of Pisa, Pisa, Italy
e-mail: riccardo.guidotti@di.unipi.it

M. Berlingerio
IBM Research, Dublin, Ireland
e-mail: michele.berlingeriog@gmail.com

H. Cherifi et al. (eds.), *Complex Networks VII*, Studies in Computational Intelligence 644, DOI 10.1007/978-3-319-30569-1_5

In this paper, we ask ourselves the question "can we enable more *enjoyable* location-based services, on the basis of the activities and interactions among people in a social network?". A possible solution to define social enjoyability for recommendations is obviously to consider people who are friends. However, by looking at only the direct (or even the two-hop) friends there could be a considerable loss in possible recommendations, especially for location-based services, as the set of friends in a given area is generally a small percentage of all the users present in that area. A trivial solution would be to put together people with similar interests, regardless of their direct connections as friends. However, not everybody may enjoy spending time with people talking about similar topics: socio-cultural diversity is often researched to increase their knowledge and enhance their enjoyable moments [12]. Hence, two persons are socially compatible according to both their interests and their willingness to be with people with either similar, or different interests.

Along these lines we tried to automatically compute a measure from available data in social network to estimate how much group of people appear to be socially compatible according to their topics of interest and to their friends. We present a measure of *enjoyability* of being together with other people, which takes into account two factors: (i) what we call *like-mindness*, i.e. a topic similarity between any two users; and (ii) what we define as our own version of *homopily*, i.e. the median of like-mindness between a person and each of his/her friends. The enjoyability is a composition between the like-mindeness and the homophily of the two persons. It reaches its maximum when they are either both homophilous and like-minded, or both heterophilous and not like-minded. A good application of enjoyability is to devise suggestions for couples of users which do not know each other and that are looking for someone to spend time together in an enjoyable way.

With the idea of producing enjoyable recommendations for location-based services in social networks as final applications, we present the analysis of a possible definition of enjoyability, computed on real world data such as Twitter. As tweets may be also geo-located, we also perform a distance based analysis of friends vs enjoyable people on a given radius, showing how enjoyable people are always at reach, for radii as little as 500 m, in the areas of Rome, and San Francisco. Our first results are encouraging, and open the way for a new class of location-based services and of recommender systems for social connections.

2 Related Work

Measures based on homphily and on common interests have long been applied in social networks for applications which try to link together people who do not know each other, or to predict their future interactions.

In general, homophily captures the similarity in social networks between individuals who share a link, or the similarity among the members of groups [10]. In the literature we can find studies analyzing how homophily influences human behavior. In [3], the authors exploited homophily in latent attribute inference to augment the users' features with information derived from Twitter profiles and from friends' posts. Their results suggests that the neighborhood context carries a substantial improvement to the information describing a user. The structure of ego-networks and homophily on Twitter is studied in [9]. The authors investigated the relations between homophily and topological features discovering an high homophily w.r.t. topics of interest. In [1], the relation between geographic and interest-based factors w.r.t social linking is analyzed. The authors found that profile similarity of users drives triangles closure in the social network and, reciprocally, closure in the social network induces profile alignment.

Homophily is generally applied in recommendation system to produce more valuable suggestions. The authors of [13] used Twitter contents with their popularity and the social network to suggest lists of users to follow. A recommender system for content is presented in [6]. The authors demonstrated that the social component improves the performances by exploiting attitudes, behaviors and preferences of the users. In [14] the authors built a personalized recommender system, based on the homophily of ego networks. Finally, in [8] the authors use friendship among Twitter users to improve the probability of acceptance for carpooling recommendations.

Also link prediction in social network is a task in which homophily have been largely exploited. Tagging and homophily is used in [2] to predict future friendships. The analysis suggests that users with similar interests are more likely to be friends, and therefore topic similarity measures among users should be predictive of social links. Finally, in [16] the authors define a set of sentiment-based features that help predicting the likelihood of two users to become "friends" based on their sentiments towards topics of mutual interest.

We propose an enjoyability measures that goes beyond the concept of simple social link and homophily, and tries to capture at the same time both differences and similarities among users. Our measure is new and, to the best of our knowledge, there are not similar studies analyzing separately the behaviors captured by the enjoyability in a comparable way in the current state-of-art.

3 Enjoyability

Given a set of users $u \in U$ we can analyze the relationships among them in terms of (i) topics of interests and kind of friendship, and (ii) mobility and spatio-temporal co-location. Every user u may consider other users in U as friends, collaborators, neighbors or direct links in general. We denote such set of users called *friends* which models the friendships as F_u. With respect to the topics of interest we take into account the fact that every user can be interested in some documents. These documents can be either generated by the user herself, or declared to be of interest for the same. We denote the set of documents of interest for a certain user u with C_u. We call *corpus* the complete set of documents for all the users $\mathscr{D}_U = \{C_u\}$. Given a user u and his/her documents C_u, we can identify a number of topics within the documents. Each topic can be weighted by their relative importance (or frequency) within the documents. Consequently, we associate every user to a vector of topics with its weights $\mathbf{t_u}$.

Definition 1 (*Like-mindness*) Given two users u, v we call *like-mindness* the cosine distance between their topic vectors:

$$lm_{u,v} = 2\frac{\mathbf{t_u} \cdot \mathbf{t_v}}{\|\mathbf{t_u}\| \, \|\mathbf{t_v}\|} - 1$$

The like-mindness expresses the similarity of interests between any two users. We say u and v are like-minded if $lm_{u,v} \approx 1$, not-like-minded if $lm_{u,v} \approx -1$.

Homophily evaluates the user's tendency to connect to people with whom (s)he has a high like-mindness, or a low one:

Definition 2 (*Homophily*) Given a user u we compute his/her *homophily* as the median of the like-mindness between him/her and all him/her friends F_u:

$$h_u = \underset{v \in F_u}{\text{median}} \, lm_{u,v}$$

where median returns the median value of a certain set of values. In social networks, the concept of homophily is well known [10], and network analysts often compute it on the degree of a node, defining the assortativity of a network the phenomenon for which most of the users tend to be homophilous by degree (i.e. they link to nodes with similar degree). If $h_u \approx 1$, we say that u tends to be *homophilous*, while if $h_u \approx -1$ we say that u tends to be *heterophilous*.

Definition 3 (*Enjoyability*) Given two users u, v, their like-mindness $lm_{u,v}$ and their homophily values h_u, h_v, we define their *enjoyability*:

$$e_{u,v} = \frac{lm_{u,v} h_u + lm_{u,v} h_v}{2}$$

Note that $e_{u,v} \approx 1$ if either both u and v are homophilous and like-minded, or u and v are heterophilus and not like-minded. In the dual case, $e_{u,v} \approx -1$. In Fig. 1 (left) users u, v are like-minded since they are interested in the same topic; user u is homophilous

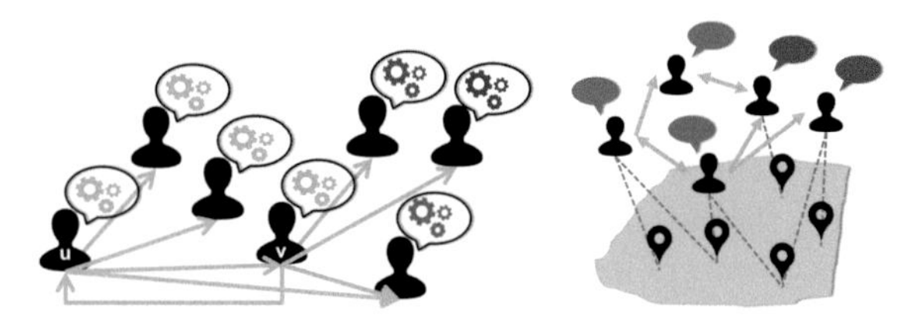

Fig. 1 (*Left*) Example of like-minded users (u, v), homophilous user (u) and heterogeneous user (v). (*Right*) Twitter data source: *social connections*, *interests* and *locations*

because also most of his/her friends are interested in the green topic, while on the other hand user v is heterophilous; finally, the enjoyability between u and v is not very high because v would not enjoy u.

4 Methodology

The proposed methodology needs three kind of input information: *social connections*, *interests* and *locations* (see Fig. 1 (right)). This is why we decided to use the Twitter social network as input. Our methodology consists of the following steps:

1. We get a set of geo-located tweets from the area of interest, using Twitter's streaming APIs[1] on a bounding box. From that set, we extract the active users U, which we filter by average ratio of distinct words used per tweet, and average tweets per day, to remove spammers and automated accounts. Then, for each user, we get the set of other users followed, i.e. *friends* (F_u).
2. From the text of the tweets $\mathscr{D}_U$, we extract for each user a topic vector by means of a Hierarchical Dirichlet Process (HDP), that is a non-parametric version of LDA. We use these vector to compute the *like-mindness* between every couple of users $lm_{u,v}$. The median of these values between each user $u \in U$ and his/her friends F_u is used as *homophily* h_u. Finally, for any two users $u, v \in U$, we can compute their *enjoyability* $e_{u,v}$.
3. In order to extract the locations we map each geo-located tweets in its corresponding cell by using a grid on the area of interest. For each user, we take the two most visited locations (typically, home and work [11]).

Topic modeling is recognized to be able to capture meaningful information in textual data [5, 7]. However, one drawback of one of the main approaches to topic modelling, i.e. Latent Dirichlet Allocation (LDA) [5], is being parametric in the number of topics to extract. To overcome this, we used a nonparametric Hierarchical Dirichlet Process (HDP) [15] algorithm on the users' tweets texts to estimate the number of topics automatically. In practice, these approaches of topic modeling, i.e. LDA and HDP, are clustering algorithms able to group together set of words related with the same topic and to produce from the clustering a vector for each user showing the interest for the various topics. Finally, to detect the two most frequently visited locations, i.e. the locations where the user is supposed to live, we map each geo-located tweets (that are about the 75 % of all the tweets) in its corresponding cell of 500 m width by using a geo-hashing function.

[1] https://dev.twitter.com/docs/streaming-apis.

5 Experiments

We conducted our tests on real Twitter social network data by following the methodology presented in the previous Section. Our goal is to study the distributions of the social measures (like-mindeness, homophily, and enjoyability), to understand how friendship and enjoyability distribute over space and which is the gain of enjoyable connections with respect to classical friendship links. Note that in this work we are not looking for an evaluation of the real effectiveness and utility of the enoyability measure. In order to estimate the reliability of the enoyability, we should perform a validation with the interested real users.. In this paper we limit ourselves to propose a new measure and to show that is able to enrich the possible recommendations in location based services. We leave for the future works its application and validations in case studies where feedback of the users will be collected with respect to suggestions provided according to the enjoyability.

5.1 *Data*

We used Twitter as data source, though the method is easily modifiable to other online sources like Flickr, or Facebook, and it can be made more general to other offline data sources as well, provided they supply information regarding *social connections*, *interests* and *locations*.

We used the Twitter's Streaming API to obtain two large datasets of geo-tagged tweets on *Rome* and *San Francisco*,[2] for 50 days from the beginning of October 2014. As a result, we collected 558,000 geo-tagged tweets from 17,600 user in Rome, and 3,286,000 geo-tagged tweets from 113,000 users in San Francisco.

For each user we retrieved historical tweets (not only geo-located) to build a larger corpus, which we cleaned by removing stopwords and performing stemming. Then we retrieve the users' friends list, together with their tweets. By applying user and tweet filtering to remove rarely active users and automated accounts, we ended up with the statistics reported in Table 1. Note that, thanks to smart filters, we lost around 93 % of users generating in total around half the tweets, i.e. we kept only users with good quality tweets which live in the bounding boxes defined, i.e. meaningful users and not users which just tweeted once in the observed area.

5.2 *Results on Social Measures*

We computed the vector of topics contained in the users' documents by using the tweets of the users resulting from the filtering. Then exploited the vector of topics to

[2]GPS coordinates bounding box: Rome (12.234498, 41.655642, 12.85576, 42.141028), San Francisco (−122.667, 36.8378, −121.2949, 38.0771).

Table 1 Statistics after filtering

Dataset	Users (%)	Tweets (%)
Rome	1,106 (06.53)	237,351 (42.19)
San Francisco	8,581 (07.60)	1,521,827 (49.32)

Numbers in brackets are the percentages over the initial unfiltered data

compute the like-mindness between any two users, and, for each user, we used the median value of it to compute the homophily. Finally with these values, we computed the enjoyability values between any two users.

Since HDP is nondeterministic, we ran it 2,000 times on our data, obtaining on average 25.48 topics ($\sigma = 1.56$) on Rome and 25.61 ($\sigma = 1.54$) on San Francisco. The results for the two datasets are reported in Fig. 2 (left). We selected the results relative to a number of topics of 25, to construct our vectors $\mathbf{t_u}$. We report in Fig. 2 (center, right) the top words for some examples of the multinomial distributions of the HDP algorithm for Rome and San Francisco.

We now report and discuss the results we obtained on Rome and San Francisco for the computation of the social measures. Figure 3 presents the distributions of like-mindness (left), homophily (center) and enjoyability between pairs of users (right) for all the users. We report no significant differences in like-mindness and homophily between Rome and San Francisco. As we can see from the first plot, computing a

Fig. 2 Number of topics distributions on 2 K runs of HDP (*left*). Top words for sample topics for Rome (*center*) and San Francisco (*right*). Colors represent different topics

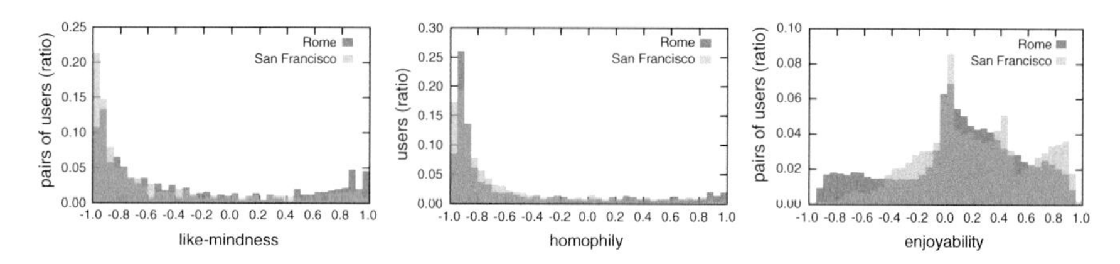

Fig. 3 The distribution of the like-mindness shows that a large majority of the pairs are not socially compatible being interested into different topics (*left*). Probably due to the number of topics extracted by HDP (i.e. 25) also most of the users result to be not homophilous (*center*). The two distributions of enjoyability for Rome and San Francisco appear to be quite different (*right*) we can observe a higher number of very enjoyable couples in San Francisco than in Rome

similarity based only on the like-mindness may end up recommending connections between a limited number of pairs of users. On the other hand, from the second plot, we learn that most of the people are heterophilous. If we combine the two things into the enjoyability, we see, in the third plot, that there is broader space for recommendations based on this measure, rather than the like-mindness. Moreover, the combination of the first two measures produces two different distributions for Rome and San Francisco, highlighting that the enjoyability is capturing a different signal than just the like-mindness.

5.3 Analysis of Tweets in Locations

As mentioned before, besides providing users' interests trough the tweets' text, and the social network modeled by the followers relationship, Twitter is also an important source of human mobility information. This is due to the fact that many users access Twitter from mobile apps and some of them choose to reveal in their tweets their current location as GPS coordinates. We used the two most frequently visited locations (i.e., cells of 500 m width over a grid) to assess the enjoyability for location-based recommendations services versus location-based recommendations services based on friendship.

Table 2 reports the aggregated (with average and standard deviations) probability of tweeting from the most frequent location (second column), the two most frequent locations combined (third column), or any other location after the second most visited one (last column). We can observe how about 80 % of the tweets are produced from these locations, typically home and work, although this analysis is out of scope for this paper. This indicates that the users live in these places and systematically tweet from there. As the first two locations cover the large majority of tweets, we only take into account these two locations for analyzing location-based services.

Since in the following we need to evaluate the distance between couples of users in order to estimate possibilities for location based services, we define the distance $d(u, v)$ between two users u and v as the minimum distance among the following four distances: (i) $d_{1,1}(u, v)$ computed between the two first top locations of u and v; (ii) $d_{1,2}(u, v)$ computed on the first top location of u and the second top location of v; (iii) $d_{2,1}(u, v)$ computed on the second top location of u and the first of v; (iv) $d_{2,2}(u, v)$ computed between the two second top locations of u and v. Hereafter, we refer to d as simply "the distance".

Table 2 Number of tweets by location: avg (std dev)

Dataset	Loc_1	$Loc_1 + Loc_2$	Other
Rome	0.70 (0.25)	0.81 (0.20)	0.19 (0.18)
San Francisco	0.62 (0.24)	0.74 (0.21)	0.26 (0.19)

Loc_1: ratio of tweets in the most visited location. Loc_1+Loc_2: ratio of tweets in the two most visited locations. Other: ratio in all other locations

5.4 Enjoyability Networks Building

To compare the improvement that the enjoyability can provide to a location-based service with respect to friendship, we build the social networks of enjoyability and friendship, named $G_E=\langle N_E, E_E\rangle$ and $G_F=\langle N_F, E_F\rangle$ respectively. In order to be comparable, their sizes in terms of number of nodes and edges must be similar. Let G_F be the Twitter social networks given by the followers. This friendship graph is quite sparse (density equals to 0.0086 and 0.0009 for Rome and San Francisco respectively) w.r.t the users in the observed bounding boxes Hence, in order to have G_E comparable with G_F, while we could create a complete graph G_E^* out of the enjoyability computed between any two nodes, we create the graph G_E representing the enjoyability with the same number of edges $|E_F|$ as found in G_F.

As enjoyability in G_E^* is computed between any two pairs, we needed to remove many edges, to end up with a graph G_E with *similar* structural properties than the friendship graph G_F. By setting a threshold on the enjoyability graph G_E to have the same number of edges of the friendship graph G_F would not work, as we would advantage edges adjacent to nodes with higher enjoyability. In fact, when we tried this strategy, we ended up with the same number of edges (this was intentional), but with roughly 10 % of the nodes.

Instead, we sampled the graph in the following way. From the friendship graph G_F, we get the degree of each node. Then we put in a multiset $\mathcal{M}$ each node N_F of the friendship graph, replicating it as many times as its degree. Finally, while the number of edges $|E_E|$ in G_E is not equal to the number of edges $|E_F|$ in G_F, we randomly pick a node u from $\mathcal{M}$, and we put in the enjoyability graph G_E only the top k edges ordered per descending enjoyability that u has in the complete enjoyability graph G_E^*, where k is the degree of u in the friendship graph G_F. Using this procedure, we can still end up with the same number of edges, but we are also preserving roughly 90 % of the nodes. The remaining nodes are nodes with very small degree, so we simply ignored those nodes.

The statistics on these two graphs are shown in Table 3. As already stated, the number of nodes between the two networks is very similar. Moreover, it is worth noting a very small intersection (5.1 % for Rome and 4.6 % for San Francisco) among the edges E_F and E_E. If the intersection would have been not present, then the location-based service based on enjoyability would have uselessly recommended users which are already friends. Furthermore, the statistics show how the two networks are indeed similar under various topological points of view. The only hard difference is the number of connected components: the enjoyability network G_E is a graph much more connected than G_F in the area observed. Also this fact is relevant for a location-based recommendation service.

Table 3 Networks statistics: $G_F = \langle N_F, E_F \rangle$ friendship, $G_E = \langle N_E, E_E \rangle$ enjoyability, $|E_{E,F}^{Rome}| = 2674$, $|E_{E,F}^{SanFrancisco}| = 19607$, $\Delta = |E_F \setminus E_E| = |E_F \setminus E_E|$, δ density, k average degree, τ triangles, C set of connected components

Dataset	$\|N_F\|$	$\|N_E\|$	Δ	$\|E_E \cap E_F\|$	δ_F	δ_E	k_F	k_E	τ_F	τ_E	$\|C_F\|$	$\|C_E\|$
Rome	785	707	2557	117	0.0086	0.0107	6.81	7.56	785	707	13	5
San Francisco	6500	5772	18708	899	0.0009	0.0011	6.03	6.79	6500	5772	152	4

5.5 Location-Based Analysis of Enjoyability and Friendship

In this section we evaluate how much the enjoyability can improve a location-based service for recommendations with respect to the friendship.

For each user u, we define three sets: F_u, E_u, and $D_u = E_u \setminus F_u$. The first contains the friends of u according to the friendship dimension of Twitter (N_F). The second contains all the people with which u is enjoyable according to the graph G_E computed above (N_E). The third set is the set difference between them, i.e. it contains those enjoyable people who are not friends with u. This last set can be used to recommend *new* people, who are enjoyable, but still not friends with the target user. Figures 4 and 5 reports two different results of the location-based analysis.

In Fig. 4 (left Rome, right San Francisco), we report different radii in km, and the ratio of users who have at least one connection in their own F, E and D (subscripts are removed as F, E and D are now computed for all the users). We used the distance

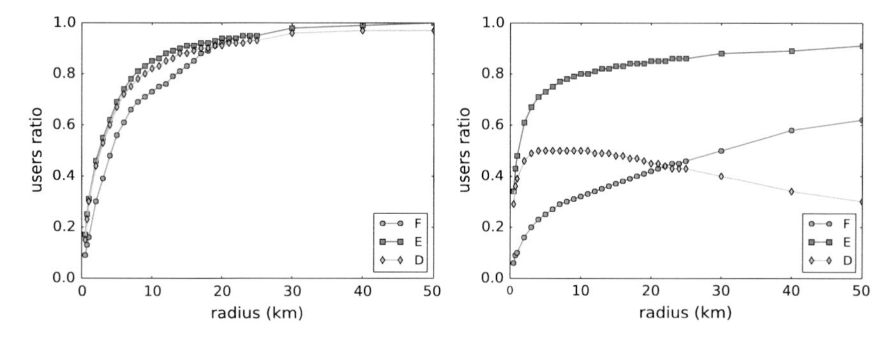

Fig. 4 Location based analysis: ratio of users who have at least one link in their own F, E, D for different radii for Rome (*left*) and San Francisco (*right*). In both the cities we can observe that 50 % or more of the users may find at least one enjoyable person who is not yet a friend within the first 10 km and a clear turning point between F and D around 20–22 km. These aspects show how it is easier to find people in D than in F, with more evidence in San Francisco

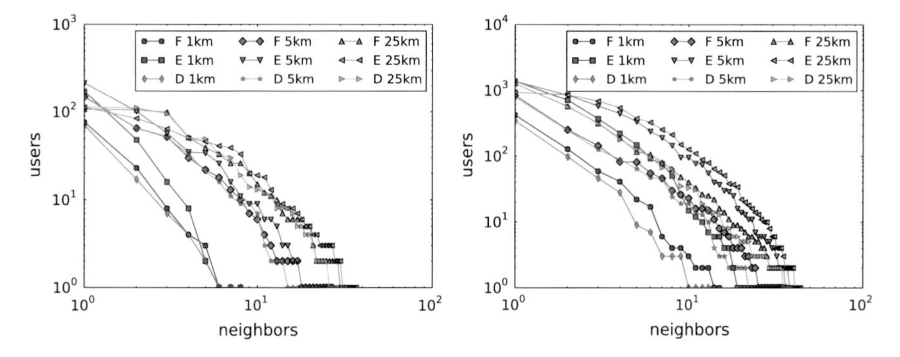

Fig. 5 Location based analysis: distribution of the number of people in F, E, D who have at least a certain number of neighbors for radii 1, 5, and 25 km for Rome (*left*) and San Francisco (*right*)

previously defined to establish if a couple of users are distant more or less than a given radius. For all the three sets, the numbers clearly increase. However, the proportion between users with at least one person in D and E change, as, generally speaking, friendship and enjoyability are distributed in a different way over space. For both the cities we see a turning point between F and D around 20 km (22 for San Francisco). As the Bay Area taken into account is wider than the area over Rome, we also see how distance has different effect in San Francisco. Nevertheless, in both the cities 50 % or more of the users may find at least one enjoyable person who is not yet a friend, within the first 10 km. For both cities it is clear how it is easier to find people in D than in F, with more evidence in San Francisco.

In Fig. 5 (left for Rome, right for San Francisco), we have the probability density function of the number of people in the three sets, for three radii: 1, 5, and 25 km (other radii were not shown to keep readability of the plots). On the x axis we have the number of neighbors in the sets, on the y axis the number of people with a given number of neighbors. For example in Rome (Fig. 5 (left)) we can find up to 10 users with two neighbors which are friends and up to 100 users with neighbors which are enjoyable in a radius of 1 km. Following this idea, this plot is particularly useful in applications where one requires to estimate how many people would be able to find, for example, another 4 soccer player for a stadium located 5 km far from the users (although this requires spatial aggregation and may be enhanced by optimization). These plots give a different perspective w.r.t the above two, as we see the full count of people in the three sets, rather than just checking for their non-emptiness.

6 Applicability

The purpose of this paper is to introduce the enjoyability on the social networks of two different geographic areas and to analyze the possibilities of its application for interest-based recommendations for location based services. The development of real location-based services and the detailed analysis of the recommendations based on enjoyability are left for future work. However, here we list a few possible use case scenarios for our methodology.

Individual recommendation. Individually recommend a new friend is the simplest application. Let think for example to someone moving to a new city where no friends are living. This approach could be used to suggest to him/her the best dinner companion or new possible friends living in the unknown neighborhood.

Group recommendation. Group recommendations are useful when there is a set of users needing to recruit another few members in the group for their activities. Some example are: team sports, group tourism, debates, social brainstorm, and so on. For example, when organizing a football match additional enjoyable players supporting the same football team (enjoyability given by homophilous users) could be found within a given radius from a soccer field and.

Location-based recommendation. Pure location-based recommendations could be useful in various mobility applications. The measure could be exploited by a carpooling service which produces proactive suggestions between couples of drivers and passengers which are feasible both in terms of mobility matches and also according to their enjoyability. This feature could help in reducing people reticence in adopting carpooling as an everyday means of transport. Finally, the users could be provided with recommendations about points of interest and the preferred time to go there according to the enjoyability of their customers.

7 Conclusions and Future Work

We have introduced the *enjoyability*, as a measure of how two people would enjoy time spent together on the basis of their interests and their homophily or heterophily. We have shown the applicability of enjoyability on data with social, textual, and geographical dimensions such as Twitter. Our approach can also be applied to other data sources. The choice for Twitter data was supported by the availability of all the components at once. One could easily replicate our experiments on Call Detail Records (CDR) data used in conjunction with phone calls transcripts. This combination would improve the data quality and quantity in all the three dimensions of our problem: better location data, better textual content, weighted friendship information. However, this type of data is not public. Alternative public data includes Flickr (whose typical usage is tourism), Foursquare (for which we would need to find an external source for textual data), or Facebook (whose APIs do not expose the same type or amount of data). Future work include the realization of an application of such new class, for individual, group or pure location-based recommendations.

Acknowledgments This work has been partially supported by the European Commission under the SMARTCITIES Project n. FP7-ICT-609042, PETRA.

References

1. Aiello, L.M., Barrat, A., Cattuto, C., Ruffo, G., Schifanella, R.: Link creation and profile alignment in the anobii social network (2010) arXiv:1006.4966
2. Aiello, L.M., Barrat, A., Schifanella, R., Cattuto, C., Markines, B., Menczer, F.: Friendship prediction and homophily in social media. TWEB **6**(2), 9 (2012)
3. Al Zamal, F., Liu, W., Ruths, D.: Homophily and latent attribute inference: Inferring latent attributes of twitter users from neighbors. ICWSM, 270 (2012)
4. Berlingerio, M., Koutra, D., Eliassi-Rad, T., Faloutsos, C.: Network similarity via multiple social theories. In: Advances in Social Networks Analysis and Mining 2013, ASONAM'13, pp. 1439–1440. Niagara, ON, Canada, 25–29 Aug 2013
5. Blei, D.M., Ng, A.Y., Jordan, M.I.: Latent dirichlet allocation. J. Mach. Learn. Res. **3**, 993–1022 (2003)
6. Carmagnola, F., Vernero, F., Grillo, P.: Sonars: a social networks-based algorithm for social recommender systems. In: UMAP, pp. 223–234. Springer (2009)
7. Chemudugunta, C., Steyvers, P.S.M.: Modeling general and specific aspects of documents with a probabilistic topic model. In: NIPS, vol. 19, p. 241 (2007)

8. Cici, B., Markopoulou, A., Frias-Martinez, E., Laoutaris, N.: Assessing the potential of ride-sharing using mobile and social data: a tale of four cities. In: Ubicomp, pp. 201–211 (2014)
9. De Choudhury, M.: Tie formation on twitter: homophily and structure of egocentric networks. In: PASSAT, pp. 465–470 (2011)
10. McPherson, M., Smith-Lovin, L., Cook, J.M.: Birds of a feather: homophily in social networks. Ann. Rev. Sociol. 415–444 (2001)
11. Oldenburg, R.: The Great Good Place: Café, Coffee Shops, Community Centers, Beauty Parlors, General Stores, Bars, Hangouts, and How They Get You through the Day. Paragon House Publishers (1989)
12. Pedreschi, D.: Big data, social mining, diversity, and wellbeing. In: SIS, pp. 1–6 (2014)
13. Rakesh, V., Singh, D., Vinzamuri, B., Reddy, C.K.: Personalized recommendation of twitter lists using content and network information. In: ICWSM (2014)
14. Sun, J., Zhu, Y.: Microblogging personalized recommendation based on ego networks. WI-IAT **1**, 165–170 (2013)
15. Teh, Y.W., Jordan, M.I., Beal, M.J., Blei, D.M.: Hierarchical dirichlet processes. J. Am. Stat. Assoc. **101**(476) (2006)
16. Yuan, G., Murukannaiah, P.K., Zhang, Z., Singh, M.P.: Exploiting sentiment homophily for link prediction. In: RecSys, pp. 17–24 (2014)

Particle Filtering as a Modeling Tool for Anomaly Detection in Networks

Ibrahima Gueye and Joseph Ndong

Abstract When linearity can be rigorously assumed for stochastic processes, the linear Kalman filter can be used as a powerful tool for anomaly detection in communication networks. However, this assumption done with a strong evidence is not generally proved in a rigorous way. So it is important to develop other methodology, for the scope of anomaly detection, which are not obliged to be based on that assumption. This paper is focused on the use of particle filtering to build a normal behavioral model for an anomaly detector. The particle filter is calibrated for entropy reduction for the scope of noise reduction in the measurements. With the help of a mixture of normal distributions, we can reuse the filtered observations to identify anomalous events in a few number of classes. Generally anomalies might be rare and thus they might happen on a few clusters. So, using a new decision process based on a hidden markov model, we can track and identify the potential abnormal clusters. We study the performances of this system by analyzing the false alarm rate vs detection rate trade-off by means of Receiver Operating Characteristic curve, and compare the results with the Kalman filter. We validate the approach to track volume anomalies over real network traffic.

Keywords Particle filter · Kalman filter · Non linear System · Entropy reduction · Anomaly detection

1 Introduction

In anomaly detection for communication networks, it is generally necessary to estimate the state of the system evolving over time, using a sequence of noisy measurements. Consequently, the initial measurements have much more noise than the

I. Gueye · J. Ndong (✉)
Department of Mathematics and Computer Science, University Cheikh Anta Diop, BP 5005, Fann Dakar, Senegal
e-mail: joseph.ndong@ucad.edu.sn

I. Gueye
e-mail: ibrahima82.gueye@ucad.edu.sn

H. Cherifi et al. (eds.), *Complex Networks VII*, Studies in Computational Intelligence 644, DOI 10.1007/978-3-319-30569-1_6

filtered measurements and this denoised process can be used to track anomalies. For example, we have techniques based on classification, clustering, nearest neighbor, information theory, statistical, spectral density, etc. The parametric statistical methods assume the existence of an underlying distribution of the state process and rely generally on the assumption of linearity of the state and measurements processes. Recently, we have proposed some anomaly detection approaches concentrated on linear Kalman filtering [13–15]. However, despite its strength, the linear Kalman filter, [12], runs with some difficulties. Generally, the innovation process is expected to be a Gaussian white noise. However, in practice, this is hardly the case as frequently the observed signals are non-gaussian/nonlinear themselves. This filter is generally set with input matrices which are often difficult to find making hard to calibrate the filter. It is also known that the linear filter performs only if the system state and the observations process are typically linear over time. In real situations, it is not obvious to demonstrate, with sampled dataset, that linearity is guaranteed for the state and the measurement processes.

Our hope in this paper is to show, for the first time in the domain of anomaly detection for communication networks, that the sub-optimal algorithm based on particle filtering (which is originally design for nonlinear/non-gaussian processes) can be view as a valuable and alternative tool for anomaly detection, in case when linearity can not be guaranteed or is hardly assumed for the state and/or measurement processes. In particle filtering, the particles represent paths through the state space, by storing the trajectory taken by each particle, so it is appropriate to study a system with dynamical states. The particle filter also performs suitably when the noise process is not assumed to be zero [17].

2 Model for Particle Filtering

The framework of particle filtering is based on the following difference equations:

$$\begin{cases} x_k = f(x_{k-1}) + v_{k-1} \\ z_k = \quad h(x_k) + n_k \end{cases} \tag{1}$$

where $f(.)$, the function which describes the internal state of the system, is used to compute the predicted state from the previous estimate and similarly, the function $h(.)$, which describes how the system state is transformed to give the observations, can be used to compute the predicted measurements from the predicted state; $x_k \in \mathbb{R}^n$ and $z_k \in \mathbb{R}^m$ are multi-dimensional vectors representing respectively the system state and the measurement. The system is assumed to be excited by an unknown process noise $v_k \sim N(O, Q_k)$ and the measurement are disturbed by unknown measurement noise $n_k \sim N(O, R_k)$.

2.1 Sub-optimal Particle Filtering

We deal with the problem of Bayesian estimation of the system state x_k at time k given the data $z_{1:k} = \{z_i, i = 1, \ldots, k\}$ up to time k. Thus, we need calculating the posterior density function (pdf) $p(x_k|z_{1:k})$. To have a filtered estimates of the x_k based on all the available measurements, we assumed that the initial pdf, i.e., the prior $p(x_0|z_0) \equiv p(x_0)$ is known. So, the pdf $p(x_k|z_{1:k})$ can be obtained recursively in a two-step prediction and update phase.

If the required pdf $p(x_{k-1}|z_{1:k-1})$ at time $k-1$ is set, the prediction step involves using the system state in Eq. 1 to determine the prior of the state at time k by means of the following Chapman-Kolmogorov equation:

$$p(x_k|z_{1:k-1}) = \int p(x_k|x_{k-1})p(x_{k-1}|z_{1:k-1})dx_{k-1}. \tag{2}$$

The probabilistic model of the state evolution $p(x_k|x_{k-1})$ is defined by the system state equation in Eq. 1) and the known statistics of v_{k-1}. At time step k, a measurement z_k becomes available, and this may be used to update the prior (update stage) via the Bayesian rule:

$$p(x_k|z_{1:k}) = \frac{p(z_k|x_k)p(x_k|z_{1:k-1})}{p(z_k|z_{1:k-1})}, \tag{3}$$

where the normalizing constant:

$$p(z_k|z_{1:k-1}) = \int p(z_k|x_k)p(x_k|z_{1:k-1})dx_k \tag{4}$$

depends on the likelihood function $p(z_k|x_k)$ defined by the measurement model in Eq. 1 and the known statistics of n_k. In the update stage 3, the measurement z_k is used to modify the prior density to obtain the required posterior density of the current state.

The recurrence Eqs. 2 and 3 form the basis for the optimal Bayesian solution. However, the solution of this recursive propagation of the posterior is intractable analytically. When the state and measurements are linear, the Linear Kalman Filter becomes a solution. Wen nonlinear/non gaussian features are assumed, sup-optimal (in the sense that approximations are necessary) estimation frameworks as the extended Kalman filter, the unscented Kalman filter or the particle filter approximate the optimal Bayesian solution. In this work we focus on the particle filtering framework.

The particle filter is based on the sequential importance sampling (SIS) algorithm which describes a Monte Carlo (MC) method [5, 6]. This algorithm has several denominations as the bootstrap filtering [7], the condensation algorithm [11], particle filtering [2] and interacting particle approximations [3, 4]. In this work we focus our attention on the denomination "particle filtering". The SIS algorithm implements a recursive Bayesian filter by MC simulations. The main idea is to represent the required pdf by a set of random samples with associated weights and to compute estimates based on these samples and weights. As the number of samples becomes very large, this MC characterization becomes an equivalent representation to the usual functional description of the posterior pdf, and the SIS filter approaches the optimal Bayesian estimate.

Consider a random measure $\{x^i_{0:k}, w^i_k\}^{N_s}_{i=1}$ which characterizes the posterior pdf $p(x_{0:k}|z_{1:k})$, where $\{x^i_{0:k}, i = 0, \ldots, N_s\}$ is a set of support points with associated weights $\{w^i_k, i = 0, \ldots, N_s\}$ and $x_{0:k} = \{x_j, j = 0, \ldots, k\}$ is the set of all states up to time k. The weights are normalized such that $\sum_i w^i_k = 1$. The posterior density at time k can then be approximated as:

$$p(x_{0:k}|z_{1:k}) \approx \sum_{i=1}^{N_s} w^i_k \delta(x_{0:k} - x^i_{0:k}). \tag{5}$$

The weights can be found by the principle of importance sampling [1, 6]. This principle supposes that $p(x) \propto \pi(x)$ is a probability density from which it is hard to draw samples but $\pi(x)$ can be found. We suppose also that $x^i \sim q(x), i = 1, \ldots, N_s$ is a set of samples generated from an *importance density* q(.). Then, we obtain a weighted approximation of the density p(.) by:

$$p(x) \approx \sum_{i=1}^{N_s} w^i \delta(x - x^i). \tag{6}$$

where

$$w^i \approx \frac{p(x^i_{0:k}|z_{1:k})}{q(x^i_{0:k}|z_{1:k})}. \tag{7}$$

For this sequential case, at each iteration, we can derive samples forming an approximation to $p(x_{0:k}|z_{1:k})$ and need to approximate with a new set of samples. To achieve this, we can write the importance density is a factorized form as:

$$q(x_{0:k}|z_{1:k}) = q(x_k|x_{0:k-1}, z_{1:k})q(x_{0:k-1}|z_{1:k-1}), \tag{8}$$

and drawn samples $x^i_{0:k} \sim q(x_{0:k}|z_{1:k})$ by augmenting each of the previous samples $x^i_{0:k-1} \sim q(x_{0:k-1}|z_{1:k-1})$ with the new states $x^i_k \sim q(x_k|x_{0:k-1}, z_{1:k})$.

The weight update equation can be then derived in the following way. First express the Eq. 3, i.e., $p(x_{0:k}|z_{1:k})$ such that:

$$p(x_{0:k}|z_{1:k}) \propto p(z_k|x_k)p(x_k|x_{k-1})p(x_{0:k-1}|z_{1:k-1}) \tag{9}$$

So, by substituting (8) and (9) to (7), the weight update equation becomes:

$$w^i_k = w^i_{k-1} \frac{p(z_k|x^i_k)p(x^i_k|x^i_{k-1})}{q(x^i_k|x^i_{0:k-1}, z_{1:k})} \tag{10}$$

If $q(x_k^i|x_{0:k-1}^i, z_{1:k})$ can be expressed as $q(x_k^i|x_{k-1}^i, z_k)$, then the importance density becomes only dependent on x_{k-1} and z_k. In this work, we focus our attention to this particular useful choice of the importance density, since we need only a filtered estimate of $p(x_k|z_{1:k})$ at each time step. In this case, only x_k^i need to be stored, so we can discard the path $x_{0:k-1}^i$ and the passed observations $z_{1:k-1}$ and the weight become finally:

$$w_k^i = w_{k-1}^i \frac{p(z_k|x_k^i)p(x_k^i|x_{k-1}^i)}{q(x_k^i|x_{k-1}^i, z_k)} \tag{11}$$

We obtained the approximated posterior filtered density as:

$$p(x_k|z_{1:k}) \approx \sum_{i=1}^{N_s} w_k^i \delta(x_k - x_k^i) \tag{12}$$

At this point, the SIS algorithm consists of a recursive propagation of the weights and support points as each observation received sequentially. For more details on how to implement this algorithm in Matlab, see [17]. The use of the SIS Algorithm as described above might leverage a difficulty, namely the degeneracy problem which have many solutions [17].

2.2 Solving the Degeneracy Problem

This problem holds when, after a few iterations, all but one particle will have negligible weight. This difficulty is due to the fact that the variance of the importance weights can only increase over time [6]. The degeneracy problem causes a large computational effort to update particles whose contribution to the approximation to $p(x_k|z_{1:k})$ is almost zero. The works in [1, 10] propose a suitable measure of degeneracy defined as the effective sample size N_{eff}:

$$N_{eff} = \frac{N_s}{1 + Var(w_k^{*i})} \tag{13}$$

where the "true weight" is $w_k^{*i} = p(x_k^i|z_{1:k})/q(x_k^i|x_{k-1}^i, z_k)$. We can only obtain an estimate of the effective sample size as:

$$\widehat{N_{eff}} = \frac{1}{\sum_{i=1}^{N_s} (w_k^i)^2} \tag{14}$$

where w_k^i is computed via (10). We always have $N_{eff} \leq N_s$, so small N_{eff} causes severe degeneracy which implies very bad results in particle filtering. One can avoid it by increasing the value of N_{eff}, but in this case the complexity will increase. Two solutions exist in the literature to solve the degeneracy problem. The first one try to build a good importance density, the second applies a resampling technique. The choice of a good importance density tries to minimize the variance of w_k^{*i} so that N_{eff} is maximized. This methods necessitates to sample from $p(x_k|x_{k-1}^i, z_k)$ which is an integral to be evaluated in a not straightforward manner. Thus, we consider the second solution which uses resampling procedure whenever significant degeneracy is observed, i.e. when N_{eff} in drastically small. The resampling operation tries to eliminate particle with small weights and consider only particles with high weights. So, the procedure involves generating a new set $\{x_k^{i*}\}_{i=1}^{N_s}$ by replacing them N_s times from an approximate discrete representation of $p(x_k|z_{1:k})$ defined as:

$$p(x_k|z_{1:k}) \approx \sum_{i=1}^{N_s} w_k^i \delta(x_k - x_k^i) \tag{15}$$

This resampling step should give $Pr(x_k^{i*} = x_k^j) = w_k^j$. The resulting sample is an i.i.d sample from the discrete density (15), so $w_k^i = 1/N_s$. The methodology described in [2, 16] and based on order statistics can be used to implement the resampling algorithm with $O(N_s)$ complexity.

3 The Model Based on the Linear Kalman Filter

In [13], the anomaly detector is built on a Linear Time State-Space (LTSS) model as shown in the following difference equations:

$$\begin{cases} x_{t+1} = A_t x_t + w_t \\ y_t = C_t x_t + v_t \end{cases} \tag{16}$$

In Eq. 1, the system state x_t and the measurable output y_t are multi-dimensional vectors of appropriate dimensions. Due to lack of place, we redirect the reader to our previous work [13] for more details for the description of this system. The system parameter $\theta = \{C_t, Q_t, R_t\}$ are found via the expectation-maximization (EM) algorithm [18, 19].

4 Normal Space Identification

After applying the particle filtering, we obtain a filtered estimate of the system state and the measurements. Thereafter, using the residual (difference between the measured and the filtered measurement) as our decision variable, we apply an unsupervised clustering technique in order to organize it into a set of clusters. Anomalous events might appear in a few number of this set of classes. We run a two-step approach to build the normal space formed by some clusters, the remaining labeled as abnormal. To see if this method is powerful enough, we apply the same detection procedure as we did in a previous paper where the system is composed with the linear Kalman filter and the two frameworks of gaussian mixture modeler (GMM) and hidden markov model (HMM).

We use GMM to reject the fact that the innovation process as output of the Kalman filter does not remain a zero mean white gaussian noise. Instead we assume that this process follows an ensemble of normally distributed processes we may identify as some clusters. We also think that temporal correlations might exist between theses classes and identifying them by means of a HMM can help us classify some cluster as "normal" of the remaining as "abnormal". Using GMM and HMM are not the scope of this paper, so we let the reader refer to our previous work, [13] for more details in the definition and method of calibration of theses techniques.

5 Discovering Normal and Abnormal HMM States

In [13], after running the GMM, we find the different clusters. Thereafter, we plug in these results to the HMM framework to separate these clusters into different states, each state having some clusters. The final step stays then on how to separate theses states in two families, the first labeled as "normal" and the second one "abnormal". Recall we have assumed that the residual (our innovation process) is assumed to be not a zero mean process. In "best conditions" where no anomaly and no heavy teals happen, it might be a zero mean process. Then, in fact, it is easy for us to think that there's a part of this process with the mean closely equal to zero. If we can find this part, we can eliminate it in the final procedure for anomaly detection, since it is potentially the place where there's no abnormal events. So after calibrating the HMM, the states containing data for which the mean are close to zero are identify as the normal family and the rest of the states where the corresponding data have their mean far from zero are labeled as the abnormal family. Finally, we can easily apply a decision based on applying thresholding to the data for the abnormal family to detect the anomalous events.

6 Model Validation

6.1 Forecasting Criterion of Accuracy

To evaluate the performances of our models, we use several accuracy measures defined to evaluate the entropy (bias) removal. Forecast accuracy can be assessed in terms of root mean square error (RMSE), mean absolute error (MAE), and mean bias error (bias) MBE. RMSE gives more weight to large errors, whereas MAE, less sensible to large errors, reveals the average magnitude of the error, and MBE indicates whether there is a significant tendency to systematically over-forecast or under-forecast. When comparing between different models, RMSE was used as the metric for minimization, that is, forecasts were trained with the goal of reducing the largest errors. These performance criterion are defined as $\text{RMSE} = \sqrt{\frac{1}{N}\sum_{i=1}^{N}(y_i - \hat{y}_i)^2}$, $\text{MAE} = \frac{1}{N}\sum_{i=1}^{N}|y_i - \hat{y}_i|$, $\text{MBE} = \frac{1}{N}\sum_{i=1}^{N}(y_i - \hat{y}_i)$, where y_i and $\hat{y}_i$ represent respectively the real and filtered measurements.

6.2 Experimental Data: Abilene Network

In this work, we used a collection of data coming from the Abilene networkbreak [13–15]. The anomalies injected in the Abilene data are small and high synthetic volume anomalies [8, 9].

6.3 Using Pre-prepared Data

We have built all the optimization process on the 41 links of the Abilene Network, but for the sake of place, we show only the results for one link (one vector of measurement) where there are six (6) true anomalous events. We have implemented the particle filter described above by considering the following system. So, we set the functions $f(x_{k-1}) = cos(x_{k-1}) + sin(x_{k-1})$ and $h(x_k) = x_k$. We suppose that system state and measurement are time invariant. We set manually the quantities $Q = 0.5$ and $R = Q \times 2.4$ and the estimation is quite perfect. The particle filter is run with a set of 200 particles. A look for the state equation of the particle filter show that the value of this state always lays in the interval $[-2; 2]$. Thus the measurement equation should evolves in the same tendency as that is the state which generate the measurement. So to make the state and the measures vary in the same level, one has to normalize the vector of measurements by its mean; this does not have any drawback for the generality. We use the same pre-prepared data for the Kalman filter to make the two methods perform with the same dataset. This situation should be avoid if the state of the particle filter should be modeled by the equation $f(x_{k-1}) = A_{k-1}cos(x_{k-1}) + B_{k-1}sin(x_{k-1})$, with convenient values for the needed parameters A and B.

6.4 Results and Discussion

The first result of our study is devoted to having a filtered estimate of the noisy measurements. The approach shows the ability of the particle filter to estimate the state of the system under noisy measurements, as can do also the Kalman filter. However, the estimates obtained by the Kalman is more smoothed, see Fig. 1. In Table 1, we show the values of the accuracy criterion which say that the Kalman filter give more error than the Particle filter. Also, we show also, on Fig. 2 the ability the particles to capture the evolution of the system state over time. If the filter is badly calibrated, there will be a severe deviation of the particles from the true state and the results will be bad. To see the dynamics on the system state, the Fig. 3 gives the evolution of the density of the system state generated by the particle filter.

After filtering for the purpose of entropy reduction, the resulting measurement is much less noisy than the original observations and thus, this filtered signal can be analyzed for the scope of anomaly detection. We do anomaly detection based on data clustering. So, we suppose that anomalies might be rare and might happen on a few number of clusters. For both methods, the GMM framework has found three (3) clusters we have plug into the HMM framework to form two (2) states. To build properly the GMM, one has to put attention on the variance for each cluster. If we calibrate a set of r models, we choose the one which generate the lowest variance fr each class. For the HMM case, we talk the model with the transition matrix for which the different states all well-separated, i.e. with the maximum likelihood estimates. Then, as we did in [13] for the Kalman technique, we have applied the detection

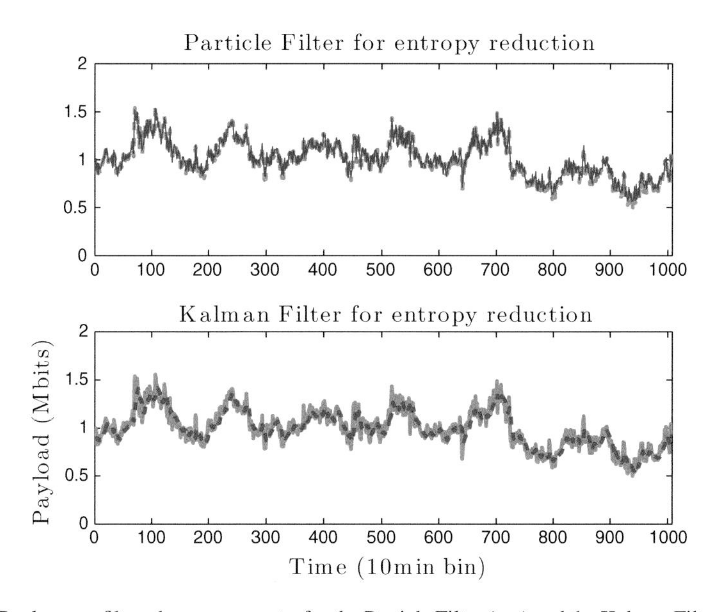

Fig. 1 Real versus filtered measurements, for the Particle Filter (*top*) and the Kalman Filter (*down*)

Table 1 Comparison of the different performance criterion for the different prediction models, Kalman Filter (KF) and Particle Filter (PF)

	RMSE (%)	MBE (%)	MAE (%)	Link
PF	0.0116	0.0050	0.0093	1
KF	0.0131	0.0094	0.0891	
PF	0.0104	0.0046	0.0089	2
KF	0.0926	0.0098	0.0709	
PF	0.0095	0.0040	0.0081	5
KF	0.0397	−0.00040	0.0294	
PF	0.0097	0.0050	0.0087	41
KF	0.0765	0.0086	0.0585	

The errors are much more larger the KF than for the PF

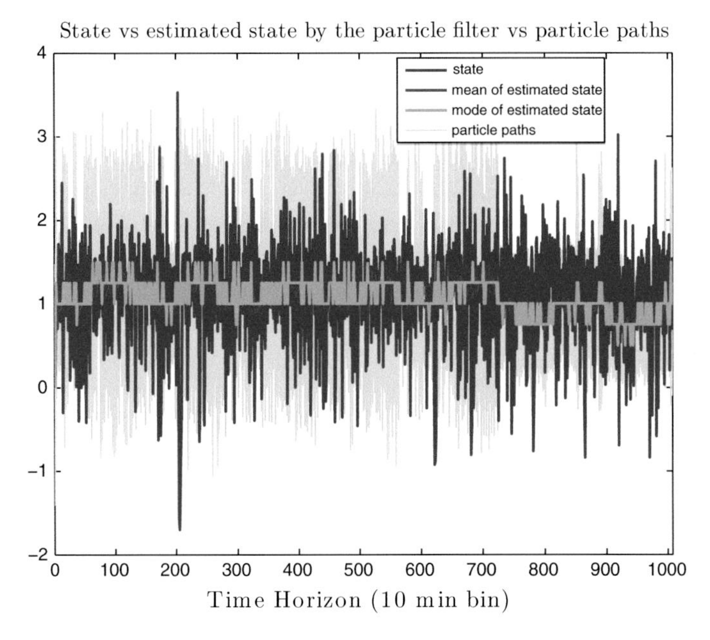

Fig. 2 Evolution of the particle filter state density

procedure in the part (data coming from some GMM classes which certainly belong to some HMM states) of the residual after removing the data which can be taken as a zero mean gaussian process. To do that, we apply the Viterbi decoder to have the best states sequence which capture the evolution of the data. Then, by observing the emission probability matrix, the state which contains data from the cluster(s) with mean closed to zero is discarded. Here, we apply the same procedure directly on the filtered observations, for the particle filter. The results are in Fig. 4. After retrieving the meaningful part of the data for anomaly detection, we apply a threshold. This threshold is built on the variance of the data and we have an interval $[a; a+\sigma]$ where

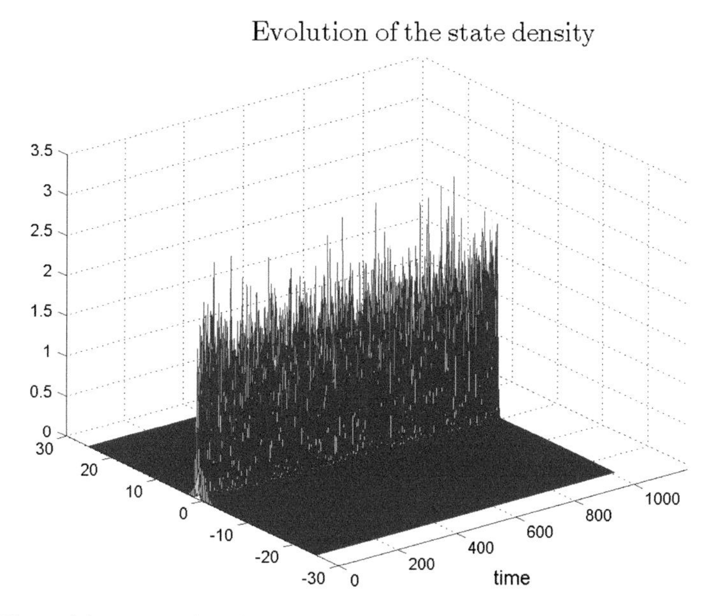

Fig. 3 The particles capture the trajectory of the estimated system state with its mean and mode

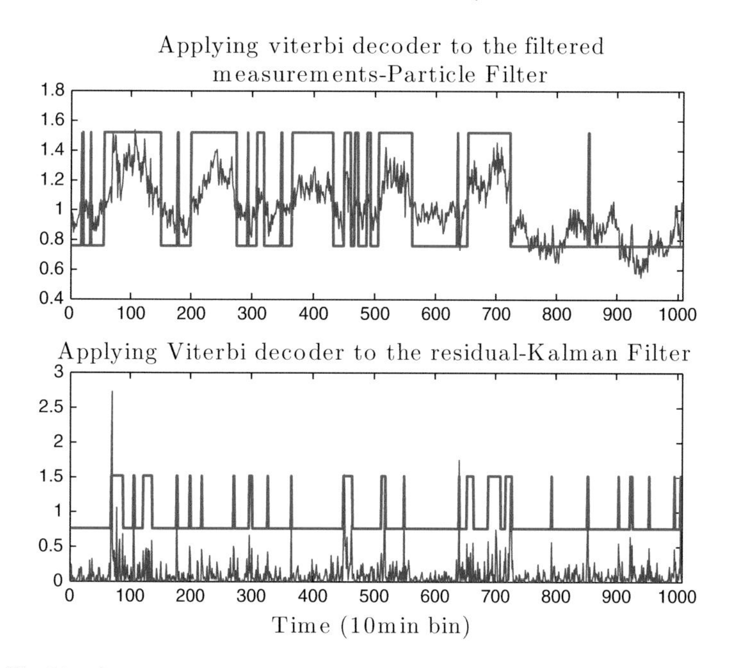

Fig. 4 Viterbi path

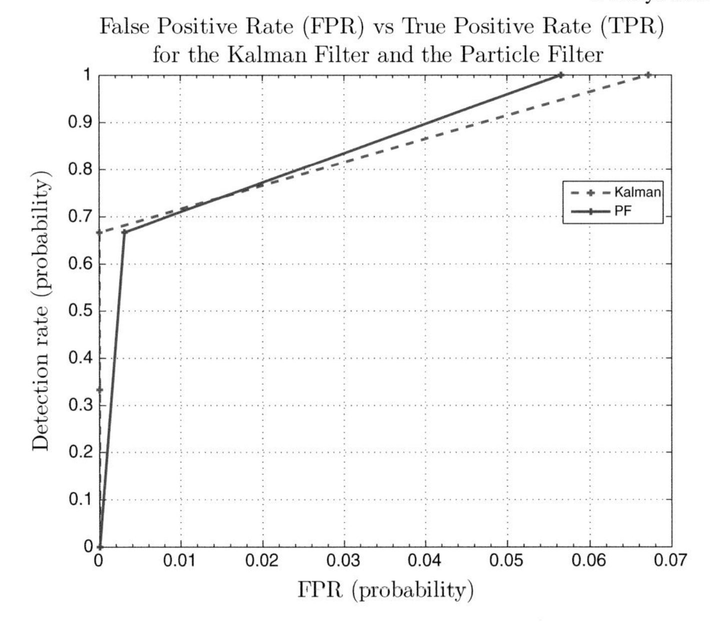

Fig. 5 ROC curve: detection rate versus false positive rate

a is the mean of the data and σ the variance of the data. All the values of the threshold are positive since we have squared the data.

The performance of an anomaly detector can be viewed in terms of detection rate and false positive percentage. A false positive is detected whenever the system leverages an alert where there's no really anomalous event. On the other side, a false negative is identified when the system is not able to detect an event which is a true anomaly. So by deriving the receiver operating characteristics (ROC) curve (with in x-axis the rate of false positive and in the y-axis the percentage of detection of true anomalies), we can judge the capacity of a anomaly detector. The ROC curves in Fig. 5 show clearly that the two methods perform with good performance. In contrast, the Kalman filter gives better results from a detection rate of 0–70 % than the particle filter. For a dection rate of 66.67 % the FPR is of 0.001 % and 0.003 % respectively for the Kalman and the Particle filters. Above 90 %, we observe the reverse tendency. as we can explore in the graph, for a detection rate of 100 %, the particle filter achieves a FPR of 0.056 % while the Kalman filter gives 0.067 %.

7 Conclusion

In this work, we have developed a particle filter to model the normal behavior of a system defined for the purpose of anomaly detection in communication networks. Many anomaly detector techniques rely on linear dynamical system for which stationarity of the underlying state system is assumed without no rigorous study of this belief. So, we believe that the particle filter can be an alternative to the Kalman filter in situations where we can not consider the system state as a linear process.

Generally, in normal conditions when we are in presence of a secure network, anomalies might be rare and targeted to local points or areas (time intervals). Techniques that use anomaly detector based on clustering are built on the assumption that the anomalous events does remain in a few number of classes. So, it is a convenient way to applying the detection procedure to a set of identified clusters to make the technique robust and fast enough to perform the calculations. The Viterbi decoder is a powerful algorithm which can be used suitably to discover these classes and separate them into normal and abnormal classes. By observing the emission probability matrix, we can extract the cluster in each HMM state and the data in the state(s) with mean close to zero are considered as "normal". An important issue when calibrating a particle filter is the initial parameter settings; one has to turn carefully the initial parameters, namely the state and measurement error covariance matrices and the number of particles needed to have the best trajectory of the particle paths in order to capture the state density evolution. We have used in this work an heuristic method based applying a threshold to the data corresponding the abnormal classes to tracking the volume anomalies. We hope to pursue this study in order to coupling the HMM-based tracking scheme with a more elaborate detection algorithm.

References

1. Bergman N., Recursive Bayesian estimation: Navigation and tracking applications, Ph.D. dissertation, Linkoping University, Linkoping, Sweden (1999)
2. Carpenter, J., Clifford, P., Fearnhead, P.: Improved particle filter for nonlinear problems. Proc. Inst. Elect. Eng, Radar Sonar Navig. (1999)
3. Crisan, D., Del Moral, P., Lyons, T.J.: Non-linear filtering using branching and interacting particle systems. Markov Process. Relat. Fields **5**(3), 293–319 (1999)
4. Del Moral, P.: Non-linear filtering: interacting particle solution. Markov Process. Relat. Fields **2**(4), 555–580
5. Doucet A., de Freitas J.F.G., Gordon N.J.: An introduction to sequential Monte Carlo methods, in sequential Monte Carlo methods in practice. In: Doucet, A., de Freitas, J.F.G., Gordon, N.J. (eds.) New York, Springer (2001)
6. Doucet, A.: On sequential Monte Carlo methods for Bayesian filtering, Department of Engineering, University of Cambridge, UK, Technical report (1998)
7. Gordon, N., Salmond, D., Smith, A.F.M.: Novel approach to nonlinear and non-Gaussian Bayesian state estimation. Proc. Inst. Elect. Eng. F **140**, 107–113 (1993)
8. Lakhina, A., Crovella, M., Diot, C.: Characterization of network-wide traffic anomalies. In: ACM Sigmetrics (2004)
9. Lakhina, A., Crovella, M., Diot, C.: Diagnosing network-wide traffic anomalies. In: SIGCOMM 2004: Proceedings of the 2004 conference on Applications. technologies, architectures, and protocols for computer communications, pp. 219–230. ACM Press, New York, NY, USA (2004)
10. Liu, J.S., Chen, R.: Sequential Monte Carlo methods for dynamical systems. J. Amer. Stat. Assoc. **93**, 1032–1044 (1998)
11. MacCormick J., Blake A.: A probabilistic exclusion principle for tracking multiple objects. In: Proceedings of the International Conference on Computer Vision, pp. 572–578 (1999)

12. Maybeck, P.: Stochastic Models, Estimation and Control, vol. 2. Academic Press (1982). Using MATLAB. Wiley Interscience (2001)
13. Ndong, J., Salamatian, K.: A robust anomaly detection technique using combined statistical methods. In: CNSR 2011, IEEE Xplore, pp. 101–108, May 2011. ISBN: 978-1-4577-0040-8
14. Ndong, J., Salamatian, K.: Signal Processing-based Anomaly Detection Techniques: A Comparative Analysis. In: The Third International Conference on Evolving Internet. INTERNET 2011. ISBN: 978-1-61208-141-0
15. Ndong, J.: Anomaly Detection: A Technique Using Kalman Filtering and Principal Component Analysis. ATAI NTC 2012 GSTF 2012
16. Ripley, B.: Stochastic Simulation. Wiley, New York (1987)
17. Sanjeev Arulampalam, M., Maskell, S., Gordon, N., Clapp, T.A.: Tutorial on particle filters for online nonlinear/non-gaussian Bayesian tracking. IEEE Trans. Signal Process. **50**(2) (2002)
18. Shumway, R.H., Stoffer, D.S.: An approach to time series smoothing and forecasting using the EM algorithm. J. Time Ser. Anal. **3**(4)
19. Sumway, R.H., Stoffer, D.S.: Dynamic linear model with switching. J. Am. Stat. Assoc. **86** (1991)

The Marginal Benefit of Monitor Placement on Networks

Benjamin Davis, Ralucca Gera, Gary Lazzaro, Bing Yong Lim and Erik C. Rye

Abstract Inferring the structure of an unknown network is a difficult problem of interest to researchers, academics, and industrialists. We develop a novel algorithm to infer nodes and edges in an unknown network. Our algorithm utilizes monitors that detect incident edges and adjacent nodes with their labels and degrees. The algorithm infers the network through a preferential random walk with a probabilistic restart at a previously discovered but unmonitored node, or a random teleportation to an unexplored node. Our algorithm outperforms random walk inference and random placement of monitors inference in edge discovery in all test cases. Our algorithm outperforms both methodologies in node inference in synthetic test networks; on real networks it outperforms them in the beginning of the inference. Finally, a website was created where these algorithms can be tested live on preloaded networks or custom networks as desired by the user. The visualization also displays the network as it is being inferred, and provides other statistics about the real and inferred networks.

1 Introduction

The exploration of complex networks is a continuously evolving study as technology progresses and networks change. In today's world, there are many networks that are unknown. How do we gain insight into these unknown networks without having to traverse every vertex and edge within the network? Is there a way to place monitors at different areas of the network to gain this insight? The objective of this paper is to explore the topic of monitor placement on network vertices in an attempt to gain insight into the true network topology.

B. Davis · R. Gera (✉) · G. Lazzaro · B.Y. Lim · E.C. Rye
Department of Applied Mathematics, Naval Postgraduate School,
Monterey, CA, USA
e-mail: RGera@nps.edu

H. Cherifi et al. (eds.), *Complex Networks VII*, Studies in Computational Intelligence 644, DOI 10.1007/978-3-319-30569-1_7

1.1 Motivation

In this paper we assume no knowledge of the true network, except for a rough approximation of the number of nodes so that the algorithm has a stopping condition. The algorithm used for network inference is tested on different synthetic and real-world complex networks of same order. The test networks are introduced in Table 1. Comparison of performance of an algorithm amongst these different test networks are normalized by looking at percentages, that is, the number of inferred nodes divided by the approximate number of total nodes, or the number of inferred edges divided by the approximate number of total edges.

In this paper we answer the following questions: As we increase the number of monitors placed up to 50 % of the nodes of the true network, what is the percent gain of new information inferred from the original network? At what percentage of monitor placement does the discovery of inferred network information begins to diminish towards a flat rate of change of the monitors discovered per monitor added? What is the minimum percentage of monitors needed to discover all nodes?

The website http://faculty.nps.edu/rgera/projects.html [4] was created where these algorithms can be tested live on preloaded networks or custom networks uploaded by the user. The visualization also displays the network as it is being inferred and that correlation to the percent edges and nodes inferred, and it provides other statistics about the real and inferred networks. Figure 1 shows two snapshots of the website, displaying the network as it is being inferred in green (top left of each figure), the leftover part of the network in white (top right), the plot of edges and nodes inferred (bottom left), and a heat-map of accuracy at each step in the inference (bottom right). Confidence intervals around the percent edges and nodes can be displayed by using multiple runs.

Table 1 Overview of the discovered data

Metrics	GR	ER	BA	FB
Node count (True network)	5242	5242	5242	4039
Edge count (True network)	14496	14496	15717	88234
Node count ($p = 0$)	4387	5083	5225	4002
Edge count ($p = 0$)	12598	12373	1418	82378
Node count ($p = 1$)	3823	5078	5223	3935
Edge count ($p = 1$)	12598	12358	14432	82179
Node count (Ideal)	5182	5201	5242	4039
Edge count (Ideal)	14348	13864	15675	85485
Node count (RW)	4491	4924	5162	3550
Edge count (RW)	12095	10976	13370	75971
Node count (RP)	4746	5090	5056	4009
Edge count (RP)	10530	10898	11747	66643

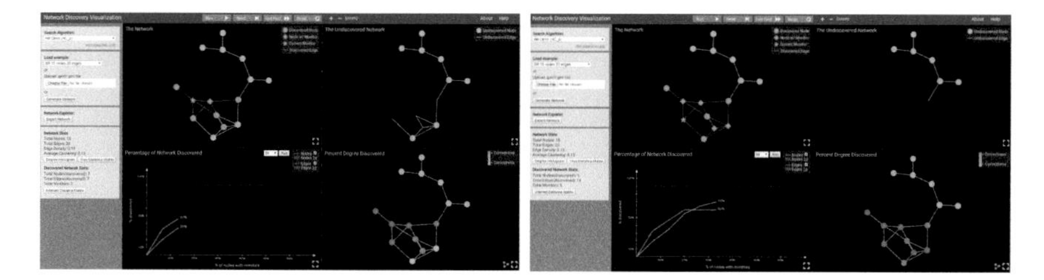

Fig. 1 Two steps in the inference of an Erdős-Rényi network from [4]

1.2 Related Work

Inferring a network can be done with no knowledge of the network at all (other than some random starting node), with partial information collected from network devices (such as knowledge of some of the nodes present in the network), or with complete information (in which case one could use the current knowledge to further monitor the network, or to re-infer an evolving network). Bliss, Danforth and Dodds [3] present recent techniques of inferring the topology of complex networks. These techniques are based on sampling nodes, sampling edges, the exploration of networks using random walks, or snowball sampling based on chain referral sampling ([2, 5]). Of course, the most relevant question is measuring the inferred network against the true network: random edge selection, depth and breath first search graph traversal, do not perform well overall; simple uniform random node selection performs surprisingly well; the best performing methods are based on random-walks starting at an arbitrary seed node (with the added probability of p at each node to teleport out of the random walk to the seed node or another arbitrary node) [5]. High degree nodes play the important role of hubs in communication and networking, and different local search strategies in power-law graphs that have costs scaling sublinearly with the size of the graph were introduced in [1]. However, the monitors in this paper infer more than just the node and edge incident with it, and thus the techniques perform differently.

Other current techniques not necessarily using complex networks are based on differential equations given one observation of one collective dynamical trajectory [11], statistical dependence between observations [14], as well as machine learning based on frequency of small subgraphs [7]. Extensions to multilayered networks have recently been published in [12, 13].

2 Preliminaries

We define a monitor to be able to see the node where it is placed, the edges incident to it, its neighbors, and possessing the ability to detect the degree and labels of its neighbors (the labels of the true topology as it is being inferred). For example, if

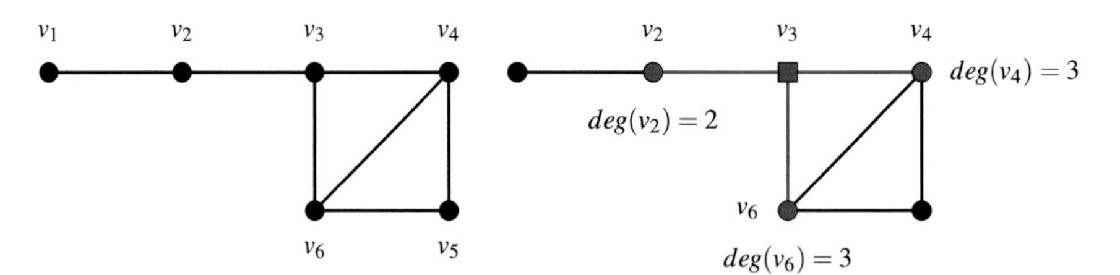

Fig. 2 A graph and a monitor placed at node v_3

each of two monitors i and j individually detect node k, they identify that it is the same k. We introduce this formally below.

Definition 1 We say that a *monitor on node i detects a node j* if (a) $d(i, j) \leq 1$, and (b) i knows the label of j and the deg j. A *monitor on node i detects an edge ij* if i and ij are incident.

Notice that a monitor always detects its closed neighborhood $N[v]$, but it infers more than just its neighbors. This is the idea used behind the domination number, $\gamma(G)$, in graphs introduced by Ore [10]. We say that a vertex *dominates* itself and its neighbors. Recall that a *dominating set* is a subset of the nodes such that each vertex of $V(G)$ is dominated by some vertex in the dominating set. The *domination number* is the cardinality of a minimum dominating set of G. The domination number could definitely be used to monitor a network, if the network is known. But in our approach of discovering the network, this is not useful since we do not assume to have much knowledge of the network (Fig. 2).

The k-Vertex Maximum Domination, introduced by Miyano and Ono in [8], is the parameter that gives the ideal placement of monitors if complete network information is known. Given a positive integer k, k-Vertex Maximum Domination (k-MaxVD) finds a subset DN of the nodes with size k that maximizes the cardinality of dominated nodes. That is, maximize $\cup_{v \in DN} N[v]$. Note that this optimization may produce a dominating set for some values of k, but does not need to, because in general not all nodes in the network are dominated. In [8], the authors show that a simple greedy strategy achieves an approximation ratio of $1 - \frac{1}{e}$ for k-MaxVD, and this approximation ratio is the best possible for k-MaxVD unless P = NP. We thus plot our inference algorithms against a greedy approach as an upper bound, and a random placement and a random walk as a lower bound on the performance of the algorithms. We refer to [9] for additional terminology not included in this paper.

3 Methodology

In this section we describe the approach used in placing monitors to infer an unknown network. We create a hill-climbing algorithm starting at some random node, with a probabilistic restart. Our algorithm first picks an initial "seed" node at random to place the first monitor. The monitor discovers the labels of its neighbors and incident edges to the monitored node. Next, the highest degree node neighbor to the monitor is chosen for the next monitor. If multiple highest degree neighbors exist, one is chosen at random.

If the process attempts to place a monitor at a node where a monitor already exists, then a stopping condition is reached. The next "seed" node could be either a previously unseen node that is discovered at random, i.e. it teleports (when $p = 0$), or the next highest degree node that was previously discovered and not used as a monitor (when $p = 1$), or a combination of both approaches (when $0 < p < 1$).

We present an initial bound on the number of monitors needed for network inference based on our algorithm. The best case is if the network topology is star. Either the first or second monitor would be placed at the center node in the best case. The worse case scenario is when the graph is a path, and the first monitor is placed at a leaf node. In that case, it would take $n - 1$ monitors to discover all n nodes of the network. We thus have the following remark, and the bounds are sharp given by the star and path described above: $1 \leq num\_monitors \leq n - 1$.

Algorithm 1 Hill-Climbing: High-degree neighbor with restart by teleportation or large seen degree

p, a given probability
$monitor \leftarrow$ randomly chosen from the network
$seen\_nodes\_list \leftarrow \emptyset$
$inferred\_graph \leftarrow \emptyset$
while 50% of the nodes in the network unmonitored **do**
  Add $monitor$ to $seen\_nodes\_list$
  Add all edges and nodes attached to the $monitor$ to the $inferred\_graph$
  Add neighbors of the $monitor$ that have not yet been discovered to the $seen\_nodes\_list$
  $highest\_deg\_node \leftarrow$ neighbor of $monitor$ with highest degree
  if $highest\_deg\_node$ does not have a monitor **then**
    $monitor \leftarrow highest\_deg\_node$
  else
    With probability $(1 - p)$, choose $monitor \leftarrow$ node randomly chosen from the complement of the $inferred\_graph$
    Otherwise, $monitor \leftarrow$ node with max degree in $seen\_nodes\_list$

4 Results and Discussion

Table 1 presents general information regarding the four data sets used in this paper: One Erdős-Rényi (ER) network, one Barabási-Albert (BA) network, one Facebook (FB) network and one General Relativity collaboration (GR) network. The real networks are from the Stanford large network data set collection [6]. The node count, edge count and number of components is shown.

The performance of each algorithm is shown and discussed for the average of the 50 trials. We plot our inference algorithms against a greedy approach as an upper bound (called *Ideal* shown in black), and two lower bounds shown in different shades of freesia representing Random Placement (RP) and Random Walk (RW). The monitors for all choices of Ideal, RW and RP are the same as we introduced for our research.

4.1 General Relativity Collaboration Network

The General Relativity collaboration network is comprised of 5242 nodes with 14496 edges, where an edge connects two nodes representing authors who have published a scholarly article together. This network consists of 355 distinct components. Figure 3 displays our inference of this network.

In this network, we achieve the best results using our algorithm with $p = 0$, outperforming all other p values, as well as the random walk (RW) and random placement (RP) strategies until about 20 % of the nodes are monitored. After the 20 % mark, RP captures a higher percentage of the nodes in the underlying network. The success of $p = 0$ initially and RP afterward is likely attributable to their preference for jumping to distinct topological components, thus capturing topology unlikely to be “seen” by inference algorithms that tend to stay within a component. RW and our algorithm when $p = 1$ tend to exhibit this “component-bound” behavior. In terms of edges, our algorithm with all p values tested ($p = 0, 0.25, 0.5, 0.75, 1$) discovered significantly more edges than the RW and RP inference algorithms as the number of monitors increased. We believe this effect is attributable to the preference for higher degree neighbors when selecting the next monitor. Neither the RW or RP algorithms

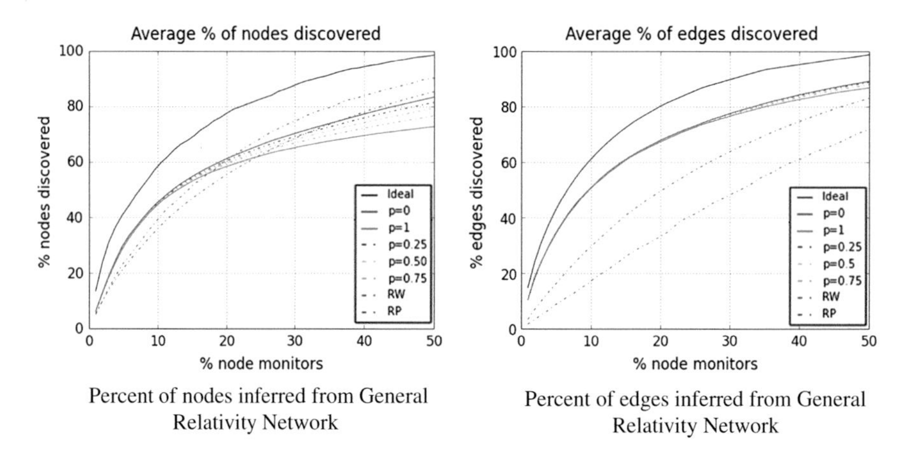

Percent of nodes inferred from General Relativity Network

Percent of edges inferred from General Relativity Network

Fig. 3 General relativity network: percent of nodes and edges in the inferred graph

prefer high-degree neighbors when selecting a successor monitor, which contributes to their under-performance in edge discovery. Figure 3 displays this difference in edge discovery.

4.2 Erdős-Rényi Random Graph

In this section, we examine the results of our inference on an Erdős-Rényi random graph, of comparable order and size to the collaboration network studied above. Of note, however, our Erdős-Rényi graph consists of only 19 connected components, compared to 355 in the General Relativity collaboration network. Figure 4 displays the results of our inference trials.

When $p = 0$, 7 of the 12 connected components are discovered, accounting for 98 % of the network nodes when 50 % of the nodes are monitored. The rate of discovery is quite high initially, with roughly 80 % of the nodes discovered after approximately 20 % of the nodes in the network are monitored. When $p = 1$, we achieve nearly identical results in terms of nodes discovered with 50 % of the nodes in the network monitored, and a slightly higher number of edges inferred (11804 vs. 12455 for $p = 0$ and $p = 1$, respectively.) Interestingly, in the $p = 1$ case, all nodes and edges discovered were contained within a single component. This reinforces the "component-bound" behavior of $p = 1$, and poses an interesting question to a potential customer of our algorithm: given an approximately equivalent amount of topological inference, is the discovery of more components within a network more or less desirable? We believe there are cases to be made for each elsewhere; here, we merely highlight this distinction.

Our investigation of variable p values are bounded by $p = 0$ and $p = 1$. Due to the tightness of the limiting p values, the variable p values do not provide much

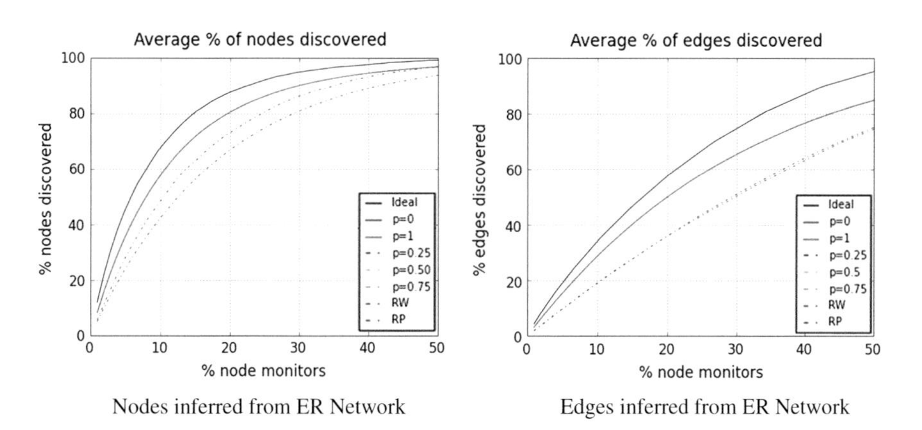

Nodes inferred from ER Network

Edges inferred from ER Network

Fig. 4 ER random networks: percent of nodes and edges in inferred graph

additional value, besides highlighting increased component discovery tendencies of low p and additional edge discovery as p increases.

4.3 Preferential Attachment Model: Barabási Albert Networks

Our graph inference trial involving a Barabási Albert-model graph was performed on a network consisting of 5242 nodes and 15717 edges, matching the number of nodes in the General Relativity example.

By construction, this network is connected, and it has hubs, unlike the General Relativity and the Erdős-Rényi Random Graphs. Due to the propensity of high degree hubs to form in the Barabási-Albert network construction model, our algorithm captures a large percentage of the nodes in the ground truth topology using relatively few monitors regardless of the choice of p value. This is evident in Fig. 5, in which the inference results for different values of p overlap throughout, significantly outperforming the random placement and random walk inferences.

This effect is due to hubs being discovered within a couple of steps from the seed, and selected for monitor placement early in the algorithm's execution. Further, we can see a diminishing return on investment as the number of monitors placed in the original graph increases, both on edges and nodes as the hubs are close to each other.

As we increase the number of monitors placed from zero up to 50 % of the nodes in the true network, the percent gain of new information per new monitor added quickly tends toward zero. In terms of nodes discovered, the derivative of the function given by our our curve in Fig. 5 decreases from a maximum of about 0.5% marginal gain at the first fifty-monitor step to about 0.1% marginal gain nodes discovered when 20% of the nodes in the graph are monitors.

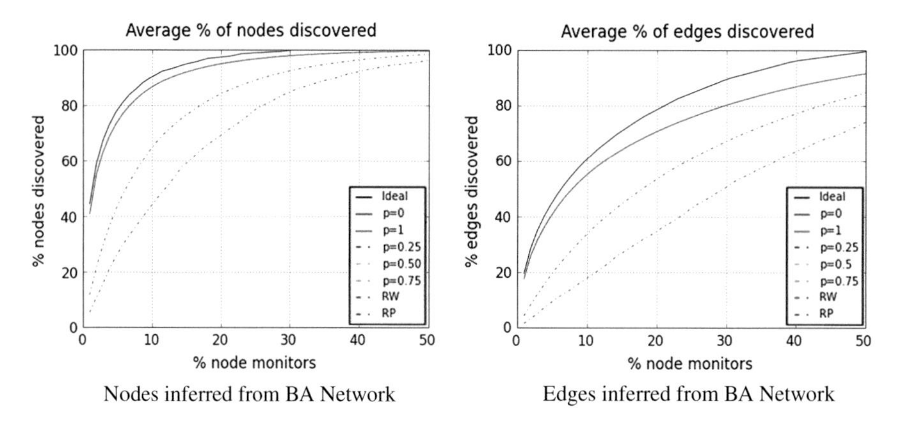

Nodes inferred from BA Network

Edges inferred from BA Network

Fig. 5 BA network: percent of nodes and edges in the inferred graph

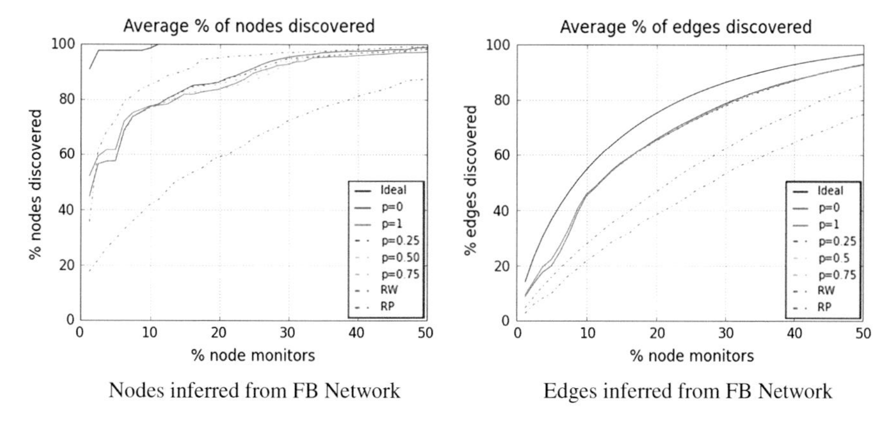

Nodes inferred from FB Network

Edges inferred from FB Network

Fig. 6 FB network: percent of nodes and edges in the inferred graph

4.4 *Facebook Network*

The final network we consider is a Facebook ego network from [6], consisting of 4039 nodes and 88234 edges. It forms a connected graph, and of note, is much more highly connected than any network studied above. The results of our inference trials are presented in Fig. 6.

In terms of edges, not much difference exists between the $p = 0$ and $p = 1$ algorithms, and our algorithms clearly outperform RW and RP. When we consider node inference, however, some variation is observed. Well-defined hubs enable node inference very quickly with few monitors; however, the existence of distinct communities tends to create "steps", due to all nodes within a particular community being exhausted as monitors before new nodes can be discovered in a disparate community. Due to this effect, out of all the choices of p, the value of $p = 0$ performs best due to its tendency to restart inference at a randomly selected node. This allows the inferences that are based just on random walks and placements to outperform our inferences after a certain point.

As a concrete demonstration of this phenomenon, consider only the vertices of degree greater than 250. Figure 7a shows the entire network, and the plot of Fig. 7b shows the graph induced on the vertices with degree greater than 250 which reveals 3 components. Consider these 3 components as communities for the entire Facebook network. One cluster has only one hub, and the second cluster has a triangle of hubs. The third cluster is centered around a star of hubs. Recall that since the entire Facebook graph is connected, we know that these clusters must be connected to each other through lower degree vertices. Thus, interconnecting paths between the clusters of the Facebook Network must contain an edge that is incident to vertices that have a low degree (since these clusters are not connected). The affinity of core hubs for each other can be measured by the Pearson correlation coefficient as mentioned in [9]. The Pearson correlation coefficient for the Facebook network is computed as 0.064,

Facebook: Overall Network Facebook: Vertices with degree above 250

Fig. 7 Facebook network visualizations

which is not indicative of either strong assortative or disassortative mixing of the graph vertices. On the Facebook network, we discovered a vertex degree of 66 to connect the entire network. Thus, we discover the nodes as a step function since many high degree nodes need to be used as monitors before getting to the lower degree node connecting the clusters.

5 Conclusions

In this research, we introduced a hill-climbing algorithm that infers a network with no knowledge of the network other than random nodes to start (or restart) the algorithm. Our algorithm has a probabilistic restart once it wants to place a monitor on a node that is occupied by a monitor: when $p = 1$ the algorithm restarts at a large degree node that has been discovered, versus when $p = 0$ the algorithm restarts at a random node of the network, and there are all the choices in between for the variable $0 \leq p \leq 1$ as expected. The value of p is chosen before the algorithm starts.

We analyzed real and synthetic networks, and present an analysis based on 50 runs of the algorithm for several values of p concluding that there is very little difference between the algorithms when we are concerned with edges being inferred. If the inference of nodes is the main goal, it is interesting to see the clear difference between the real networks and the synthetic networks. On the synthetic networks, there is no difference between any of our algorithms and they outperform the random placement and random walk, being extremely close to the ideal case in the presence of hubs. On the real networks, we see lots of variance between the algorithms, our inferences outperforming the random walks and placement towards the beginning of the inference, at which point random placement performs better since it does not choose nodes in the same clusters or component as our algorithms do. This suggests that the current algorithms should be used for quick inferences with a few monitors. Also, on the real networks, we observed that if there are no random restarts, our algorithm infers the denser part of the graph in more detail.

A user that desires to infer an unknown complex network with this algorithm needs to know a rough estimate of the size of the network to define the budget of total monitors, which was set to $n/2$ in this research. Secondly, the user needs to have a goal of inferring nodes or edges. If the inferred nodes are the goal, then select $p = 0$, and for edges edges select $p = 1$. The variable parameter (probability of staying within the current component) p of the algorithm combines the two different kinds of search methodologies, namely edge-finding or node-finding allowing a better discovery of both nodes and edges on average.

6 Further Studies

It is assessed that the real world networks could be accurately correlated to

Open question 1: One possible extension for the detection algorithm would be to increase the capability of the detection monitor. Future work could consider a monitor that has the capability to detect a triangle, that is, the ability to detect neighboring vertices, the edges to neighboring vertices, and edges between those neighbors; or nodes at further steps from the monitors.

Open question 2: Another possible improvement is to combine algorithms after a certain number of steps, or to add restarts more often. This will avoid the step increases observed in the Facebook network due to the clusters of hubs.

Open question 3: The biggest improvement that the authors see is finding a way of comparing the topology of inferred networks to the true network that uses other metrics besides the percent nodes and percent edges discovered. This requires a different type of analysis complementing this article.

Acknowledgments The authors would like to thank the DoD for partially sponsoring the current research. We would also like to thank and acknowledge the Naval Postgraduate School's Center for Educational Design, Development, and Distribution (CED3) for creating the live visualization [4] for this project.

References

1. Adamic, L.A., Lukose, R.M., Puniyani, A.R., Huberman, B.A.: Search in power-law networks. Phys. Rev. E **64**(4), 046135 (2001)
2. Biernacki, Patrick: Waldorf, Dan: Snowball sampling: problems and techniques of chain referral sampling. Sociol. Methods Res. **10**(2), 141–163 (1981)
3. Bliss, C.A., Danforth, C.M., Dodds, P.S.: Estimation of global network statistics from incomplete data. PloS One **9**(10), e108471 (2014)

4. Gera, R.: Network discovery visualization: an analysis of network discovery (2015). http://faculty.nps.edu/rgera/projects
5. Leskovec, J., Faloutsos, C.: Sampling from large graphs. In: Proceedings of the 12th ACM SIGKDD international conference on Knowledge discovery and data mining, pp. 631–636. ACM (2006)
6. Leskovec, J., Krevl, A.: SNAP Datasets: Stanford large network dataset collection (2014). http://snap.stanford.edu/data
7. Middendorf, M., Ziv, E., Wiggins, C.H.: Inferring network mechanisms: the drosophila melanogaster protein interaction network. Proc. Natl. Acad. Sci. U.S.A. **102**(9), 3192–3197 (2005)
8. Miyano, E., Ono, H.: Maximum domination problem. In: Proceedings of the Seventeenth Computing: The Australasian Theory Symposium, vol. 119, pp. 55–62. Australian Computer Society Inc (2011)
9. Newman, Mark: Networks: An Introduction. Oxford University Press Inc, New York (2010)
10. Ore, O.: Theory of graphs. Am. Math. Soc. Colloq. Publ. **38**, (1962)
11. Shandilya, S.G., Timme, M.: Inferring network topology from complex dynamics. New J. Phys. **13**(1), 013004 (2011)
12. Sharma, R., Magnani, M., Montesi, D.: Missing data in multiplex networks: a preliminary study. In: Third International Workshop on Complex Networks and their Applications (2014)
13. Sharma, R., Magnani, M., Montesi, D.: Investigating the types and effects of missing data in multilayer networks. In: IEEE/ACM International Conference on Advances in Social Networks Analysis and Mining (2015)
14. Tieu, K., Dalley, G., Grimson, W.E.L.: Inference of non-overlapping camera network topology by measuring statistical dependence. In: Tenth IEEE International Conference on Computer Vision, 2005. ICCV 2005, vol. 2, pp. 1842–1849. IEEE (2005)

Part V
Community Detection

Analysis of the Temporal and Structural Features of Threads in a Mailing-List

Noé Gaumont, Tiphaine Viard, Raphaël Fournier-S'niehotta, Qinna Wang and Matthieu Latapy

Abstract A link stream is a collection of triplets (t, u, v) indicating that an interaction occurred between u and v at time t. Link streams model many real-world situations like email exchanges between individuals, connections between devices, and others. Much work is currently devoted to the generalization of classical graph and network concepts to link streams. In this paper, we generalize the existing notions of intra-community density and inter-community density. We focus on emails exchanges in the Debian mailing-list and show that threads of emails, like communities in graphs, are dense subsets loosely connected from a link stream perspective.

1 Introduction

Exchanges in a mailing-list are often studied as complex networks: there is a link between two individuals if they exchange emails. In particular, communities in such complex networks capture groups of friends or close colleagues (individuals that exchange many more messages within the group than outside the group, typically) [2]. However, removing all time information has important consequences if one wants to study the dynamics of email exchanges.

N. Gaumont (✉) · T. Viard · Q. Wang · M. Latapy
Sorbonne Universités, UPMC Univ Paris 06, CNRS, LIP6 UMR 7606,
4 Place Jussieu, 75005 Paris, France
e-mail: noe.gaumont@lip6.fr

T. Viard
e-mail: tiphaine.viard@lip6.fr

Q. Wang
e-mail: qinna.wang@lip6.fr

M. Latapy
e-mail: matthieu.latapy@lip6.fr

R. Fournier-S'niehotta
CNAM, CEDRIC, Paris, France
e-mail: fournier@cnam.fr

H. Cherifi et al. (eds.), *Complex Networks VII*, Studies in Computational Intelligence 644, DOI 10.1007/978-3-319-30569-1_8

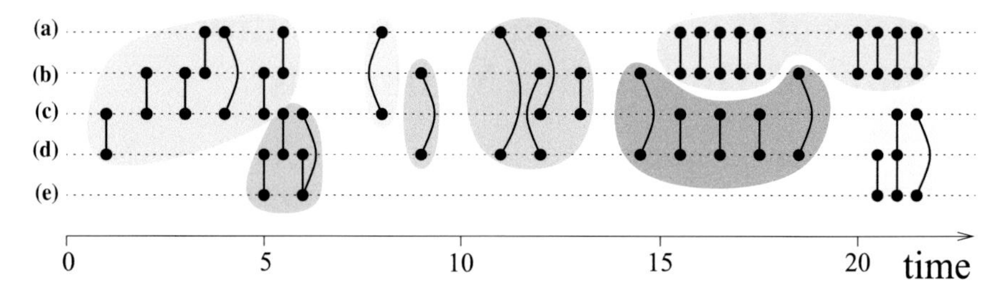

Fig. 1 An example of link stream representing email exchanges between individuals a, b, c, d and e, with threads represented by *colored areas*. For instance, at time 5, b and c exchange an email, as well as d and e. Threads are a priori dense series of exchanges involving a limited group of nodes during a limited period of time

In order to study those dynamics, one may label each link with the frequency of exchanges or the times at which they occur [5], but capturing both the structure and the dynamics of exchanges remains challenging. In particular, studying threads calls for methods that capture the temporal nature of interactions more accurately, without loosing the power of network analysis.

We propose here to model email exchanges directly as link streams, i.e. series of triplets (t, a, b) meaning that individuals a and b exchanged an email at time t. We then introduce notions that capture both the temporal and structural nature of these exchanges. We use a typical dataset obtained from a public mailing-list archive to illustrate our approach. We analyze this dataset using our model, with a special focus on the properties of threads within the whole archive. Our goal is to understand how the now classical concept of communities in complex networks may translate to threads in link streams representing email exchanges. Indeed, we expect the exchanges of a given thread to involve a specific set of individuals for a specific period of time, thus being dense from both structural and temporal points of view. This is illustrated in Fig. 1.

2 Dataset

Archives of exchanges in various mailing-lists are readily available on the web, and studying them provides very rich insights on various issues. They have the advantage of being publicly available in many cases, and some involve large amounts of users over long time periods.

A typical example is provided by Debian mailing-list [4]: it contains emails sent from over 51753 email addresses, over almost 20 years. In addition, exchanges in this mailing-list have been studied in the past [1, 3, 7]. Finally, this dataset provides the thread information for each message, that we can use as a ground truth. For all these reasons, we use in this paper the Debian mailing-list to illustrate and validate our approach.

More precisely, we crawled the Debian mailing-list web archive [4]. For each message m, we extract its author $a(m)$, the date $t(m)$ at which it was posted (converted into UTC time), and the message it is replying to $p(m)$ (through the IN- REPLY- TO entry), which has a corresponding author $a(p(m))$. This corresponds to an interaction between $a(m)$ and $a(p(m))$ at time $t(m)$ in the link stream. Some messages are not answers to any other message (they are directly sent to the mailing-list), and in this case we state that $p(m) = m$. Such messages are called *root* messages.

We capture the mailing-list from January 1st, 1996 to December 31st, 2014. We obtain a dataset $\mathscr{D}$ of $n = 722716$ emails sent from 51753 distinct email addresses.

Each root message m naturally induces a thread: it is the set $\mathscr{T}(m)$ of messages such that m belongs to $\mathscr{T}(m)$ and if a message m' is in $\mathscr{T}(m)$ then all messages m'' such that $p(m'') = m'$ also belong to $\mathscr{T}(m)$. In other words, $\mathscr{T}(m)$ contains exactly m, the answers to m, the answers to these answers, and so on. The focus of this paper is the study of structural and temporal features of these threads.

Our data contains incomplete threads: the ones that have an email in our dataset but began before and/or continued after the data collection period. Some threads also exhibit inconsistencies, for instance a reply has a smaller timestamp than the message it replies to. We remove those threads, as well as all threads that last for more than 2 years, or that start 2 years before the end of our data collection.

After this bias correction procedure, we obtain $n = 554233$ emails, involving 34648 distinct authors over a duration of 598532269 s (18 years, 11 months and 19 days) and 116999 threads.

3 Framework and Notations

Our goal is to study the structural and temporal properties of threads within a mailing-list archive. In order to do so, we propose a model of the data that captures both its temporal and structural nature, and allows for easy manipulation of threads.

We model our mailing-list archive as the link stream $D = (T_D, V_D, E_D)$ with $T_D = [\alpha, \omega]$, $V_D = \{a(m) : m \in \mathscr{D}'\}$ and $E_D = \{(t(m), a(m), a(p(m))) : m \in \mathscr{D}'\}$ where $\mathscr{D}'$ is the set of emails in our dataset after cleaning. In other words, a triplet (t, u, v) in E_D indicates that individual u answered to an email of individual v at time t.

Such a link stream naturally contains sub-streams: $L' = (T', V', E')$ is a sub-stream of $L = (T, V, E)$ if and only if $T' \subseteq T$, $V' \subseteq V$ and $E' \subseteq E$. In other words, all the interactions of L' also appear in L. Given a set of nodes S, we define the sub-stream $L(S)$ of L induced by S as the largest sub-stream of L such that all the links in $L(S)$ are between nodes in S.

Any link stream $L = (T, V, E)$ also induces a graph $G = (V_G, E_G)$ where $V_G = \{u : \exists t \in T, v \in V \text{ s.t. } (t, u, v) \in E\}$ and $E_G = \{(u, v) : \exists t \in T \text{ s.t. } (t, u, v) \in E\}$. In our case, the whole mailing-list archive induces the graph $G(D)$ among authors of emails, and each thread induces a sub-graph of $G(D)$.

In a graph $G = (V, E)$, a community structure is defined by a partition $C = \{C_i\}_{i=1..k}$ of V into k communities. In other words, $\bigcup_i C_i = V$ and $C_i \cap C_j = \emptyset$ whenever $i \neq j$. In a similar way, one may consider a link stream $L = (T, V, E)$ and a partition of its links into k sub-streams $P = \{P_i = (T_i, V_i, E_i)\}_{i=1..k}$. In other words, for any $(t, u, v) \in E$, there exists a unique j between 1 and k such that (t, u, v) is a link of E_j.

The threads in our email dataset are exactly a partition of the whole stream, which we denote by $\mathscr{T} = \{P_i\}_{i=1..k}$ where k is the number of threads and each P_i is a sub-stream representing a thread (with our notations above, there exists a message m such that $P_i = \mathscr{T}(m)$). See Fig. 1.

Notice that, although the threads are a partition of the whole stream, their induced graphs may overlap: some nodes and links of $G(D)$ belong to several sub-graphs $G(P_i)$. As a consequence, threads do not induce a partition of $G(D)$ into communities. Instead, one may see the partition of D into threads as a community structure, and this is the focus of our work.

Notice finally that we consider that links are undirected (i.e. $(t, u, v) = (t, v, u)$) and happen at an instant in time (regardless, for instance, of when the message is read). Taking into account the direction and duration of links is out of the scope of this work.

4 Basic Statistics

In this section, we present the basic statistics describing the threads in our dataset and the whole archive.

The most basic description of our data certainly is the number of links (i.e. emails) they contain, the number of distinct nodes (i.e. authors) involved, the number of distinct links they contain (distinct pairs of authors in direct interaction), and their duration (time from the first email to the last one). Figure 2 display the distribution of these values for each thread.

Although the largest thread lasts more than a year, most threads are contained within a few days (100000 s is a bit more than 24 h). Similarly, the largest thread involves 100 messages, though all intermediate sizes are represented in the dataset. Most threads are very short and involve less than 3 messages.

In order to gain more insight, we observe correlations between some of these basic statistics. Figure 3 (left) shows that thread duration and size are correlated (the larger a thread is, the longer it is likely to be); notice however that for small-sized threads, all types of durations are represented. Looking at the correlations between the size of threads and the number of distinct authors involved shows that threads nearly always involve more messages than authors. This is a typical feature of mailing-lists [1] and as such is dataset-dependent.

In a link stream $L = (T, V, E)$ with $T = [\alpha, \omega]$, we define, for all $(u, v) \in V \times V$, the maximal sequence $t_{uv} = (\alpha, t_0, \ldots, t_k, \omega)$ such that for all i between 0 and k,

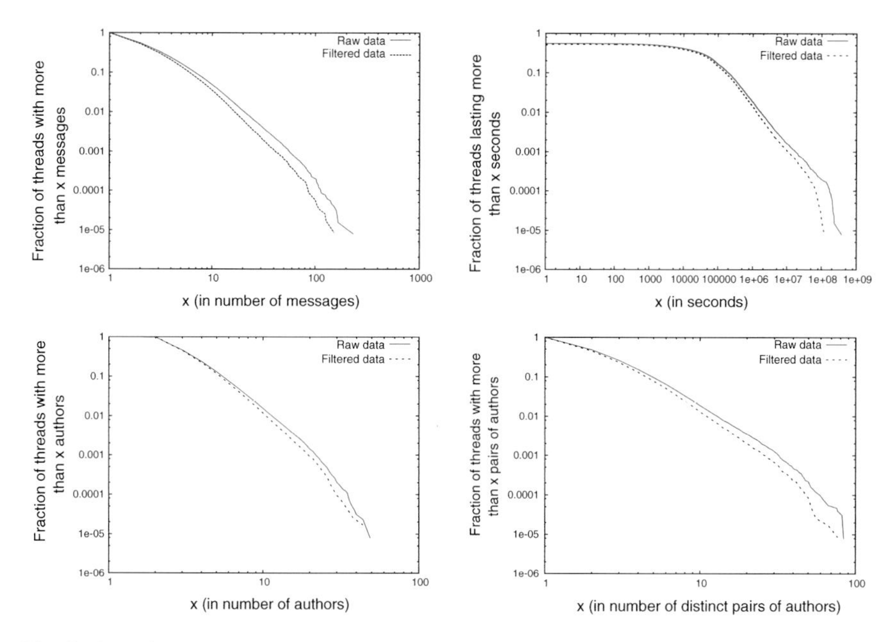

Fig. 2 Complementary cumulative distributions for basic statistics of our raw (*solid line*) and filtered (*dotted line*) datasets. *Top left* thread sizes (number of messages per thread); *top right* thread durations (time elapsed between the first and the last message of the thread); *bottom left* number of distinct authors; *bottom right* number of distinct pairs of authors

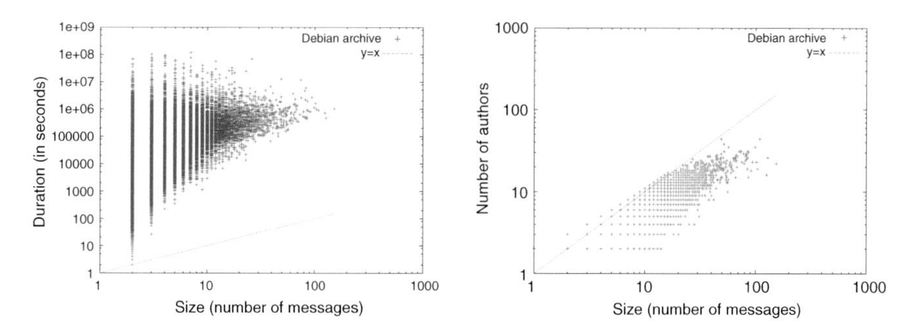

Fig. 3 *Left* Correlations between size and duration of threads. *Right* Correlations between size of threads and the number of authors involved

there exists $(t_i, u, v) \in E$, and for all i between 0 and $k-1$, $t_i \leq t_{i+1}$. In other words, t_{uv} is the ordered sequence of apparitions of the link (u, v) to which we add α and ω.

We further define $\tau(u, v) = (t_{i+1} - t_i)_{i=0..k+1}$ the sequence of intercontact times of a pair of u and v in V. In other words it is the series of times elapsed between two consecutive occurrences of a link between them. Figure 4 (left) shows the intercontact times distribution in the Debian mailing-list for all pairs of nodes (u, v).

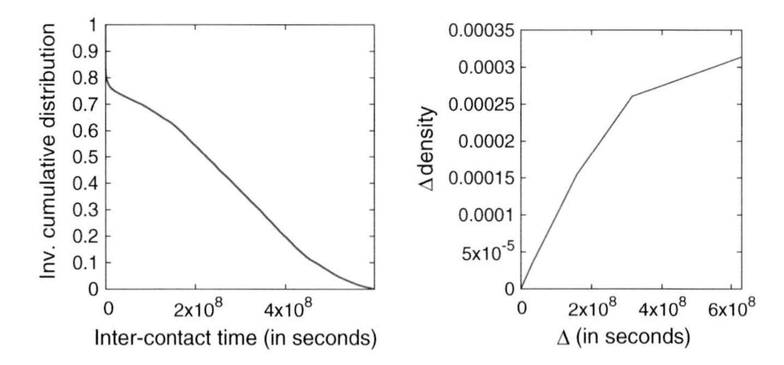

Fig. 4 *Left* Inter-contact times distribution in the Debian mailing-list dataset. *Right* Evolution of the Δ-density of the link stream for Δ from 1 s to 20 years

5 Interactions Within Threads

The key feature of communities is the fact that they form dense subgroups. This section is therefore devoted to the study of density of interactions within threads, from both structural and temporal point of views.

5.1 Density of Threads

In a graph, the density is the probability that two randomly chosen nodes are linked together. In other words, it captures the extent at which *all* nodes are directly connected to each other. The density of the graph $G(D)$ induced by our dataset is 3.139×10^{-4}.

In [6], we introduced the notion of Δ-density to capture a similar intuition in link streams, involving both structure and time. Indeed, given a duration Δ, the Δ-density of link stream L is the probability that a link appears between two randomly chosen nodes during a randomly chosen time interval of duration Δ. It captures the extent at which all nodes are directly connected to each other at least every Δ time units. Formally, it is defined as:

$$\delta_\Delta(L) = 1 - \frac{2 \cdot \sum_{u,v \in V, u \neq v} \sum_{t \in \tau(u,v)} \max(0, t - \Delta)}{|V| \cdot (|V| - 1) \cdot \max(0, \omega - \alpha - \Delta)}$$

where $\tau(u, v)$ denotes the inter-contact times between u and v, and α and ω are the start and end time of the link stream.

In order to study the Δ-density in our data, we first have to choose an appropriate Δ. We use here several values which capture email dynamics at different scales: $\Delta = 1$ min, 1 hour, 1 day, 1 week, 1 month, 1 year and 20 years (the whole duration of the dataset). Figure 4 (right) displays the evolution of the Δ-density of the stream for all theses values of Δ. It shows that the Δ-density is small for small Δs, and converges to the density of the graph induced by the email exchanges (in our case, 3.139×10^{-4}).

In Fig. 4 (right), the inflexion points give information on the values of Δ where the dynamics change. Still, looking at the density of the whole stream is very coarse and yields little information. A finer approach consists in looking at the Δ-density of relevant sub-streams. In our case, the threads between authors are a natural object to study.

5.2 Intra-thread Density

More globally, given a graph $G = (V, E)$ and a partition $C = \{C_i\}_{i_1..k}$ of V into k communities, the density within communities of C is captured by the *intra-community density*:

$$\frac{2 \cdot \sum_i |\{(u, v) \in E, u \in C_i \text{ and } v \in C_i\}|}{\sum_i |C_i| \cdot (|C_i| - 1)}$$

In other words, intra-community density is the probability that two nodes chosen at random in the same community are linked together.

In our case, this notion does not directly make sense: as already noticed, we do not have communities defined on $G(D)$ since the graphs induced by threads overlap. However, we extend the notion of intra-community density to link streams as follows. The intra-thread Δ-density is the probability that two randomly chosen authors contributing to the same thread are linked together within a randomly time interval of duration Δ, for a given Δ:

$$1 - \frac{2 \cdot \sum_i \sum_{u,v \in V_i, u \neq v} \sum_{t \in \tau_i(u,v)} \max(0, t - \Delta)}{\sum_i |V_i| \cdot (|V_i| - 1) \cdot \max(0, \omega_i - \alpha_i - \Delta)}$$

where V_i is the set of authors involved in thread P_i, α_i is the time of the first message in the thread (i.e. the minimal t such that there exists a $(t, u, v) \in E_i$), ω_i is the time of the last message in the thread (i.e. the maximal t such that there exists a $(t, u, v) \in E_i$) and $\tau_i(u, v)$ denotes the inter-contact times in P_i.

In our data, the inverse cumulative distribution of intra-thread Δ-densities are in Fig. 5 (left) for several values of Δ ranging from 1 min to 1 year. For each point on the x-axis, the plot gives the proportion of threads in the mailing-list that have an intra-thread Δ-density higher than x. As expected, the higher the Δ used, the higher the density is. However, there is no significant change between a Δ of 7 days and a Δ of 1 year.

Moreover, these distributions confirm that the interactions within threads are much denser (both structurally and temporally) than in the global mailing-list. Indeed, the median intra-thread Δ-density ranges from 2.69×10^{-4} to 0.28 while the link stream

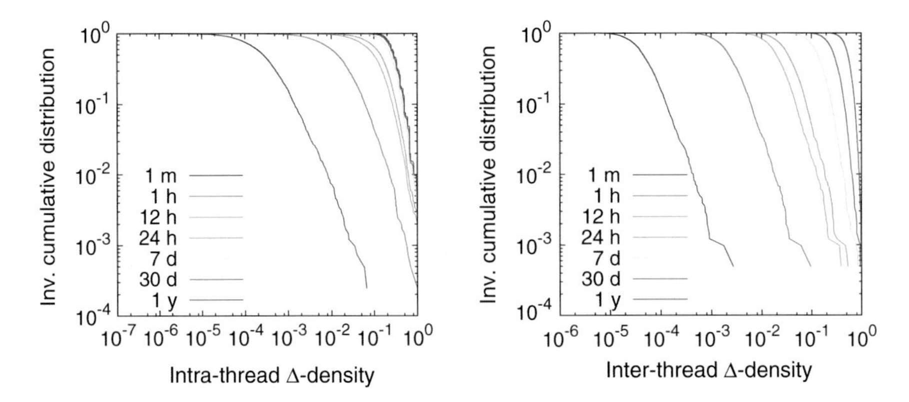

Fig. 5 *Left* Inverse cumulative distributions of values of intra-thread Δ-density for different Δs. *Right* Inverse cumulative distributions of values of inter-thread Δ-density for different Δs

Δ-density ranges from 1.05×10^{-10} to 3.42×10^{-5}. The intra-thread Δ-density typically is 10^5 times larger than the global Δ-density.

This shows that threads are indeed dense substreams in our link streams.

6 Relations Between Threads

In the previous section, we focused on structural and temporal properties *inside* threads, compared to the whole link stream. We now turn to the study of relations *between* threads.

6.1 Inter-thread Density

Let us first study the density of relations between threads in a way similar to above. Given a graph $G = (V, E)$ and a partition $C = \{C_i\}_{i_{1..k}}$ of V into k communities, the inter-community density is the probability that two nodes chosen at random in two different communities are linked together:

$$\delta^{inter}(C_i) = \frac{1}{|C|} \sum_{j,i \neq j} \frac{|\{(u, v) \in E \text{ s.t. } u \in C_i \text{ and } v \in C_j\}|}{|C_i| \cdot |C_j|}$$

Again, this notion does not directly make sense in link streams, as threads do not induce a partition of nodes. As a consequence, we introduce the inter-thread Δ-density as the probability that two randomly chosen nodes in different communities are linked together during a time interval of duration Δ chosen at random during the time duration of both threads.

Let us define the inter-thread substream between a thread P_i and a thread P_j: $L_{ij} = (T_{ij}, V_{ij}, E_{ij})$, with $T_{ij} = [\min(\alpha_i, \alpha_j), \max(\omega_i, \omega_j)]$, $V_{ij} = V_i \cup V_j$ and $E_{ij} = \{(t, u, v) : t \in T_{ij}, u, v \in V_{ij}, (t, u, v) \in E \backslash E_i \cup E_j\}$. In other words, this is the substream containing the links between nodes of P_i or P_j that are not involved in threads P_i and P_j. The inter-thread density between P_i and P_j is the Δ-density of L_{ij}. In order to obtain the inter-thread Δ-density of P_i to all other threads, we simply average the inter-threads Δ-densities of P_i and all other threads. More precisely:

$$\delta_{\Delta}^{inter}(C_i) = \frac{1}{|C|} \sum_{j, i \neq j} \delta_{\Delta}(L_{ij})$$

In our data, the inverse cumulative distribution of inter-thread Δ-densities are displayed in Fig. 5 (right) for different values of Δ. For each point on the x-axis, the plot gives the proportion of threads in the mailing-list that have an intra-thread Δ-density higher than x. Again, larger Δ correlates with larger Δ-densities. However, the inter-thread Δ-density does not plateau, even for large values of Δ. This is natural, since the number of links considered in the computation of the inter-thread Δ-density naturally grows with Δ.

In Fig. 6, the correlations between the inter- and intra-thread Δ-density are plotted for some values of Δ. As expected, intra-threads are denser than inter-threads. This relation holds as Δ is bigger, even though the difference between inter and intra thread Δ shrinks. Further experimentation shows that for $\Delta = 20$ years, the difference is non-existent. The figure is omitted for brevity. This is due to the fact that the bigger the Δ, the less the temporal characteristics of threads are important.

6.2 Graphs Between Threads

Relations between sub-streams L_i, $i = 1..k$, may have different forms, and in particular they have a temporal and a structural nature. In order to capture the temporal

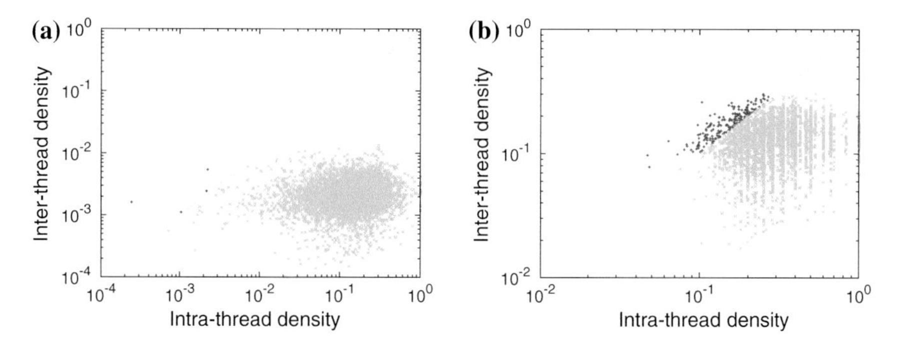

Fig. 6 Correlations between inter- and intra-thread densities for different values of Δ. **a** $\Delta = 1$ day **b** $\Delta = 1$ year

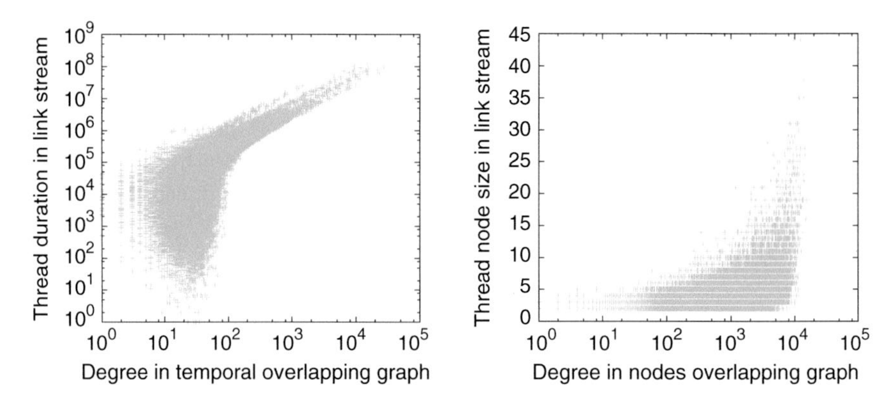

Fig. 7 *Left* Correlation between the degree in the time overlap graph X and the thread size. *Right* Correlation between the degree in the node overlap graph Y and the thread duration

relations between sub-streams, one may define the temporal overlap graph as follows: $X = (V, E)$ with $V = \{i, i = 1..k\}$ and there is a link (i, j) in E whenever P_i and P_j have a temporal intersection (i.e. $[\alpha_i, \omega_i] \cap [\alpha_j, \omega_j] \neq \emptyset$). Likewise, one may define the node overlap graph as follows: $Y = (V, E)$ with $V = \{i, i = 1..k\}$ again and there is a link (i, j) in E whenever there is a node v involved in both P_i and P_j (i.e. there exists a t, a t'n a u and a u' such that there is a link (t, u, v) in P_i and a link (t', u', v) in P_j.

The graphs contain 116999 nodes (the number of threads) and about 2 million edges for the temporal overlap graph and 63 millions for the node overlap graph. These graphs encode much information about relations between threads. For instance, the degree of node i in X is the number of threads active at the same time as P_i.

We display in Fig. 7 (left) the correlations between the degree in X and the thread size. There is a clear correlation between the thread duration and the degree in temporal overlap graph when threads have a duration of at least 10^5s. Also, it appears that some time up to 10^4 threads are present simultaneously as reflected by the maximal degree.

Figure 7 (right) shows the correlations between the degree in Y and the thread duration. The correlation is less clear between the thread node size and the degree in the node overlap graph. However, the trends appears: threads with a lot of participants have a high degree in the graph.

6.3 Quotient Stream

The *quotient* graph is another key notion for studying the relations between communities in a graph $G = (V, E)$. Given a partition $C = \{C_i\}_{i=1..k}$ of V into communities, in the quotient graph $\overline{G}$ each node i, $i = 1..k$, represents community C_i and there

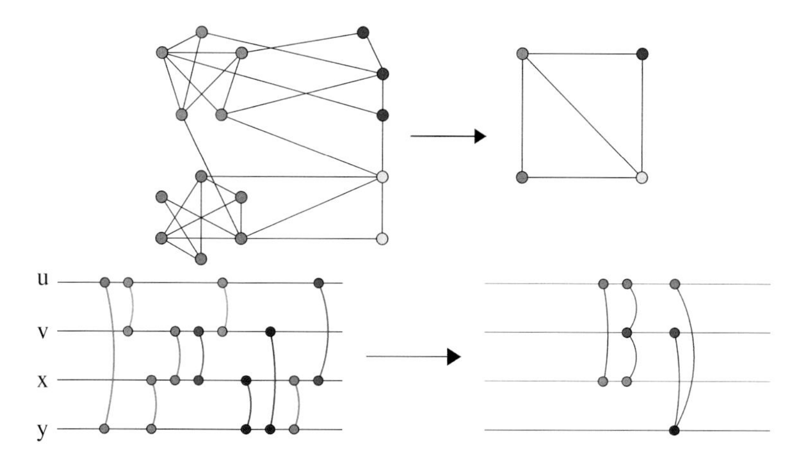

Fig. 8 *Top* An example of graph exhibiting communities and its corresponding graph quotient. *Bottom* An example of link stream with communities and its corresponding quotient stream

is a link between two nodes i and j, $i \neq j$, if there is a link between a node in C_i and a node in C_j in G. See Fig. 8 for an illustration. One may add on each link a weight indicating the number of links between communities. Clearly, the quotient graph captures relations between the communities under concern; for instance, its density indicates up to what point all communities have links between them.

To deepen our understanding of our data, we capture here both temporal and structural nature of relations between sub-streams. We define the quotient stream induced by a partition $P = \{P_i = (T_i, V_i, E_i)\}_{i=1..k}$ of link stream L as the stream $Q = (T_Q, V_Q, E_Q)$ such that $(P_i, P_j, t) \in E_Q$ if and only if there exists (u, v, t_1) in E_i, (u, v', t_2) in E_i and (u, v'', t) in E_j with $t_1 \leq t \leq t_2$. In other words, there is a

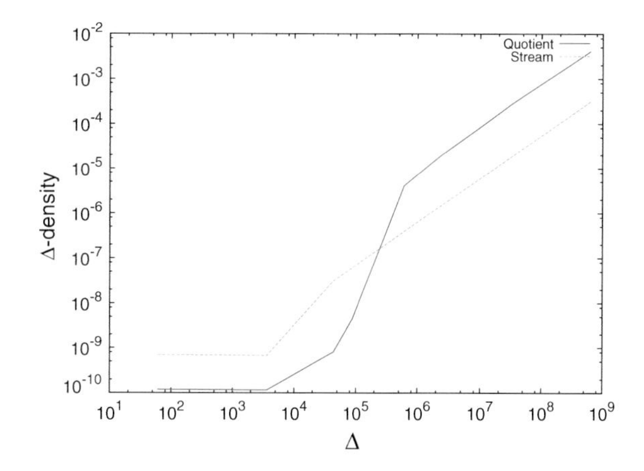

Fig. 9 Δ-density of the link stream and the quotient stream as a function of Δ, for $\Delta = 1mn$, $1h$, $12h$, $1d$, $37d$, $30d$, $1y$ and $20y$

node u that has a link within P_j occurring between two of its links in P_i. This means that u is involved in the two streams during the same time period.

The quotient stream induced by the threads in our dataset has 12281269 links and involves 68524 distinct nodes (i.e. threads). Since our dataset contains 116999 threads, this implies that 48475 threads are not in relation with any others.

Figure 9 shows the Δ-density of the quotient stream and the Δ-density of the original stream for different values of Δ. The quotient is not very Δ-dense, i.e. threads are not densely connected together, though it is slightly denser than the stream for large values of Δ. This is comparable to graphs.

Acknowledgments This work is supported in part by the French *Direction Générale de l'Armement* (DGA), by the Thales company, by the CODDDE ANR-13-CORD-0017-01 grant from the *Agence Nationale de la Recherche*, and by grant O18062-44430 of the French program *PIA—Usages, services et contenus innovants*.

References

1. Dorat, R., Latapy, M., Conein, B., Auray, N.: Multi-level analysis of an interaction network between individuals in a mailing-list. In: Annales des télécommunications, vol. 62, pp. 325–349. Springer (2007)
2. Fortunato, S.: Community detection in graphs. Phys. Rep. **486**(3), 75–174 (2010)
3. Sowe, S., Stamelos, I., Angelis, L.: Identifying knowledge brokers that yield software engineering knowledge in OSS projects. Inf. Softw. Technol. **48**(11), 1025–1033 (2006)
4. SPI. Debian mailing-list archive: https://lists.debian.org/debian-user/
5. Sun, J., Faloutsos, C., Papadimitriou, S., Yu, P.S.: Graphscope: parameter-free mining of large time-evolving graphs. In: Proceedings of the 13th ACM SIGKDD International Conference on Knowledge Discovery and Data Mining, KDD '07, pp. 687–696, New York, NY, USA. ACM (2007)
6. Viard, J., Latapy, M.: Identifying roles in an ip network with temporal and structural density. In: 2014 IEEE Conference on Computer Communications Workshops (INFOCOM WKSHPS), pp. 801–806. IEEE (2014)
7. Wang, Q.: Link prediction and threads in email networks. In: 2014 International Conference on Data Science and Advanced Analytics (DSAA), pp. 470–476. IEEE (2014)

Incorporation of Social Features in the Naming Game for Community Detection

Thais Gobet Uzun and Carlos Henrique Costa Ribeiro

Abstract The organization of individuals in groups or communities is an observed property of complex social networks and this structural organization emerges naturally due to the relationships built between people on a daily basis. We believe that the opinion exchange among individuals is a key factor to this community construction, given that sharing opinions bounds people together, and disagreeing constantly would probably weaken a relationship. In this work, we analyse three models of opinion exchange that uncover the community structure of a network, based on the Naming Game (NG), a classic model of linguistic interactions of agreement. The NG-based models applied in this work insert time-changing social features to the NG dynamics in order to form communities of nodes sharing different language conventions. For this matter, we explore the models NG-AW—that incorporates trust—, NG-LEF—that incorporates uncertainty—and NG-SM—finally incorporating opinion preference. We test the algorithms in LFR networks and show that the separate addition of each social feature in the Naming Game results in improvements in community detection. Our simulations show that opinions coexist at the end of the game in non-convergent executions, each name tagging a different community, identifying, by a socially guided language dynamics, the topological communities present on the network. Moreover, the resulting trust in edges and uncertainty in nodes classify them according to role and position in the network, respectively. We observed this behavior in large networks with disjoint communities generated using LFR benchmark, and we compared our results with existing results from the literature, focusing on the quality of the community detection per se. Our model with secondary memory has shown accuracy comparable with algorithms designed specifically for topological community detection, while modeling social features that reveal communities as an emergent property, as observed in real-world social systems.

Keywords Naming game · Community detection · Complex social systems

T.G. Uzun (✉) · C.H.C. Ribeiro
ITA - Aeronautics Institute of Technology,
Sao José dos Campos, SP, Brazil
e-mail: thaisgobet@gmail.com

H. Cherifi et al. (eds.), *Complex Networks VII*, Studies in Computational Intelligence 644, DOI 10.1007/978-3-319-30569-1_9

1 Introduction

Most complex systems in nature and society can be analysed as networks, sets of nodes joined together in pairs by links, to capture the intricate web of relations among the units they are made of. One property that many real networks share is community structure, when nodes organize themselves in sub-units, where nodes are highly connected among each other and connections between communities are much more rare [5]. Many research has been done for detection of the communities in these systems, mainly focusing on the topological features of the network, and every year more algorithms are presented in the literature, some of them having great performance and accuracy in detecting communities [4].

However, one can notice that, in real social networks, communities are formed in a natural emergent way. Those groups, however, tend to be bound closely or dissolve through time naturally due to changes in the aspects they share. One of these aspects, that will be tackled in this work, is the linguistic interaction or opinion exchange between people in a network.

For this matter, we will use as base the Naming Game (NG), a widely known model of language exchange, that, when played by agents in networks with extremely high community structure can enter a meta-stable state where communities are revealed in an emergent way [8]. Our aim in this work is to show that, the revelation of communities can happen for an arbitrary network when modeling agents' social features in the NG dynamics. For this purpose, we will analyse three variations of the Naming Game, each incorporating a new social feature, trying to show that the insertion of each social attribute leads to a better emergence of the present community structure.

2 The Naming Game

The Naming Game [1] is a model for the emergence of a common vocabulary in a multi-agent system without central control, using only peer-to-peer interactions. The game is played by N identical agents, trying to agree in the name of a subject. Every agent has a memory, that starts empty and the game ends when a consensus or convergence state, where all agents have only one and the same word in their memories, is reached.

The Naming Game can be shown to converge for various network topologies, and the main difference in behavior is the time to reach convergence and the memory necessary to store all words. Interestingly, in [8] the authors apply NG to a real high-school friendship network with extremely high community structure, where NG does not converge but reaches a state with coexisting meta-stable word clusters. This *non-convergent state* evidences existing communities in the friendship network, separating students from middle-school and high-school and among different ethnic lines. Figure 1 [8] shows the resulting words in memories after the Naming Game

for *each time step* **do**
  One agent is randomly chosen as a speaker;
  Among its neighbors, one agent is chosen to be the hearer;
  if *the speaker's memory is empty* **then**
    the speaker invents a word;
  else
    the speaker selects one of the words in his inventory;
  end
  The speaker transmits the word to the hearer;
  if *the hearer has the word in his memory* **then**
    the communication is a *success* and both speaker and hearer delete all the words from their memories except the transmitted word;
  else
    the communication is a *failure* and the hearer adds the transmitted word to his inventory
  end
end

Algorithm 1: Naming Game

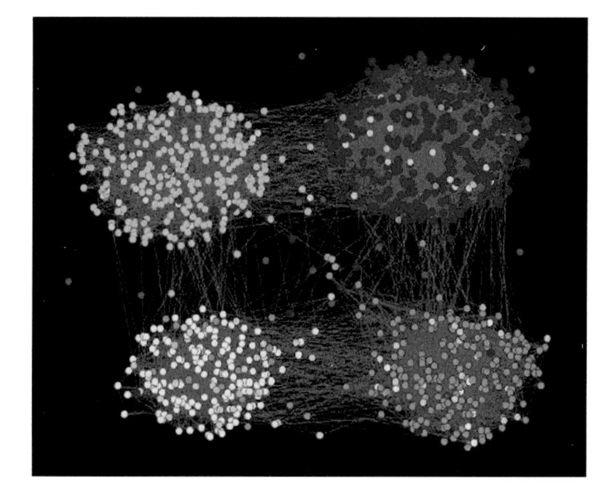

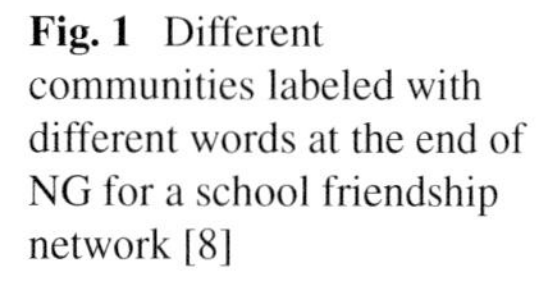

Fig. 1 Different communities labeled with different words at the end of NG for a school friendship network [8]

(different colors mean different words) for the mentioned network, where we can observe the separation of the communities.

However, due to the non-determinism of this game, this behavior does not happen in all executions, but only in a few, because of the initial choices of hearer: If in the first steps many communications happen in edges between communities—here called *external edges*—the probability of convergence increases drastically (not shown). In other words, if external edges are chosen with more frequency to communicate in the early beginning of the game, convergence is likely. On the other hand, the existence of the meta-stable of non-convergence suggests that for some instances the Naming Game can unravel the communities present in a network.

Several variations of the Naming Game have been proposed in the literature [8–10], being used sometimes as an agreement model and an opinion spreading model [14]. One of these variations is the NG with Adaptive Weights and will be presented next.

3 Naming Game with Adaptive Weights

The Naming Game with Adaptive Weights (NG-AW) was proposed by Lipowska et al. [7], as a model that results in an adaptive weighted network, based on the assumption that a person talks preferably with those with whom he already have agreed with. The weight of each edge is the ratio between its number of successes and its number of attempts, and changes over time. It is originally applied to fully connected networks and works as in Algorithm 2.

input : ϵ value;
for *each time step:* **do**
  One agent i is randomly chosen as a speaker;
  The agent i chooses his hearer j among his neighbors with probability

$$p_{ij} = \frac{w_{ij} + \epsilon}{\sum_{k=neighbors}(w_{ik} + \epsilon)} \tag{1}$$

  where $w_{ij} = \frac{Successes_{ij}}{Attempts_{ij}}$;
  if *the speaker's memory is empty* **then**
    the speaker invents a new word;
  else
    the speaker selects randomly one from his inventory;
  end
  The speaker transmits the word to the hearer;
  if *the hearer has the word in his memory* **then**
    the communication is a success and both speaker and hearer delete all the words from their memories except the transmitted word;
  else
    the communication is a failure and the hearer adds the transmitted word to his inventory
  end
end

Algorithm 2: Naming Game with Adaptive Weights

As suggested by the work in [7], it is easy to see that in some way sharing conventions and therefore developing a high trust brings individuals closer, while being unable to communicate creates a distance between them. This motivates us to interpret the edges' weights as *trust*, reinforcing and weakening the connections among individuals as communication interactions take place. In this way, a communication will be more probable to happen through a high trust relationship. The calculation of an edge weight can be seen as a special case of the definition of trust in [13] where trust is represented as $\alpha = \frac{r}{r+s}$, where r and s are real numbers and represent respectively the positive and negative experiences a truster has with a trustee.

The ϵ value is given a priori and is the responsible for the randomness in the communications: a small ϵ indicates that the agent will prefer edges that had many successes in the past interactions. As we increase ϵ the game approaches the original Naming Game, behaving more randomly. In the Community Detection (CD) scope, as the weight is the edge's success rate, when applied to networks with community structure, external edges have small weight during the game, since transmits words from different clusters. If the communication flows through high weight edges, interactions will take place mostly inside communities, avoiding convergence.

In this work we will test the detection of communities using the LFR benchmark [6], one of the most used CD benchmarks. This benchmark is based on the fact that real-world networks are characterized by heterogeneous distributions of node degree, with tails that often decay according to a power law. Likewise, it also takes into account that different communities are not always the same size: in fact, the distribution of community sizes of real networks can also be approximated by a power law [6]. LFR benchmark generates networks with power law distributions of degree and community size, with exponents τ_1 and τ_2, respectively. The *mixing parameter* μ is responsible for the community structure in the network, and indicates the fraction of external links each node will have. Consequently, each node will have $(1 - \mu)$ of its edges with nodes from its community.

Figure 2 shows the weights of edges after NG-AW with $\epsilon = 1$ for a LFR network with N = 1000, $\langle k \rangle = 20$, $\tau_1 = 2$, $\tau_2 = 1$, $k_{max} = 50$, size of communities $C_{min} = 10$, $C_{max} = 50$ and $\mu = 0.1$. As we can see, the weights of internal edges tend to be higher than weights for external ones, meaning that the algorithm weakens or reinforces a connection if it is external or internal, respectively. If we relate each different word to a different community, we have a Naming Game-based model that detects communities as groups of agents sharing the same word after the game, in non-convergent executions.

We will use the *Normalized Mutual Information* (NMI) [3], that is a measure of similarity of partitions based on Information Theory, as the standard partition measure. If the found and correct partitions are identical the resulting NMI will

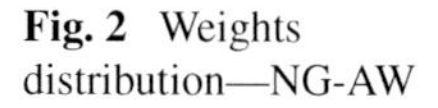
Fig. 2 Weights distribution—NG-AW

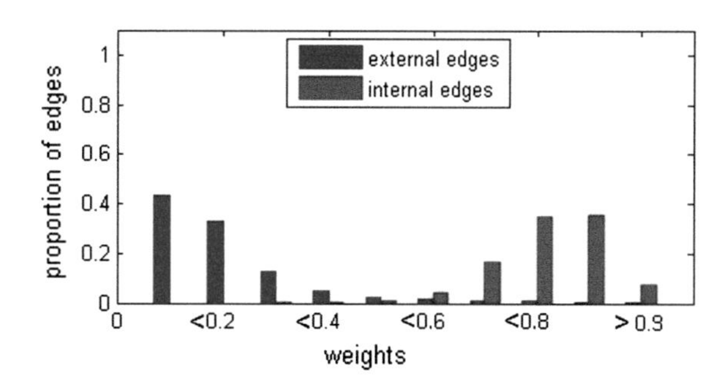

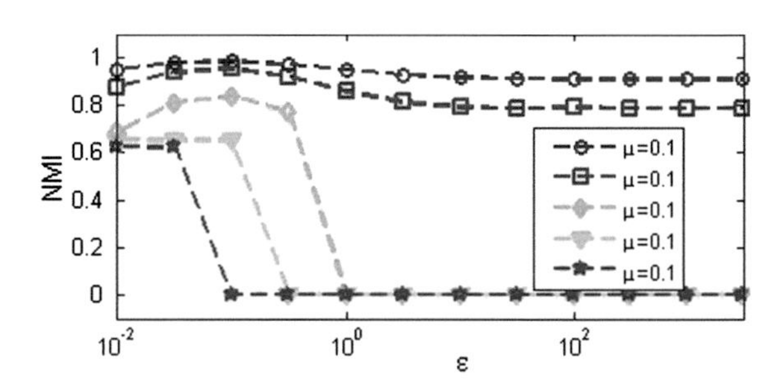

Fig. 3 NMI for NG-AW

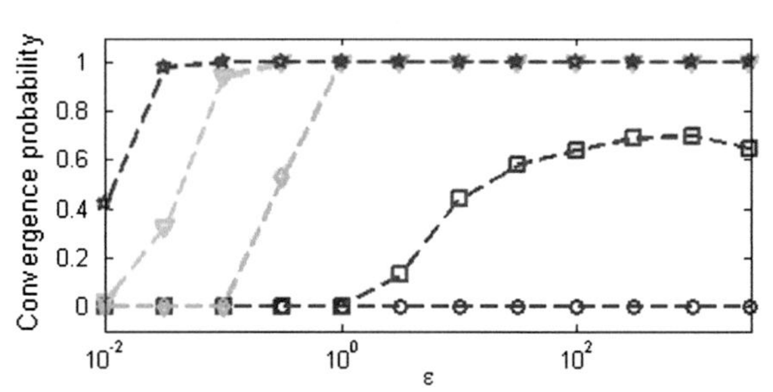

Fig. 4 NG-AW convergence probability

be 1, the maximum possible value. If having complete knowledge of partition A we would still know nothing about partition B and vice versa, NMI tends to 0. Figure 3 shows the resulting NMI for LFR networks with the same parameters varying the parameter μ, responsible for the community structure of the network. For networks with high community structure ($\mu = 0.1$ and 0.2) we have a value of NMI close to 1. When the community structure of the network is a little less obvious, the algorithm only detects communities for a small interval of low ϵ input values and with worse accuracy. This interval becomes smaller as the network has more external edges. As we see in Fig. 4, this happens because for larger ϵ values convergence is reached with higher probability, as the communications happen more randomly and the communities for such networks are less redundant. However, these results verify that an opinion dynamics model, like the Naming Game or variations, can be considered as a dynamical CD algorithm, in which the uncovering of communities occur in an emergent way through pairwise interactions.

Trying to establish a NG-based approach that naturally detects community structure for a more extensive range of networks and with higher probability, next we will explore an algorithm based on [7], the NG-LEF, relating ϵ to a time-changing local exploration factor, that is socially interpreted as the agent's uncertainty.

4 Naming Game with Local Exploration Factor

The Naming Game with Local Exploration Factor (NG-LEF) [11] incorporates *uncertainty* in the dynamics, by turning the global ϵ in NG-AW in a *local* ϵ_i that decays over time. We interpret ϵ_i as a measure of uncertainty of the agent i in its knowledge about the environment, that would be the agents' trust during the game. Thus, an agent chooses the hearer based on a deterministic part—his built trust w—biased by a random factor—the ϵ value—, that is the uncertainty of the agent in such trust.

Relating this local ϵ in a social scope to an uncertainty factor for *each* agent, we produce an antropomorphic analogy with people meeting people randomly in a social network. A person would communicate with higher uncertainty as he/she knows very little about the environment (early stages), and with higher certainty as he/she learns about it, as the person exchange information with other people. In NG-LEF the idea is that, in the beginning of the game, the random exploration of the Naming Game holds, satisfying the need for redundancy. In later times, communication happens more often between nodes with a history of successes. Algorithm 3 shows NG-LEF. The value of 10 % of ϵ decay was obtained empirically.

In NG-LEF, as the node has more failures, its ϵ will decay faster, decreasing the randomness of this node's choice of hearer. Then, nodes with more external edges, having smaller ϵ, will tend to communicate more preferably through edges with high trust, making the node behave like in NG-AW. Nodes with less or none external edges, however, typically have larger values of ϵ, since they have more successes, and would communicate more randomly with their neighbors. So the behavior of the node will depend on its relative position in the network. This is shown in Fig. 5, where we divide the nodes in three classes of proportion of external edges: low, medium and high, for the LFR network with $\mu = 0.1$ and $\epsilon(0) = 10$. As can be seen, different nodes get different values for ϵ_i along the game, making a node communicate more randomly as it has more internal connections. Also, as Fig. 6 shows, NG-LEF weakens external edges and reinforces internal ones with more precision, therefore, better classifying the edges. If we take both of these values as outputs of the algorithm, we can have not only the resulting partition but also a classification of nodes by their proportion of external edges and classification of edges in internal or external by their weights.

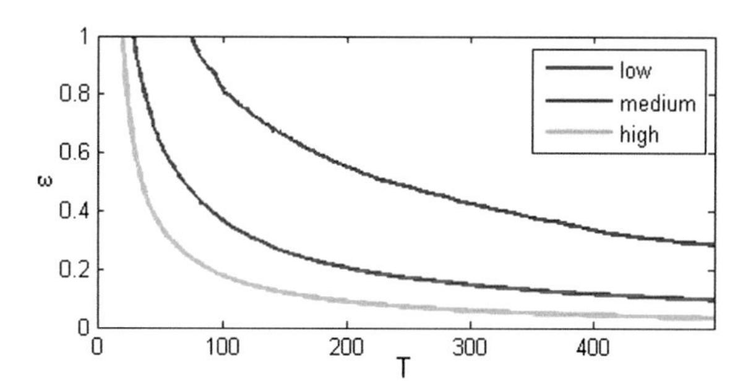

Fig. 5 NG-LEF ϵ evolution

input : Initial value of random factor $\epsilon(0)$;
$\epsilon_k = \epsilon(0)$ for all nodes k;
for *each time step:* **do**
  One agent i is chosen randomly to be the speaker;
  The speaker i chooses a hearer j with probability proportional to

$$p_{ij} = \frac{w_{ij} + \epsilon_i}{\sum_{k=neighbors}(w_{ik} + \epsilon_i)} \qquad (2)$$

  if *the speaker's memory is empty* **then**
    the speaker invents a new word;
  else
    the speaker selects randomly one from his inventory;
  end
  The speaker transmits the word to the hearer;
  if *the hearer has the word in his memory* **then**
    the communication is a success and both agents erase their memories keeping only the transmitted word;
  else
    the communication is a failure, the hearer adds the word to his inventory and both agents decrease their ϵ in 10 %
  end
end

Algorithm 3: Naming Game with Local Exploration Factor

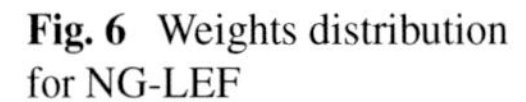
Fig. 6 Weights distribution for NG-LEF

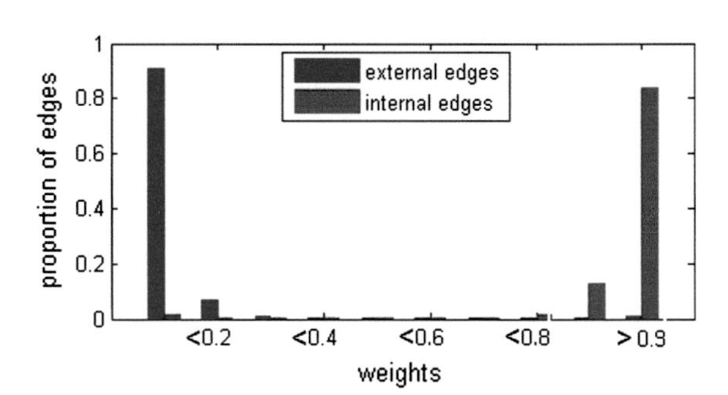

As we see in Fig. 7, for networks with high community structure NG-LEF presents a more stable behavior, having plateau of NMI for values of ϵ larger than 1, approximately. However, for networks with higher values of μ the behavior is similar to in NG-AW, even though we had better construction of edges' weights. This happens because, as the networks have lower proportion of internal connections, external edges have a slightly higher chance to be chosen to communicate , creating paths of high weight between different communities, making easier to a competing word to reach them. Once the word enters the community, many times through more than one link, fewer internal connections helps the competing word to disseminate, leading to poor detection of communities, and often to convergence. As we see in Fig. 8, for networks with strong communities the convergence rate is lower than for NG-AW.

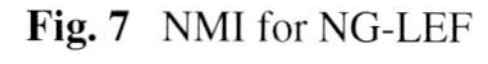

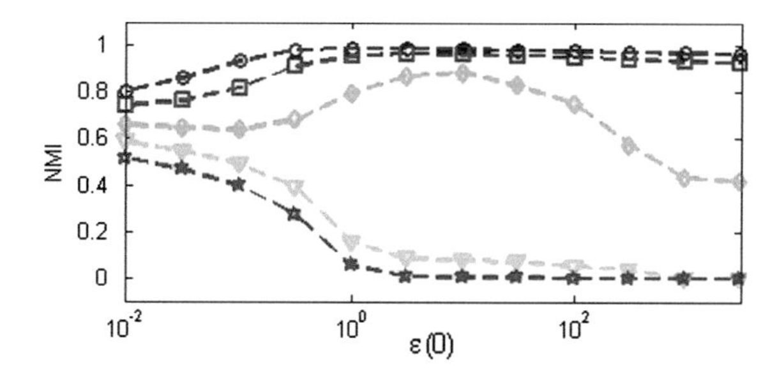

Fig. 7 NMI for NG-LEF

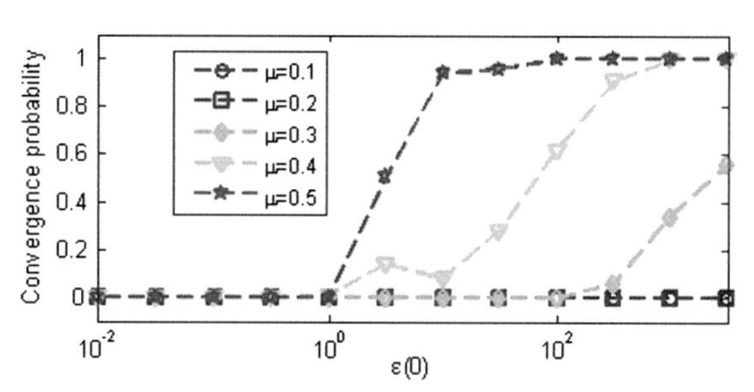

Fig. 8 Convergence probability NG-LEF

However, for networks with less community structure, NMI is poor for small ϵ and convergence is reached with high probability for larger ϵ values.

Even though the incorporation of trust and then uncertainty have led us to some improvement in community detection, we see that this is not enough to reveal or form the community structure of an arbitrary network. In the next section we explore a model that incorporates a third social feature: opinion preference.

5 Naming Game with Secondary Memory (NG-SM)

Still on the search for a model where the utterance dynamics reveals the community structure, we analyse a recently introduced model, the Naming Game with Secondary Memory [12], where agents have two inventories: the main memory, that behaves just like in NG; and the *secondary memory*, that keeps count of every word the agent has ever been exposed to. Our motivation is, even if one's mind is made out about a given subject, one do not properly forget other opinions had in the past. In both NG-AW and NG-LEF, agents are flighty, forgetting their opinions based on a single successful communication, regardless of how many communications the agent had with the forgotten words. A word that is massively present in the agent's neighbourhood, and therefore would be in the agent's inventory most of the time, would be erased at any moment by few communications with agents from a different community (with a different word). In a social sense, it would be equivalent to a person sharing an opinion with a close group for most of his/her life, and changing it completely when hearing a different one twice in a row. In real life, however, a person could even change his/her mind after talking a few times with an outsider, but would have a predisposition of believing again in a previously well agreed opinion. With the preference for transmission of the word with more occurrences, the spread of outsider words inside communities would be minimized and the state of non-convergence should be more stable, leading to a better detection of the existing communities (Figs. 9 and 10).

input : Initial value of random factor $\epsilon(0)$;
$\epsilon_k = \epsilon(0)$ for all nodes k;
for *each time step:* **do**
    One agent i is chosen randomly to be the speaker;
    The speaker i chooses a hearer j with probability proportional to

$$p_{ij} = \frac{w_{ij} + \epsilon_i}{\sum_{k=neighbors}(w_{ik} + \epsilon_i)} \tag{3}$$

    if *the speaker's memory is empty* **then**
        the speaker invents a new word;
    else
        the speaker selects one word k from his principal memory with probability proportional to the success of the word S_{i_k} in the secondary memory;
    end
    The speaker transmits the word to the hearer;
    if *the hearer has the word in his principal memory* **then**
        the communication is a success and both agents increase the number of occurrences of the transmitted word in 1 and delete all other words from the principal memory;
    else
        the communication is a failure, the hearer adds the word to his principal memory and secondary memory and both agents decrease their ϵ in 10 %
    end
end

Algorithm 4: NG with Secondary Memory

Fig. 9 NG-SM ϵ evolution

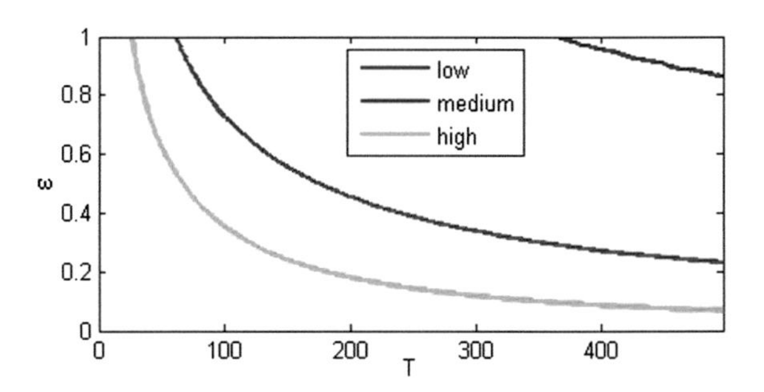

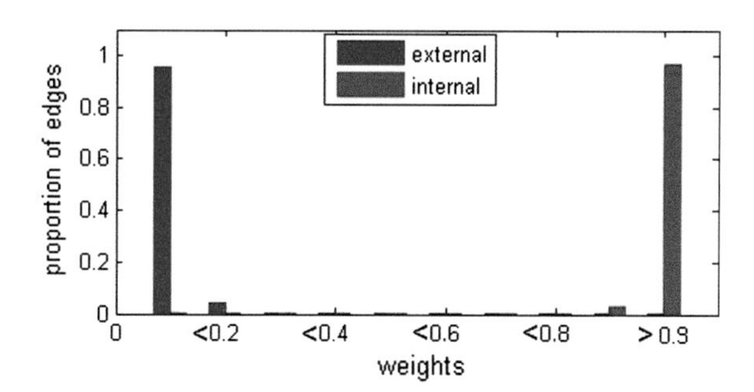

Fig. 10 Weights distribution—NG-SM

Figures 11 and 12 presents respectively the Normalized Mutual Information and the convergence rate for the same LFR networks after NG-SM. We observe a high NMI value for all networks with $\mu < 0.5$, only observing a few convergences in the network with $\mu = 0.5$. For more evident communities, the game always reaches non-convergence, so emergence of communities will happen in all executions. Figures 9 and 10 show the ϵ evolution and the histogram of edge weights for NG-SM. We see that, not only NG-SM maintains those classifications, but makes them more accurate: we see that the numeric difference between classes averages is larger through the game. Also, the values for all classes are mainly larger, indicating more uncertainty in the conversations, without leading the game to convergence. This is important because the random part of the communications is fundamental for gathering intrinsic information about the network. For the edges' weights we can also notice some improvement, on the misclassified internal edges, and a larger proportion of external

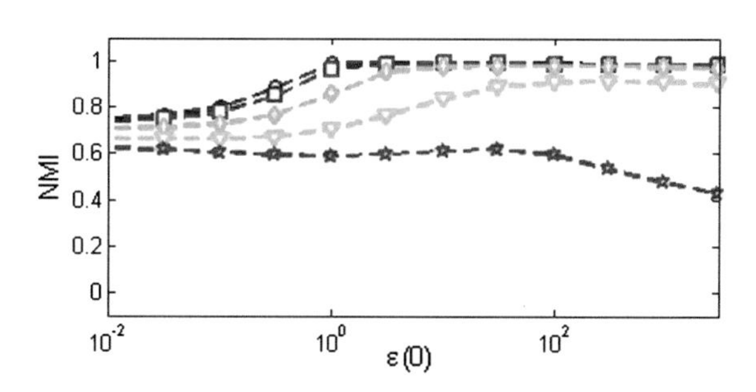

Fig. 11 NMI for NG-SM

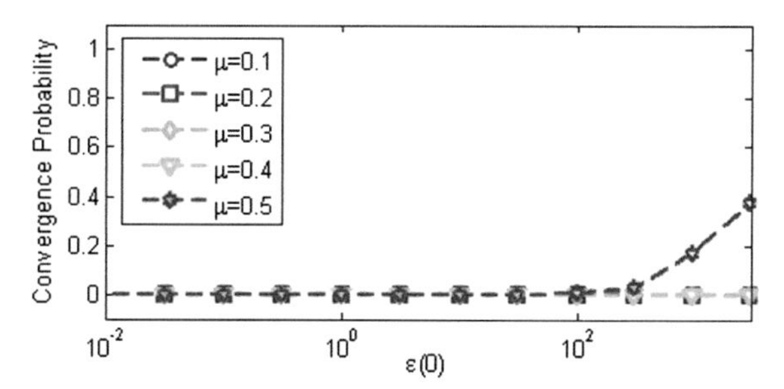

Fig. 12 Convergence rate of NG-SM

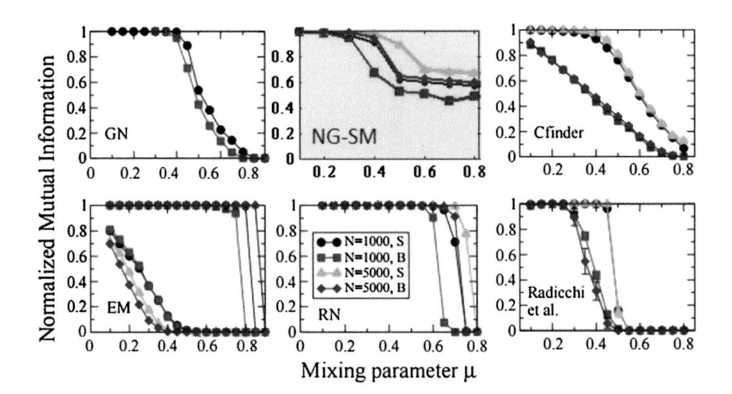

Fig. 13 NG-SM ϵ evolution

edges with weight smaller than 0.1 and internal edges with weight larger than 0.9. Figure 13 shows the NMI values for NG-SM (highlighted) with $\epsilon(0) = 10$ and also—from [4]—for different existing CD algorithms, in LFR networks with the same previous parameters, N = 1000 and N = 5000. The community sizes also vary as small (S), from 10 to 50 nodes or big (B), from 20 to 100 nodes. We can see that NG-SM has better accuracy for larger networks and it seems to work better when the communities in the network are smaller. Also, we see that NG-SM is suited for CD, comparable and in some cases better than existing algorithms.

That way, with the incorporation of opinion preference, uncertainty and trust, a model like the Naming Game can reveal the community structure of an arbitrary network with considerable accuracy.

6 Conclusions

In this work we use NG-variations to detect disjoint communities. We showed that with the insertion of trust in the NG—using the model in [7]—the system reaches more often a non-convergent meta-stable state, where nodes in the same communities share the same words. Also, NG-AW results in a classification of edges in external or internal, due to the resulting weights of these edges. This happens for networks with obvious community structure, as for other networks, convergence happens with high probability. We interpret the local exploration faction in NG-LEF [11] as uncertainty, and show that this algorithm, that models trust and uncertainty, has better accuracy than NG-AW. Moreover, the resulting uncertainty values of nodes are related to their proportion of external edges, or how connected they are in the community they belong to. The resulting edges weights—or the connections trust values—are better divided in internal and external. The incorporation of opinion preference, the third social feature, happens with the Naming Game with Secondary Memory [12]. NG-SM showed improvement in uncovering the communities in the networks, where networks that were intractable with NG-LEF have high NMI and zero convergence probability. Also, both classifications of edges with trust and nodes with uncertainty are more well divided and accurate, with NG-SM.

We tested the models in networks generated with the well known LFR community detection benchmark [6], that presents real-world characteristics and we showed that the found communities are more similar to the topological communities as we increase the number of modeled social features. With NG-SM, that incorporates all three features, for all LFR networks with $\mu < 0.5$ we have accurate detection of communities with NMI > 0.9 and guaranteed non-convergence. The classification of nodes, edges and the unravel of the communities all happen on an emergent way. In this way, each insertion of social feature results in a partition closer to the known a priori partition.

Thaís Uzun thanks CNPq (143356/2011-9) and Carlos Ribeiro thanks CNPq (303738/2013-8) and Fapesp (2013/13447-3) for financial support.

References

1. Baronchelli, A., Felici, M., Caglioti, E., Loreto, V., Steels, L.: J. Stat. Mech. P06014 (2006)
2. Brigatti, E.: Phys. Rev. E **78**, 046108 (2008)
3. Danon, L., Diaz-Guilera, A., Duch, J., Arenas, A.: J. Stat. Mech, P09008 (2005)
4. Fortunato, S.: Phys. Rep. **486**, 75–174 (2010)
5. Girvan, M., Newman, M.E.J.: Proc. Natl. Acad. Sci, USA (2002)
6. Lancichinetti, A., Fortunato, S., Radicchi, F.: Phys. Rev. E **78**, 046110 (2008)
7. Lipowska, D., Lipowski, A.: Artif. Life **18**(3), 311–323 (2012)
8. Lu, Q., Korniss, G., Szymanski, B.K.: J. Econ. Interact. Coord. **4**, 221 (2009)
9. Silva-Filho, R., Brust, M., Ribeiro, C.H.: CoRR, vol. Abs/0912.4553 (2009)
10. Tang, C.-L., Lin, B.-Y., Wang, W.-X., Hu, M.-B., Wang, B.-H.: Phys. Rev. E **75**, 027101 (2007)
11. Uzun, T. G., Ribeiro, C.H.C.: In: ECAI, pp. 1115–1116 (2014)
12. Uzun, T.G., Ribeiro, C.H.C.: In: Proceedings of CISIS (2015). doi:10.1109/CISIS.2015.21
13. Wang, Y., Singh, M.P.: ACM TAAS, **5**, 14:1–14:28 (2010)
14. Xie, J. Rensselaer Polytechnic Institute, Ph.D. thesis (2012). http://www.cs.rpi.edu/szymansk/theses/xie.phd.2012.pdf

A Novel Approach to Evaluate Community Detection Algorithms on Ground Truth

Giulio Rossetti, Luca Pappalardo and Salvatore Rinzivillo

Abstract Evaluating a community detection algorithm is a complex task due to the lack of a shared and universally accepted definition of community. In literature, one of the most common way to assess the performances of a community detection algorithm is to compare its output with given ground truth communities by using computationally expensive metrics (i.e., Normalized Mutual Information). In this paper we propose a novel approach aimed at evaluating the adherence of a community partition to the ground truth: our methodology provides more information than the state-of-the-art ones and is fast to compute on large-scale networks. We evaluate its correctness by applying it to six popular community detection algorithms on four large-scale network datasets. Experimental results show how our approach allows to easily evaluate the obtained communities on the ground truth and to characterize the quality of community detection algorithms.

1 Introduction

Evaluating the results provided by a community detection algorithm is one of the most difficult tasks of complex network analysis, since there is no a shared and universally accepted definition of what a community is [1, 2]. Each approach hence defines its own idea of community and maximizes a specific quality function (e.g. modularity, density, conductance, etc.). Even though the communities identified by a given algorithm on a network are consistent with its community definition, it is not guaranteed that they are able to capture the real sub-topology of the network. For this reason, the common way to state the quality of a community detection algorithm is

G. Rossetti (✉) · L. Pappalardo · S. Rinzivillo
Institute of Information Science and Technologies (ISTI),
National Research Council of Italy (CNR), Pisa, Italy
e-mail: Giulio.Rossetti@isti.cnr.it

L. Pappalardo
e-mail: Luca.Pappalardo@isti.cnr.it

S. Rinzivillo
e-mail: Salvatore.Rinzivillo@isti.cnr.it

H. Cherifi et al. (eds.), *Complex Networks VII*, Studies in Computational Intelligence 644, DOI 10.1007/978-3-319-30569-1_10

to evaluate the similarity between the communities it produces and the ground truth communities of the network. Generally, the communities produced by the algorithm are compared to the ground truth communities specified in the network dataset using metrics like the Normalized Mutual Information score (NMI) [3]. Unfortunately the computational complexity of this metric is quadratic in the number of communities of the network, which makes it unsuitable on large-scale complex networks where a large number of communities emerge.

In this paper we propose a novel community evaluation approach that leverages ground truth communities and copes with the computational issues that arise when calculating NMI on large community sets. To do that we define two measures, namely *community precision* and *community recall*, which provide information about how much the nodes of a given community tend to be in the same ground truth community. In particular, community precision quantifies the level of label homophily between a community and a ground truth community, while the community recall quantifies the ratio of nodes in the ground truth community covered by a given algorithm community. To validate our methods we apply six popular community detection algorithms on four large-scale networks with ground truth communities. We then compute the proposed community precision and community recall metrics on the produced community sets in order to compare them on the ground truth. We show how the evaluation can be easily performed through density scatter plots, where the presence and position of visual clusters well identify the properties of the community sets in terms of precision and recall. The evaluation can be also summarized into a single number using the $F1$-measure (the harmonic mean of community precision and community recall), which provides a clear and concise evaluation of the quality of a community set.

The paper is organized as follows. Section 2 revises the main works in community detection and community evaluation. Section 3 introduces the community precision and the community recall metrics and Sect. 4 describes our experiments, the community detection algorithms used (Sect. 4.1), the network datasets (Sect. 4.2) and the results obtained (Sect. 4.3). Finally, Sect. 5 concludes the paper illustrating some possibile improvements of the proposed metrics.

2 Related Works

Community detection has become during the last decade one of the most challenging and studied problems in complex network analysis, due to its relevance for a wide range of applications such as the study of information and disease spreading [4, 5], the prediction of future interactions and activities of individuals [6, 7], and even the analysis of the patterns of human mobility [8, 9]. Two surveys by Fortunato [1] and Coscia et al. [2] explore all the most popular techniques to find communities in complex networks, highlighting that several algorithms have been proposed in literature to detect different definitions of network community. The plethora of many community definitions makes the evaluation of a community detection algorithm a difficult task. In literature, the most used evaluation method is to compare the community set produced by an algorithm on a network with ground truth communities of the same network. Due to the scarse availability of real networks with ground truth communities, the evaluation of an algorithm is often performed using synthetic network generators that also provide ground truth communities (such as the LFR benchmark [10]). In such scenario, the comparison is generally done by the Normalized Mutual Information score (NMI) a measure of similarity borrowed from information theory [3, 11, 12], defined as:

$$NMI(X, Y) = \frac{H(X) + H(Y) - H(X, Y)}{(H(X) + H(Y))/2} \quad (1)$$

where $H(X)$ is the entropy of the random variable X associated to an algorithm community, $H(Y)$ is the entropy of the random variable Y associated to a ground truth community, and $H(X, Y)$ is the joint entropy. NMI ranges in the interval [0, 1] and is maximized when the algorithm community and ground truth community are identical. One drawback of NMI is that, assuming that the algorithm community set and the ground truth community set have approximately the same size n, the overall NMI computation requires $O(n^2)$ comparisons, making it unsuitable for large-scale networks.

3 Approach Definitions

The computation of NMI on large community sets is often prohibitive: following Eq. (1) given the algorithm community set X of size m and ground truth community set Y of size n, to compute NMI we need to identify the communities best matches with cost $O(mn)$. Assuming $m \simeq n$ the NMI computation requires $O(n^2)$ comparisons thus making it often unsuitable for large-scale networks.

To overcome this drawback, we propose a novel approach that provides valuable insights on the quality of the community sets produced by a community detection algorithm. Given a community set X produced by an algorithm and the ground truth community set Y, for each community $x \in X$ we label its nodes with the ground truth community $y \in Y$ they belong to. We then match community x with the ground truth community with the highest number of labels in the algorithm community. This procedure produces (x, y) pairs having the highest homophily between the node labels in x and all the ground truth communities. We then measure the quality of the mappings by the two following measures:

- *Precision*: the percentage of nodes in x labeled as y, computed as

$$P = \frac{|x \cap y|}{|x|} \in [0, 1] \quad (2)$$

- *Recall*: the percentage of nodes in y covered by x, computed as

$$R = \frac{|x \cap y|}{|y|} \in [0, 1]. \quad (3)$$

Given a pair (x, y) the two measures describe the overlap of their members: a perfect match is obtained when both *precision* and *recall* are 1. We thus have a many-to-one mapping: multiple communities in X can be connected to a single ground truth community in Y. This policy enables the adoption of the proposed methodology both in case of algorithms producing crisp partitions or algorithm producing overlapping communities. Moreover, analyzing the *precision* and *recall* of each pair we are able to detect both underestimations and overestimations made by the adopted algorithm.

We can combine *precision* and *recall* into their harmonic mean obtaining the $F1$-measure, a concise quality score for the individual pairing:

$$F1 = 2\frac{precision * recall}{precision + recall}. \quad (4)$$

Given a network, the $F1$ score can be averaged among all the identified pairs in order to summarize the overall correspondence between the algorithm community set and ground truth community set. The mean $F1$, along with its standard deviation, makes possible to compare the performances of different algorithms on the same network with ground truth communities. The proposed approach as complexity $O(|V| + |C|) \simeq O(|V|)$ since it is composed by two steps: (i) node labeling (linear in the number of nodes $|V|$) and (ii) communities F1-computation (linear in the number of identified communities $|C|$). The averaging of community F1s has constant cost.

4 Experiments

In this section we evaluate the proposed methodology on the community sets produced by popular community detection algorithms on large-scale real-world networks with ground truth communities. In Sect. 4.1 we introduce the algorithms used and in Sect. 4.3 we evaluate the quality of the algorithms by using the proposed approach.[1]

4.1 Community Detection Algorithms

We use six different community detection algorithms designed to maximize different functions: LOUVAIN, INFOHIERMAP, CFINDER, DEMON, ILCD and EGO-NETWORK.

[1] A Python implementation of our approach is available at: http://goo.gl/kWIH2I.

LOUVAIN is an heuristic method based on modularity optimization [13] and it is proven to be fast and scalable on large-scale networks. The modularity optimization is performed in two steps. First, the method searches for "small" communities by optimizing modularity locally. Second, it aggregates nodes belonging to the same community and builds a new network whose nodes are communities. These steps are repeated iteratively until a maximum modularity is obtained, producing a complete non-overlapping partitioning of the graph. As most of the approaches based on modularity optimization, it suffers from a "scale" problem that causes the extraction of few huge communities and an high number of tiny ones.

INFOHIERMAP is one of best performing hierarchical non-overlapping clustering algorithms for community detection [14] studied to optimize community conductance. The graph structure is explored with a number of random walks of a given length and with a given probability of jumping into a random node. The underlying intuition is that random walkers are trapped in a community and exit from it very rarely. Each walk is described as a sequence of steps inside a community followed by a jump. By using unique names for communities and reusing a short code for nodes inside the community, this description can be highly compressed, in the same way as re-using street names (nodes) inside different cities (communities). The renaming is done by assigning a Huffman coding to the nodes of the network. The best network partition will result in the shortest description for all the walks.

CFINDER is an algorithm for finding dense overlapping groups of nodes in networks, based on the Clique Percolation Method (CPM) [15]. Its community definition is based on the observation that a typical member in a community is linked to many other members, but not necessarily to all other nodes in the community. In other words, a community can be interpreted as a union of smaller complete subgraphs that share nodes. These complete subgraphs are called k-cliques, where k is the number of nodes in the subgraph, and a k-clique-community is defined as the union of all k-cliques that can be reached from each other through a series of adjacent k-cliques. Two k-cliques are said to be adjacent if they share $k - 1$ nodes.

DEMON is an incremental algorithm that uses an approach based on the extraction of ego networks, that is, the set of nodes connected with a certain ego node u [16]. The communities are extracted by using a bottom-up approach: each node gives the perspective of the communities surrounding it and then all the different perspectives are merged together in an overlapping structure. In practice, the ego network of each node is extracted and a label propagation is performed on this structure ignoring the presence of the ego itself, since it will be judged by its peer neighbors. Then, with equity, the vote of everyone in the network is combined. The result of this combination is a set of overlapping modules, the guess of the real communities in the global system, made not by an external observer, but by the actors of the network itself.

ILCD is an algorithm for the detection of overlapping communities in dynamic networks. It can also be used on static networks and works on large-scale networks. It is not based on the modularity, but, on the contrary, on the idea that communities are defined locally (intrinsic communities) [17].

Table 1 Networks statistics of the four large-scale real-world networks analyzed

Network	Nodes	Edges	Clustering	Diameter	Ground truth com.
Amazon	334,863	925,872	0.3967	44	75,149
DBLP	317,080	1,049,866	0.6324	21	13,477
Youtube	1,134,890	2,987,624	0.0808	20	8,385
LiveJournal	3,997,962	34,681,189	0.2843	17	287,512

EGO-NETWORKS is a naive algorithm that models the communities as the set of induced subgraphs obtained considering each node with its neighbors. This approach provides the highest overlap among the considered approaches: each node u belongs exactly to $|\Gamma(u)| + 1$ communities, where $\Gamma(u)$ identifies its neighbors set.

4.2 Network Data

We use four large-scale network datasets in our experiments: DBLP, Youtube, Amazon and LiveJournal [18], filtering them on the nodes covered by the ground truth partition (network statistics shown in Table 1).[2]

The DBLP network is a co-authorship network where two authors of computer science papers are connected if they publish at least one paper together. The ground truth communities are defined by the publication venue, e.g. journal or conference, hence authors who published to a certain journal or conference form a community.

Youtube is a popular video-sharing website where the users form friendships each other and can create groups which other users can join. The user-defined groups are the ground truth communities of the network.

The Amazon network has been collected by crawling Amazon website. It is based on Customers-Who-Bought-This-Item-Also-Bought feature of the Amazon website. If a given product i is frequently co-purchased with product j, the graph contains an undirected edge from i to j. Each product category provided by Amazon defines each ground truth community.

LiveJournal is a free online blogging community where users can declare friendships to each other. It also allows users to form a group which other members can then join. Each of these user-defined groups is a ground truth community.

[2]The network datasets are available at: https://snap.stanford.edu/data/.

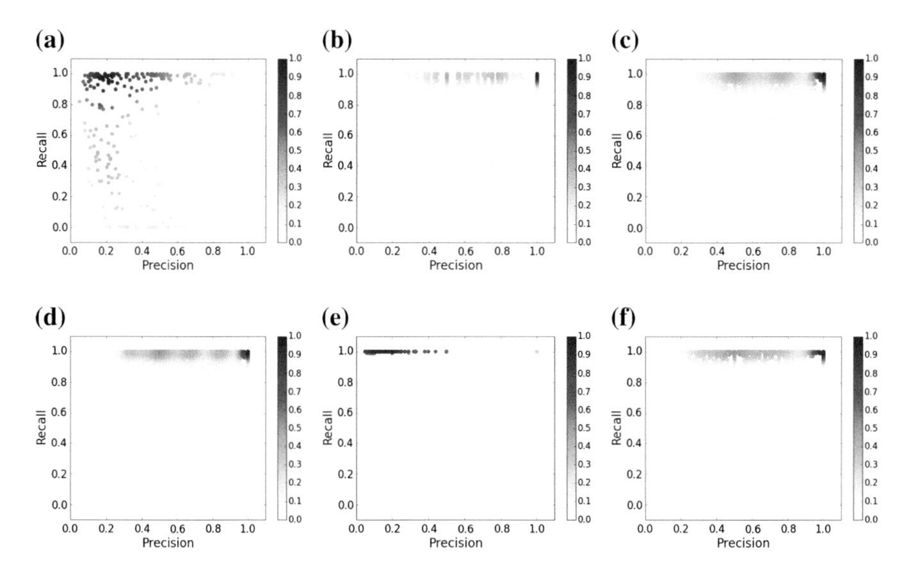

Fig. 1 Density scatter plots describing community precision and community recall on the six community sets extracted from the Amazon network. **a** Amazon—LOUVAIN. **b** Amazon—ILCD. **c** Amazon—CFINDER. **d** Amazon—DEMON. **e** Amazon—INFOHIERMAP. **f** Amazon—EGO-NETWORK

4.3 Results

We apply the six algorithms introduced in Sect. 4.1 to extract communities from the four large-scale network datasets described in the Sect. 4.2. We then use the proposed evaluation approach to compare the obtained community sets and rank the tested algorithms. Figures 1, 2, 3 and 4 show the density scatter plots describing community precision and community recall computed on the six community sets produced on the Amazon, DBLP, Youtube and LiveJournal networks respectively. In this representation, we report the community precision on the x-axis and the community recall on the y-axis: the color of a point (x, y) in a scale from yellow to red indicates the number of community matchings having precision x and recall y: the more red is the color the higher is the volume. We have a perfect match when both precision and recall are 1 (top-right corner of the plot): in this scenario, the algorithm community is identical to the corresponding ground truth community. The proposed visualization also allows an intuitive identification of the community scale:

- pairings having maximal recall and low precision (i.e. points that clusters close to the upper left corner of the plot) identifies network substructures that overestimate the ground truth;
- pairings having low recall and maximal precision (i.e. points that clusters close to the lower right corner) identifies network substructures that underestimate the ground truth.

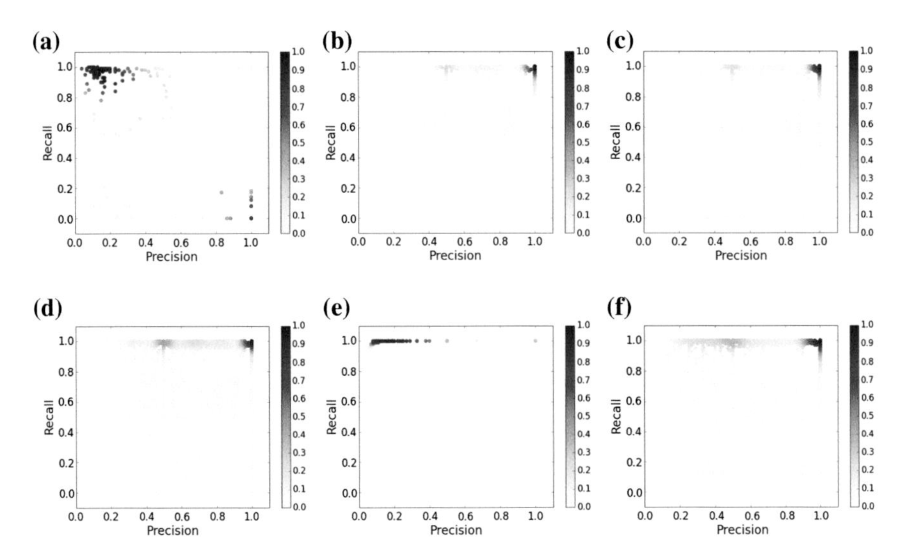

Fig. 2 Density scatter plots describing community precision and community recall on the six community sets extracted from the DBLP network. **a** DBLP—LOUVAIN. **b** DBLP—ILCD. **c** DBLP—CFINDER. **d** DBLP—DEMON. **e** DBLP—INFOHIERMAP. **f** DBLP—EGO-NETWORK

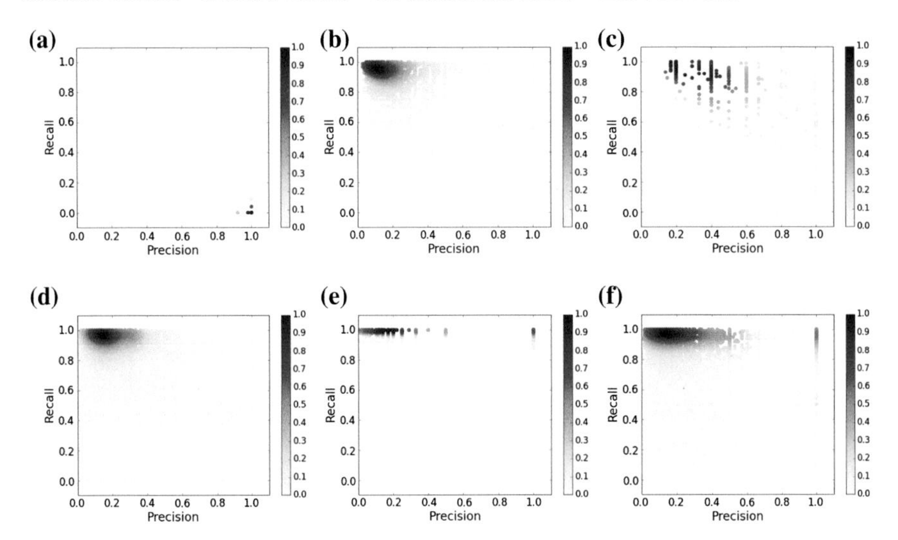

Fig. 3 Density scatter plots describing community precision and community recall on the six community sets extracted from the Youtube network. **a** Youtube—LOUVAIN. **b** Youtube—ILCD. **c** Youtube—CFINDER. **d** Youtube—DEMON. **e** Youtube—INFOHIERMAP. **f** Youtube—EGO-NETWORK

The former scale tells us that the algorithm produces communities that group together more nodes than it should, while in the latter case the ground truth communities are fragmented in smaller communities.

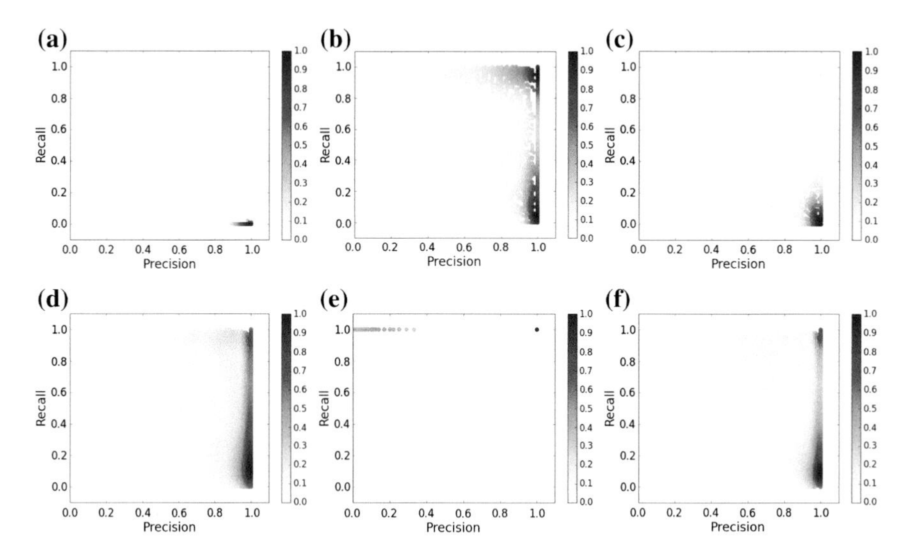

Fig. 4 Density scatter plots describing community precision and community recall on the six community sets extracted from the LiveJournal network. **a** LiveJournal—LOUVAIN. **b** LiveJournal—ILCD. **c** LiveJournal—CFINDER. **d** LiveJournal—DEMON. **e** LiveJournal—INFOHIERMAP. **f** LiveJournal—EGO-NETWORK

Table 2 The average $F1$-measure for the four networks

Network	LOUVAIN	INFOHIERMAP	CFINDER	DEMON	ILCD	EGO-NETWORKS
Amazon	0.40 (0.26)	0.46 (0.29)	0.72 (0.27)	0.70 (0.24)	**0.74 (0.23)**	0.72 (0.22)
DBLP	0.26 (0.24)	0.45 (0.31)	**0.82 (0.24)**	0.75 (0.24)	0.81 (0.23)	0.81 (0.22)
Youtube	0.16 (0.05)	**0.59 (0.32)**	0.50 (0.20)	0.36 (0.10)	0.35 (0.20)	**0.58 (0.28)**
LiveJournal	0.01 (0.006)	0.66 (0.30)	0.21 (0.30)	0.56 (0.29)	**0.71 (0.04)**	0.52 (0.30)

Each row shows the average $F1$-measure (standard deviation within brackets) achieved when matching the communities identified by the algorithms and the ground truth communities of a specific network. In bold the best score for each network

From the plots, for the Amazon and DBLP networks a difference among the algorithms clearly emerges: while DEMON, ILCD, CFINDER and EGO-NETWORKS produce community sets with high precision and high recall denoting a high correspondence to the ground truth communities, LOUVAIN and INFOHIERMAP produce community sets with low precision (low label homophily) and high recall (they cover a large fragment of the corresponding ground truth community). On the Youtube network LOUVAIN shows very high precision and very low recall, while the other algorithms behave the opposite producing communities with low precision and high recall. On the LiveJournal network all the algorithms produce communities with high precision, while the recall varies a lot across the communities. Table 2 summarizes all

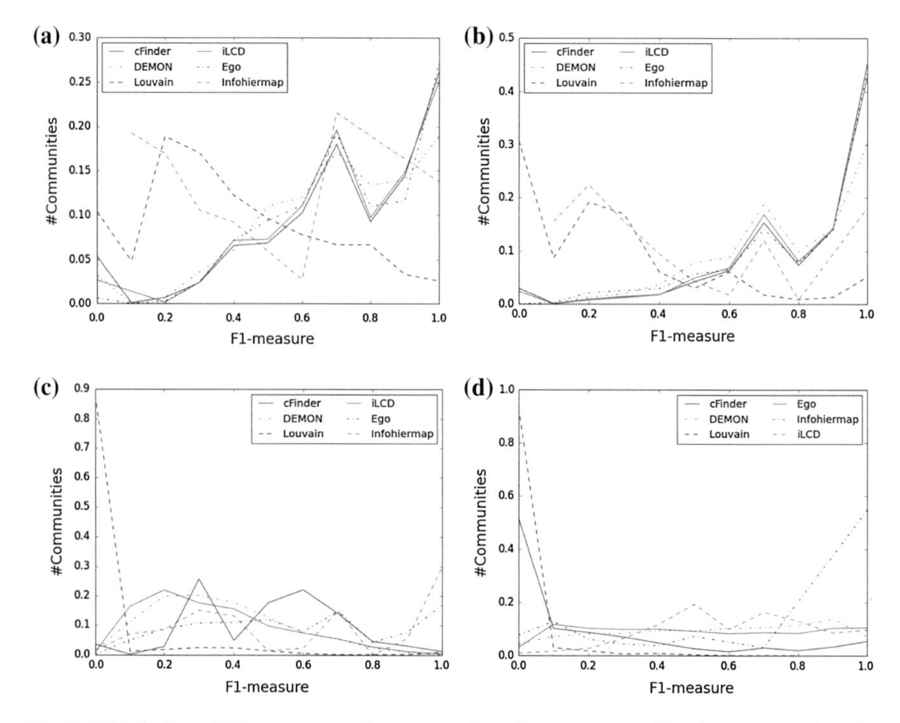

Fig. 5 Distribution of F1-measure on the community pairings generated by the six algorithms the Amazon (**a**), DBLP (**b**), Youtube (**c**) and LiveJournal (**d**) networks

these observation reporting, for each algorithm and dataset, the average $F1$-measure computed on the identified pairings. We observe how the average $F1$-measure is useful to understand two main aspects of community evaluation:

- First, it summarizes how well the communities produced by an algorithm corresponds to the ground truth communities. For instance, from our experiments is clear how LOUVAIN shows lower correspondence with the ground truth than all the other algorithms: this result is clearly due to the so called *scale* problem of modularity based approaches. Indeed, as shown from all the density scatter plots, LOUVAIN produces either huge or tiny communities thus providing respectively an overestimation (high recall, low precision—i.e. Amazon and DBLP) or a underestimation (low recall, high precision—i.e. LiveJournal and Youtube) of the ground truth communities;
- Second, the $F1$-measure helps also in evaluating the quality of the ground truth itself: on the Youtube dataset for example no algorithm produces communities with high correspondence with the ground truth ones, denoting either a low quality of the ground truth communities or that the community definition underlying the ground truth radically differs from the community definition of the tested algorithms.

However the $F1$-measure indicator is computed as an average of community-pairs $F1$s and it can show a high standard deviation (Table 2). For this reason we report in Fig. 5, for each network and algorithm, the complete distribution of $F1$ across the community pairs. We observe how the distributions endorse the validity of the proposed indicator even in presence of high standard deviation.

5 Conclusion

Evaluating the quality of community detection algorithms is a hard task, especially because the problem itself is ill-posed: each algorithm optimizes a different quality metric introducing its own community definition. In this paper we tackled the problem of estimating the correspondence between algorithm communities and ground truth communities. When available, the information about ground truth communities of a network can be used to compare the results provided by a set of algorithms: so far the NMI has been the common way to perform this task. However, NMI has a major drawback: its computational complexity is quadratic in the number of communities. For this reason we introduced a novel and fast approach to estimate the quality of the communities produced by an algorithm that can be applicable to large-scale networks. With the support of visual tools, our methodology provides a reliable index that captures the quality of a community set and describes if the adopted algorithm underestimates or overestimates the ground truth community structure. As future works, we plan to use the proposed approach to identify and characterize sub-profiles among the communities extracted: by applying clustering techniques using precision and recall as features, we can group communities according to their degree of correspondence to the ground truth and then study their network features.

Acknowledgments This work was partially funded by the European Community's H2020 Program under the funding scheme "FETPROACT-1-2014: Global Systems Science (GSS)", grant agreement #641191 CIMPLEX "Bringing CItizens, Models and Data together in Participatory, Interactive SociaL EXploratories", https://www.cimplex-project.eu. Our research is also supported by the European Community's H2020 Program under the scheme "INFRAIA-1-2014-2015: Research Infrastructures", grant agreement #654024 *"SoBigData: Social Mining & Big Data Ecosystem"*, http://www.sobigdata.eu.

References

1. Fortunato, S.: Community detection in graphs, Physics Reports, vol. 486, no. 3–5, pp. 75–174 (2010). http://www.sciencedirect.com/science/article/B6TVP-4XPYXF1-1/2/99061fac6435db4343b2374d26e64ac1
2. Coscia, M., Giannotti, F., Pedreschi, D.: A classification for community discovery methods in complex networks. Stat. Anal. Data Min., **4**(5), 512–546 (2011). http://dx.doi.org/10.1002/sam.10133

3. Lancichinetti, A., Fortunato, S., Radicchi, F.: Benchmark graphs for testing community detection algorithms. Phys. Rev. E, **78**(4), 046110 (2008). http://pre.aps.org/abstract/PRE/v78/i4/e046110
4. Bhat, S., Abulaish, M.: Overlapping social network communities and viral marketing. In: International Symposium on Computational and Business Intelligence, pp. 243–246, Aug 2013
5. Wu, X., Liu, Z.: How community structure influences epidemic spread in social networks. Phys. A: Stat. Mech. Appl. **387**, 623–630 (2008)
6. Rossetti, G., Pappalardo, L., Kikas, R., Pedreschi, D., Giannotti, F., Dumas, M.: Community-centric analysis of user engagement in skype social network. In: Proceedings of the 2015 ACM/IEEE International Conference on Advances in Social Network Analysis and Mining (2015)
7. Rossetti, G., Guidotti, R., Pennacchioli, D., Pedreschi, D., Giannotti, D.: Interaction prediction in dynamic networks exploiting community discovery. In: Proceedings of the 2015 ACM/IEEE International Conference on Advances in Social Network Analysis and Mining (2015)
8. Rinzivillo, S., Mainardi, S., Pezzoni, F., Coscia, M., Giannotti, F., Pedreschi, D.: Discovering the geographical borders of human mobility. KI - Künstliche Intelligenz (2012)
9. Bagrow, J.P., Lin, Y.-R.: Mesoscopic structure and social aspects of human mobility. PLoS ONE **7**(5), p. e37676, 2012. http://dx.doi.org/10.1371/journal.pone.0037676
10. Lancichinetti, A., Fortunato, S.: Benchmarks for testing community detection algorithms on directed and weighted graphs with overlapping communities. Phys. Rev. E **80**(1), 016118 (2009)
11. McDaid, A.F., Greene, D., Hurley, N.J.: Normalized mutual information to evaluate overlapping community finding algorithms. CoRR, arXiv:1110.2515 (2011)
12. Detecting the overlapping and hierarchical community structure in complex networks. New J. Phys. (2009)
13. Blondel, V.D., Guillaume, J.-L., Lambiotte, R., Lefebvre, E.: Fast unfolding of communities in large networks. J. Stat. Mech.: Theory Exp. **2008**(10), P10008 (2008)
14. Rosvall, M., Bergstrom, C.T.: Maps of random walks on complex networks reveal community structure. Proc. National Acad. Sci. **105**(4), 1118–1123 (2008)
15. Palla, G., Derényi, I., Farkas, I., Vicsek, T.: Uncovering the overlapping community structure of complex networks in nature and society. Nature **435**(7043), 814–818 (2005)
16. Coscia, M., Rossetti, G., Giannotti, F., Pedreschi, D.: Demon: a local-first discovery method for overlapping communities. In: Agarwal, D., Pei, J. (eds.), KDD, Q. Y. 0001, pp. 615–623. ACM (2012)
17. Cazabet, R., Amblard, F., Hanachi, C.: Detection of overlapping communities in dynamical social networks. In: SocialCom, pp. 309–314 (2010)
18. Jaewon, Y., Leskovec, J.: Defining and evaluating network communities based on ground-truth. Knowl. Inf, Syst (2015)

Improving Network Community Structure with Link Prediction Ranking

Mingming Chen, Ashwin Bahulkar, Konstantin Kuzmin and Boleslaw K. Szymanski

Abstract Community detection is an important step of network analysis that relies on the correctness of edges. However, incompleteness and inaccuracy of network data collection methods often cause the communities based on the collected datasets to be different from the ground truth. In this paper, we aim to recover or improve the network community structure using scores provided by different link prediction techniques to replace a fraction of low ranking existing links with top ranked predicted links. Experimental results show that applying our approach to different networks can significantly refine community structure. We also show that predictions of edge additions and persistence are confirmed by the future states of evolving social networks. Another important finding is that not every metric performs equally well on all networks. We observe that performance of link prediction ranking is correlated with certain network properties, such as the network size or average node degree.

Keywords Community structure · Link ranking · Network dynamics

This work was supported in part by the ARL under Cooperative Agreement Number W911NF-09-2-0053 and by the Office of Naval Research Grants No. N00014-09-1-0607 and N00014-15-1-2640, and by the EU's 7FP Grant Agreement No. 316097 and by the Polish National Science Centre, the decision no. DEC-2013/09/B/ST6/02317.

M. Chen · A. Bahulkar · K. Kuzmin · B.K. Szymanski (✉)
Department of Computer Science, Rensselaer Polytechnic Institute, Troy, NY 12180, USA
e-mail: szymab@rpi.edu

M. Chen
e-mail: chenm8@rpi.edu

A. Bahulkar
e-mail: bahula@rpi.edu

K. Kuzmin
e-mail: kuzmik@rpi.edu

B.K. Szymanski
Wroclaw University of Technology, 50-370 Wroclaw, Poland

H. Cherifi et al. (eds.), *Complex Networks VII*, Studies in Computational Intelligence 644, DOI 10.1007/978-3-319-30569-1_11

1 Introduction

Detecting and characterizing network community structure are among the fundamental techniques of network science. Community detection reveals latent but meaningful structures in a wide range of networks [1], yet the results often do not represent the reality. The primary reason is that available network datasets are often incomplete or inaccurate because of lost, incorrect. or misrepresented data, especially when gathered from massive networks. Consequently, the networks derived from such data may have some edges missing while some edges present in the network dataset may not exist in reality.

In this paper we introduce and evaluate methods of recovering or improving the network community structure by removing extraneous (or transient) edges and restoring (or creating) the missing ones.

We start by setting the fraction of edges to be replaced which defines the number of added and deleted edges. Next, we rank all the edges by the chosen link prediction method. Then, we complement the network with non-existing highest ranked edges and remove the same number of existing lowest ranking links using three popular link prediction metrics. We evaluate this approach on seven real-world network datasets, including two friendship networks, two collaboration networks, and a co-purchasing network. After enhancing the networks with our link improvement procedure, we first run community detection algorithms to find community structure. Then we measure the quality of the discovered community structure with two global and six local metrics. The results show that the community structure of five out of seven real-world networks is significantly refined.

The rest of the paper is organized as follows. The related work is presented in Sect. 2. The detailed description of the approach is provided in Sect. 3. The analysis of the results using community structure metrics is given in Sect. 4. Section 5 includes the results for an evolving network with the known ground truth data. The conclusion and future work are outlined in Sect. 6.

2 Related Work

In [2] the authors focus on finding missing edges and communities. First, they classify the reasons for missing edges but do not consider extraneous edges. They use a partial network which is simulated by deleting a certain fraction of edges from the input dataset. The quality of the resulting communities is compared to their true versions which are assumed to be available. Normalized mutual information (NMI) [3] is used as a measure of community quality.

In [4] the authors use the results of community detection to guide the addition of missing links. In their approach, intra-community edges suggested by a link predictor are added to the network first, followed by inter-community edges. Experimental verification was performed on the LFR benchmark [5] and six very small real-world networks using several link predictors and two community detection algorithms.

A more detailed analysis of the relation between community structure and link formation is provided in [6]. Given an array of communities, the density of links inside a community and between any two communities determines the probability of adding a particular link. A further development [7] uses the network's local structural information for improved performance.

A common approach to enhancing the quality of community detection methods using link prediction techniques is to introduce a preprocessing step to ameliorate the network by reinforcing its community structure. An example is a method [8] in which link prediction is applied to assign weights to the existing edges of a network and then a community detection algorithm is applied to the weighted network. This approach uses five different community detection methods. The community quality is measured using NMI for synthetic networks and modularity for real-world ones.

A more complicated solution has been proposed recently in [9]. It involves running link prediction multiple times on the same input network thus creating a family of enhanced networks. Community detection is then performed for each network in this family. The final result is constructed by aggregating community detection results of each individual network. This approach was implemented only for disjoint community detection.

3 Link Replacing Methodology

Algorithm 1 defines our approach. First, every possible edge (whether existing or potential) in the network is assigned a rank based on the score returned by a link predictor $\mathscr{L}$. We use LPmade library [10] for unsupervised link prediction and analysis selecting three local computationally efficient metrics described below.

The Number of Common Neighbors (*CN*) is simply the count of the number of neighbors that any two given nodes have in common. The computational complexity to calculate this score for a network is $O(E)$, where E denotes the number of edges in the network.

Adamic-Adar (*AA*) [11] is a refinement of the CN in which each common neighbor of the two nodes for which the metric is evaluated adds the inverse of the logarithm of its degree to the result rather than adding a constant of 1. The computational complexity to calculate this score for a network is $O(E)$.

PropFlow (*PF*) [12] measures the geodesic proximity of the two nodes for which the metric is computed. We restrict the degree of the considered neighborhood to four, so the complexity of computing this metric is $O(d^4N)$ where d is the average node degree and N is the number of nodes in the network. For a sparse network, the complexity is linear in the number of nodes.

Edges and their corresponding rank scores are kept separately for existing and potential edges.

Algorithm 1 : Link ranking and replacement

Input: Graph $G = (V, E)$, link predictor $\mathscr{L}$, fraction of edges to be replaced f
Output: Graph $G' = (V, E')$ with improved community structure

```
E' ← E
Ē ← {{u, v} : ∀u ∈ V, ∀v ∈ V, s.t. u ≠ v} \ E
R_E ← ()
R_Ē ← ()
for all {u, v} ∈ E s.t. (deg(u) > 1 and deg(v) > 1 do
    Add ({u, v}), L({u, v}) to R_E
end for
Sort R_E in the order of ascending rank values
for all e ∈ Ē do
    Add (e, L(e)) to R_Ē
end for
Sort R_Ē in the order of descending rank values
n ← ⌊f · |E|⌋
for i = 1 to n do
    e ← edge from the i^th top tuple of R_Ē
    E' ← E' ∪ {e}
end for
for i = 1 to n do
    e ← edge from the i^th top tuple of R_E
    E' ← E' \ {e}
end for
```

During the second phase of the algorithm, edge replacements take place. First, a number of edges (denoted by f) with the highest rank among the non-existing edges are added to the network. Next, the same number of the lowest ranked existing edges are removed from the network. In order to prevent the formation of isolated nodes (i.e., nodes with a degree of 0), an edge is not considered for removal if one or both of its endpoints have a degree of 1. Then, we use the community detection algorithms SpeakEasy [13] and GANXiS [14] to detect the community structure of the modified network.

SpeakEasy is a label propagation community detection algorithm which identifies communities using top-down and bottom-up approaches simultaneously. Specifically, nodes join communities based not only on the nodes' local connections but also on the global information about the network structure. It adopts consensus clustering to get robust community structure. In our experiments, we choose to make 50 label propagation iterations with no node receiving a new label before terminating. We conduct 20 replicate runs for consensus clustering to get more robust and deterministic results.

GANXiS is a fast algorithm using a general speaker-listener information propagation process. It spreads one label at a time between nodes according to the interaction rules. The worst-case time complexity of GANXiS is $O(E)$.

Both GANXiS and SpeakEasy can detect overlapping communities, but for our experiments we configure them to detect only disjoint communities. Once the community structure is found, we measure its quality using several metrics to check the performance of our approach. The impact of selecting the value of parameter f on performance is discussed in Sect. 4.3. Our approach differs from [2] since we do not know the ground truth networks. Instead, we consider different metrics (see Sect. 4.1 for details) and if the majority of them agree on the improvement of communities after the replacement, we accept the results.

4 Evaluation and Analysis

4.1 Community Quality Metrics

To evaluate our approach without ground truth, we adopt two global community quality metrics, modularity (Q) [15] and modularity density (Q_{ds}) [16], and the following six local community quality metrics: *Intra-density ID*, *Contraction CNT*, *Expansion EXP*, *Conductance CND* [16], *Fitness F* [17], and the *Modularity Degree D* [18]. For the sake of space, we list the metrics above, and refer the reader to the cited references for their formal definitions and descriptions.

4.2 Dataset Descriptions

We consider seven real-world network datasets, including two friendship networks, two collaboration networks, and a co-purchasing network. Below we describe the basic properties of these datasets and provide the number of nodes (N) and edges (E) for each.

Gowalla was collected from a location-based social networking provider. There are 391,222 users with public profiles (friends and check-ins) that were active from the middle of September 2011 to late October of the same year [19]. There are 2,176,188 edges in this network that indicate friendships between users.

Amazon is a product co-purchasing network of the Amazon website [20]. There are 334,864 nodes in the network that represent products and 925,872 edges that link commonly co-purchased products.

DBLP is a scientific collaboration network with 317,080 nodes representing authors and 1,049,866 edges connecting authors that have co-authored a paper [20].

Santa Fe is the largest connected component of the collaboration network of scientists at Santa Fe Institute during the years 1999 and 2000 [21]. It has 118 nodes and 200 edges.

Football is a network that represents the schedule of games between college football teams in a single season with 115 nodes and 613 edges [21].

Dolphin is a social network of frequent associations between 62 dolphins connected by 150 edges and living in a community off Doubtful Sound, New Zealand [22].

Karate is a small network representing the friendships between 34 members of a karate club at a US university during two years [23]. It has 78 edges.

4.3 Experimental Results

In this part, we present the quality metrics for the community structure in which the percentage $f = [0, 1, 2, 5, 10, 15, 20, 25, 30, 40, 50]$ of edges were replaced. $f = 0$ means that there is no change to the original networks.

Figure 1 shows the results for the Gowalla dataset. The horizontal lines show the quality metric of the community structure detected in the original unchanged network. All three link predictors improved the networks according to eight community quality metrics. PropFlow performs extremely well on Gowalla, except for the Expansion measure. The improvement goes beyond values of $f \leq 50$ reported here. We can observe a limit (varying for different link prediction metrics) of how many links could be replaced for the purpose of improving the community structure of the

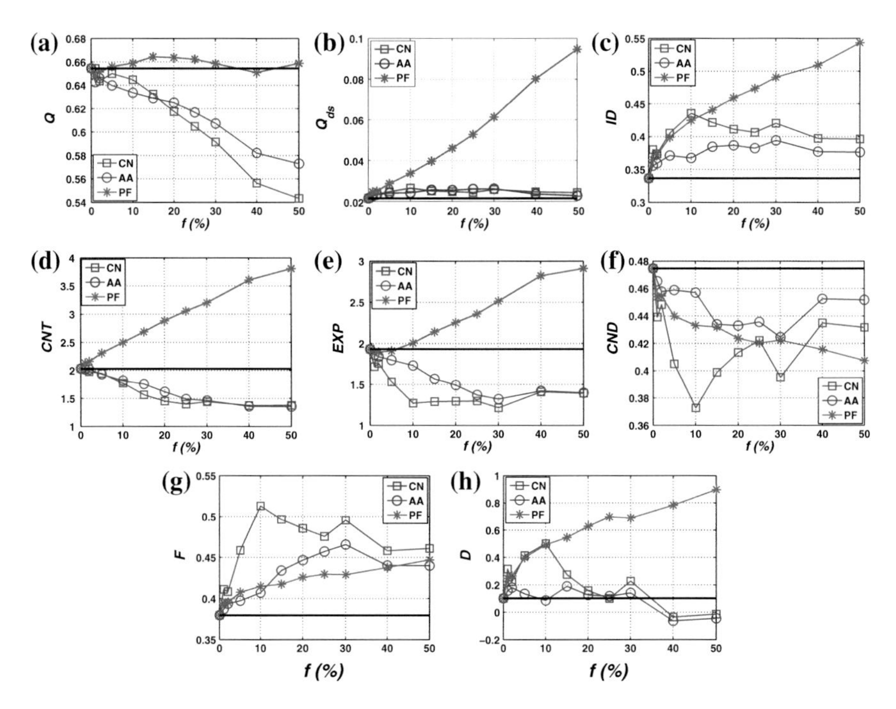

Fig. 1 The quality metrics for the community structure that SpeakEasy discovers on the networks generated from Gowalla using our link improvement method. **a** Q. **b** Q_{ds}. **c** Intra-density. **d** Contraction. **e** Expansion. **f** Conductance. **g** Fitness. **h** Modularity degree

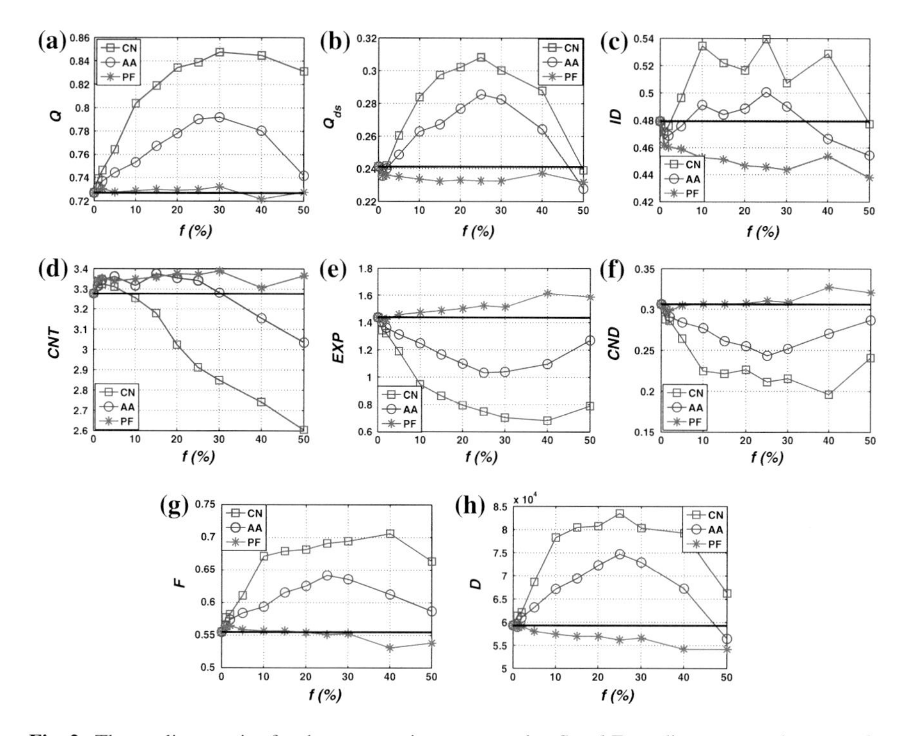

Fig. 2 The quality metrics for the community structure that SpeakEasy discovers on the networks generated from Amazon using our link improvement method. **a** Q. **b** Q_{ds}. **c** Intra-density. **d** Contraction. **e** Expansion. **f** Conductance. **g** Fitness. **h** Modularity degree

network. Figure 1b shows that the value of modularity density grows by an order of magnitude compared to the original value with PropFlow.

Figure 2 presents the results of the Amazon dataset. We can observe that CN and Adamic-Adar metrics work well on this network.

Figure 3 displays the results of DBLP. With *CN*, the values of Q_{ds}, *ID*, and *CNT* generally decrease as the replacing percentage f increases. While Q achieves its maximum at $f = 30$, *EXP* and *F* reach their optima at $f = 40$, and *CND* and *D* attain theirs at $f = 15$. With Adamic-Adar, the values of *Q*, *EXP*, *CND*, and *F* reach their optima at $f = 20$, while Q_{ds} and *D* achieve theirs at $f = 15$.

Results for smaller networks are summarized in Table 1. This excludes Santa Fe and Karate datasets which, although small, have a well-evolved edge structure and no need for improvement.

Table 1 presents link improvement results for CN and Adamic-Adar metrics on the Amazon, DBLP, Football, and Dolphin datasets. The results for PropFlow are omitted because it works well only on the Gowalla dataset. The cells in the table that contain the fraction of replaced links f show the best f for the corresponding community quality metric. *RI* stands for the relative improvement

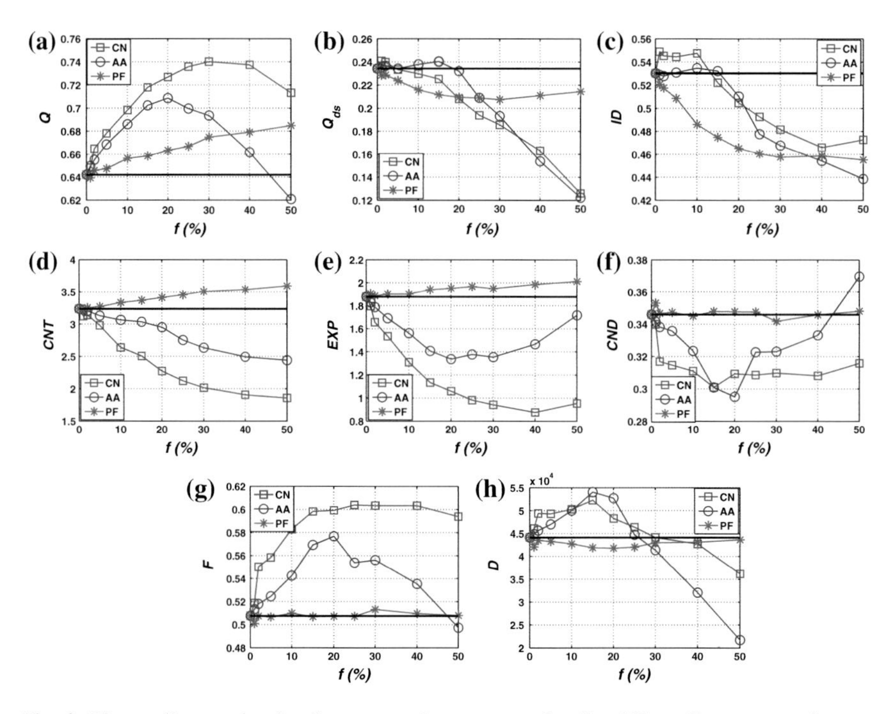

Fig. 3 The quality metrics for the community structure that SpeakEasy discovers on the networks generated from DBLP using our link improvement method. **a** Q. **b** Q_{ds}. **c** Intra-density. **d** Contraction. **e** Expansion. **f** Conductance. **g** Fitness. **h** Modularity degree

of the corresponding community quality metric. It is the percentage of improvement of the metric attained with the best f compared to its original value for the unchanged network ($f = 0$). The ✗ in the table indicates that our link prediction method results in no improvement ($RI < 0$) in that row. This table and all the above figures show that our link improvement procedure is able to significantly refine the community structure of five out of seven networks.

Generally, our method performed best when the number of common neighbors were used as the link prediction metric, followed by the Adamic-Adar metric. Yet for the Gowalla network, the best performing link prediction metric was PropFlow. Therefore, we conclude that a single link prediction metric cannot perform well on all networks. The basic reason for this is that each metric performance depends on the meaning of the relationships which define links in the network. Another reason is that networks also differ in their size, structure, and dynamics. The impact of these factors on link prediction is often unclear. Thus, our method can be used to evaluate the performance of link prediction metrics. If the highly ranked predicted links do not improve quality of communities, they are unlikely to be formed quickly. We also observe that there is a threshold (which varies for different link prediction metrics)

Table 1 The best replacing percentage *f* and the corresponding relative improvement (*RI*) of the community quality metric achieved using the two link improvement method: the number of common neighbors and Adamic-Adar on Amazon, DBLP, Football, and Dolphin

		Datasets							
		Amazon		DBLP		Football		Dolphin	
		f	*RI*	*f*	*RI*	*f*	*RI*	*f*	*RI*
CN	*Q*	30	16.6	30	15.3	***15***	20.5	20	11.7
	Q_{ds}	25	27.8	✗	✗	***15***	34.1	10	25.7
	ID	25	12.6	✗	✗	10	9.8	10	9.4
	CNT	✗	✗	✗	✗	***15***	20.4	25	3.9
	EXP	***40***	***52.6***	***40***	***53.4***	***15***	43.6	***30***	52.3
	CND	***40***	36.0	15	13.0	***15***	41.9	25	42.1
	F	***40***	27.2	***40***	18.8	***15***	35.0	***30***	40.5
	D	25	40.8	15	18.5	***15***	***69.2***	25	***73.0***
AA	*Q*	***30***	8.9	***20***	10.4	15	16.9	15	7.1
	Q_{ds}	25	18.4	15	2.6	15	33.2	10	20.8
	ID	25	4.4	✗	✗	15	9.1	✗	✗
	CNT	15	3.0	✗	✗	***20***	19.4	✗	✗
	EXP	25	***28.2***	***20***	***28.8***	***20***	36.0	20	33.4
	CND	25	20.5	***20***	14.7	***20***	36.5	***30***	20.2
	F	25	15.6	***20***	13.6	***20***	29.1	***30***	18.8
	D	25	26.0	15	22.5	15	***60.1***	15	***40.4***

of how many links could be replaced for the purpose of improving community structure of a network. Going beyond this threshold may lead to higher cost and lower performance although the quality of the community structure may still be better than that of the original unchanged network.

4.4 Impact of the Community Detection Algorithm

To test how much our outcomes depend on the choice of the community detection algorithm, we present here the results of the experiments on one of the largest datasets, Amazon, using GANXiS algorithm [14]. Figure 4 shows the experiment outcomes that are qualitatively similar to those reported in Sect. 4. The scale of improvements and the range of percentages *f* over which the improvements are seen are very similar, while the order of the link prediction methods sorted by their performance remains the same. This is a clear indication that switching to a different community detection algorithm did not impact our conclusions about the use of link prediction methods on the Amazon dataset.

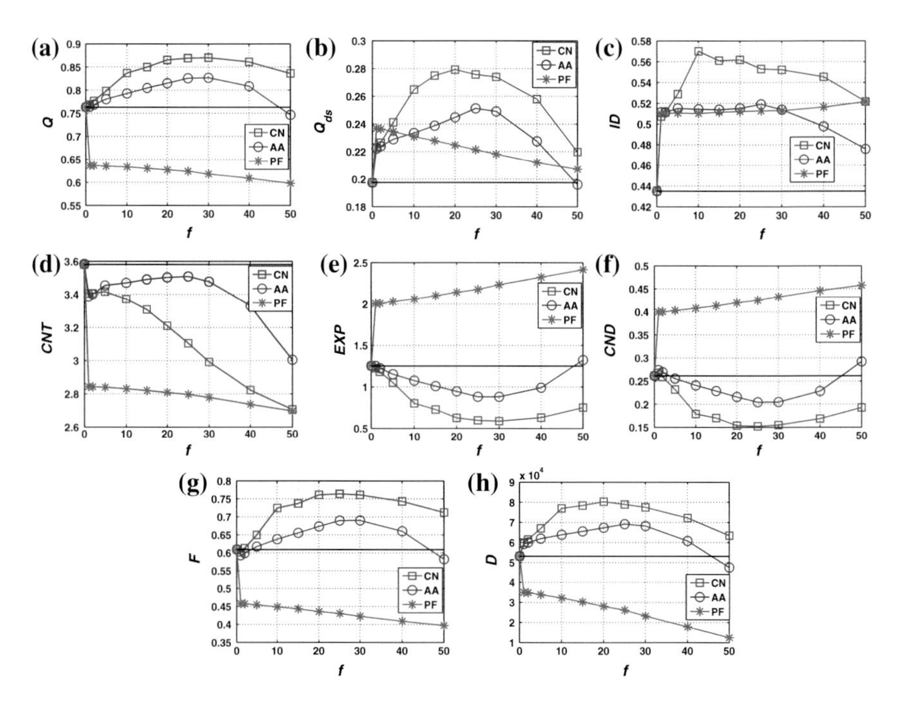

Fig. 4 The quality metrics for the community structure that GANXiS discovers on the networks generated from Amazon using our link improvement method. **a** Q. **b** Q_{ds}. **c** Intra-density. **d** Contraction. **e** Expansion. **f** Conductance. **g** Fitness. **h** Modularity degree

5 Evolving Networks

In search for ground truth examples to further verify our methodology, we apply it to social networks evolving from their initial state. This evolution happens when a group of people is put together and makes initial links which are being refined later based on increasing members knowledge about each other. Examples are when middle school students move to high school or later when they join universities. Links from the initial period that are a miss dissolve and are replaced by other links while some initial links persists. The final stage is a stable community, like researchers in Santa Fe or Karate club members. This is distinct from truly dynamic networks where change is constant. So here we are not trying to detect changing points in evolving networks or evolving community structure in dynamic networks. The networks that we consider evolve from an initially suboptimal state to a final stable state. By observing evolving networks at different stages, we are able to measure how our predictions of edge creation and persistence based on the past network states compare to the future state. We use the ArnetMiner paper citation and author collaboration dataset [24]. We restrict the network to a subgraph whose nodes represent authors only in the United States, since collaboration requires language and location affinity.

We divide the dataset into three subsets defined by time periods containing collaborations between 1995–2004, 2005–2009, and 2010–2014. To compare collaborations within any two time periods, we first find the subgraphs for each period that contain an intersection of nodes of both graphs. For the chosen periods, the subgraphs have $N = 18{,}382$ nodes and $E_o = 44{,}182$, $E_n = 43{,}267$ edges in the older and newer subgraph, respectively. The newer subgraph has $EN = 28{,}996$ existing new edges and $EP = 14{,}271$ existing edges persisting from the older subgraph. With these values, the fraction of new edges in a random sample of edges non-existing in the older graph is $\frac{2*EN}{N(N-1)-2*E_o}$ which is 0.017 % for our graph.

Next, non-existing but highly ranked edges are added as new edges to the older subgraph while low ranking existing edges are removed from it. Then using the newer graph as ground truth, we count how many new edges in the older subgraph actually exist in the newer one. Creating new edges successfully requires a metric consistent with the meaning of the links, which in our case is co-authorship. Hence, in addition to the CN metric, we also introduce a new metric, *Complete Recent Triangles* (CRT) for ranking edges. CRT first identifies all new triangles that are created by adding a new edge to the network. For each such newly created triangle, the CRT metric increases the score of the new added edge by the sum of weights of the two previously existing edges of the triangle. The weight of each such edge is the sum of the recency values of papers co-authored by the authors represented by the edge endpoints. Four age categories are set; less than 2, 2–4, 4–6 years, and older than 6 years. The corresponding values of recency are 1, 0.8, 0.65, and 0.5.

To measure the accuracy and coverage of edges selected as new or persistent by our method, we vary f from 1 to 50. The results are computed with the older period set to 1995–2004, and the newer one set to 2005–2009. The results based on other periods are qualitatively similar. Then, for each f, we compute the numbers of all edges selected as new by the link prediction SN(Selected and New), and the number of such edges that actually exist in the newer subgraph is denoted SEN. The ratio of SEN to SN measures the quality of selection of new edges while the ratio of SEN to EN tells us what percentage of the new edges is covered by the selected new edges. Table 2 shows the results. For the ArnetMiner networks generated for the periods selected for the reported experiments and for CRT, the first ratio varies between 12.22–17.91 %, thus it is up to 1,000 times greater than the fraction of new edges in a random sample of edges. The second ratio shows that the coverage of new edges reaches up to 9.3 % for $f = 50$. The results for CN are worse, the first ratio peaks at $f = 5$ and yields 13.8 % while the coverage peaks at 8.4 % for $f = 50$. For the middle range of f the two metrics perform similarly.

For ArnetMiner network, we used edge persistence selection which is complementary to edge deletion considered for the other network. Like previously, we first rank all existing edges using link prediction method, here CN and CRT. Then we remove $100 - f$ percentage of the lowest ranking edges, thus preserving f percentage of existing edges as persistent. SP denotes the number of existing edges selected as persistent, while SEP is the number of those edges that actually exist in the newer subgraph.

Table 2 Results of predicting added and deleted edges in evolving networks

Measurements	*f*									
	1	2	5	10	15	20	25	30	40	50
SEN/SN for CN	9.30	12.91	13.81	13.67	13.47	13.07	12.13	11.54	10.68	11.08
SEN/SN for CRT	18.14	17.10	16.03	14.24	13.29	12.78	12.97	12.86	13.02	13.42
SEN/EN for CN	0.14	0.39	1.05	2.08	3.08	3.98	4.62	5.28	6.51	8.44
SEN/EN for CRT	0.28	0.52	1.22	2.17	3.04	3.89	4.94	5.88	7.94	10.23
SEP/SP for CRT	79.86	72.51	64.16	58.68	54.01	51.92	49.32	47.11	44.27	42.71
SEP/EP for CRT	1.18	2.16	4.87	9.23	13.33	17.88	22.41	27.27	39.00	54.68

All ratios are represented as percentages

The results with CN for edge persistence are at the level of random chance, and they are clearly impacted by the massive number of deleted edges. Therefore, we omitted those results here. However, using CRT we again observe the improvements. When persistent edges are selected randomly, the success rate is 32.3 %. At the same time, using CRT with the two smallest values of f, yields the success rate of over 70 %. The best success rate of 79.9 % is achieved with $f = 1$ which is 2.5 times greater than at random. The coverage of persistent edges reaches 54.7 % for $f = 50$.

6 Conclusion and Future Work

In this paper, we introduce an approach for improving the network community structure by removing a certain fraction of low ranking existing links and replacing them with highly ranked new links. The proposed method significantly improves the community structure of the networks we considered. However, there is a threshold of how many links can be replaced in order to refine the community structure of a network. Going beyond this threshold may lead to higher cost and lower performance.

Generally, the link improvement method using the number of common neighbors for link prediction has the best performance, followed by Adamic-Adar, while PropFlow performs extremely well only on Gowalla dataset. We conclude that a single link prediction method cannot perform uniformly well on every network. Some metrics are more suitable than others for a particular network depending on the nature of the links. This was confirmed by our study of the evolving network in which a new link prediction metric for co-authorship, Complete Recent Triangles, delivered the improvement of three orders of magnitude over randomly selecting new edges. Finally, we observe that there is a correlation between the performance of link prediction improvement and certain network properties. Two influential factors are the network size and the degree to which nodes possess global knowledge about the network structure. To confirm that our conclusions do not depend on the use of a specific community detection method, we processed the Amazon dataset with two community detection algorithms, obtaining similar results.

In the future, we plan to design and adopt more link prediction metrics for our approach to explore their performance on different types of networks. We also plan to explore how much our link improvement method could refine the quality of overlapping community structure.

References

1. Fortunato, S.: Community detection in graphs. Phys. Rep. **486**, 75–174 (2010)
2. Yan, B., Gregory, S.: Finding missing edges and communities in incomplete networks. J. Phys. A **44**, 495,102 (2011)
3. Chen, M., Kuzmin, K., Szymanski, B.: Community detection via maximization of modularity and its variants. IEEE Trans. Comput. Soc. Syst. **1**(1), 46–65 (2014)

4. Yan, B., Gregory, S.: Finding missing edges in networks based on their community structure. Phys. Rev. E **85**(5), 056,112 (2012)
5. Lancichinetti, A., Fortunato, S., Radicchi, F.: Benchmark graphs for testing community detection algorithms. Phys. Rev. E 78, 046,110 (2008)
6. Liu, Z., He, J.L., Kapoor, K., Srivastava, J.: Correlations between community structure and link formation in complex networks. PLoS ONE **8**, 72,908 (2013)
7. Liu, Z., Dong, W., Fu, Y.: Local degree blocking model for missing link prediction in complex networks (2014). arXiv:1406.2203
8. Yan, B., Gregory, S.: Detecting community structure in networks using edge prediction methods. J. Stat. Mech: Theory Exp. **2012**(09), P09,008 (2012)
9. Burgess, M., Adar, E., Cafarella, M.: Link-prediction enhanced consensus clustering for complex networks (2015). arXiv:1506.01461
10. Lichtenwalter, R.N., Chawla, N.V.: LPmade: link prediction made easy. J. Mach. Learn. Res. **12**, 2489–2492 (2011)
11. Adamic, L., Adar, E.: Friends and neighbors on the Web. Soc. Netw. **25**(3), 211–230 (2003)
12. Lichtenwalter, R.N., Lussier, J.T., Chawla, N.V.: New perspectives and methods in link prediction. In: Proceedings of the 16th ACM SIGKDD International Conference Knowledge Discovery and Data Mining, pp. 243–252. New York, NY, USA (2010)
13. Gaiteri, C., Chen, M., Szymanski, B.K., Kuzmin, K., Xie, J., Lee, C., Blanche, T., Neto, E.C., Huang, S.C., Grabowski, T., Madhyastha, T., Komashko, V.: Identifying robust clusters and multi-community nodes by combining top-down and bottom-up approaches to clustering. Scientific Reports 5 (2015)
14. Xie, J., Szymanski, B.K., Liu, X.: SLPA: Uncovering overlapping communities in social networks via a speaker-listener interaction dynamic process. In: Proceedings of the Data Mining Technologies for Computational Collective Intelligence Workshop, pp. 344–349. IEEE (2011)
15. Newman, M.E.J., Girvan, M.: Finding and evaluating community structure in networks. Phys. Rev. E **69**, 026,113 (2004)
16. Chen, M., Nguyen, T., Szymanski, B.K.: A new metric for quality of network community structure. ASE Human J. **2**(4), 226–240 (2013)
17. Lancichinetti, A., Fortunato, S., Kertész, J.: Detecting the overlapping and hierarchical community structure in complex networks. New J. Physics **11**(3), 033,015 (2009)
18. Li, Z., Zhang, S., Wang, R.S., Zhang, X.S., Chen, L.: Quantitative function for community detection. Phys. Rev. E **77**, 036,109 (2008)
19. Nguyen, T., Chen, M., Szymanski, B.: Analyzing the proximity and interactions of friends in communities in Gowalla. In: Proceedings of the IEEE 13th International Conference Data Mining Workshops (ICDMW), pp. 1036–1044 (2013)
20. Yang, J., Leskovec, J.: Defining and evaluating network communities based on ground-truth. In: Proceedings of the ACM SIGKDD Workshop on Mining Data Semantics, pp. 3:1–3:8. ACM, New York, NY, USA (2012)
21. Girvan, M., Newman, M.E.J.: Community structure in social and biological networks. Proc. Natl Acad. Sci. USA **99**(12), 7821–7826 (2002)
22. Lusseau, D., Schneider, K., Boisseau, O.J., Haase, P., Slooten, E., Dawson, S.M.: The bottlenose dolphin community of Doubtful Sound features a large proportion of long-lasting associations. Behav. Ecol. Sociobiol. **54**(4), 396–405 (2003)
23. Zachary, W.: An information flow model for conflict and fission in small groups. J. Anthropol. Res. **33**, 452–473 (1977)
24. Tang, J., Zhang, J., Yao, L., Li, J., Zhang, L., Su, Z.: ArnetMiner: Extraction and mining of academic social networks. Phys. Rev. E **78**, 046,110 (2008)

A Subgraph-Based Ranking System for Professional Tennis Players

David Aparício, Pedro Ribeiro and Fernando Silva

Abstract This paper introduces a novel ranking system for competitive sports based around the notion of subgraphs. Although the system is targeted specifically to professional tennis it could be applied to any dominance network due to its generality. The results of about 140,000 tennis matches played between Top-100 players are used to create a colored directed network where colors represent different surfaces and edge direction depends on head-to-read results between players. The main contribution of this work is a ranking system which relies on the occurrences of 4-node directed subgraphs and the positions (or orbits) where the players appear on them. Since the concept of orbit is intrinsically connected with node dominance, appearing frequently in dominant orbits indicates that the player himself is dominant. Even in a very sparse network and without any background knowledge on the tournaments or stages of the matches, our proposal is able to extract meaningful rankings which capture the intricate competitive relationships between players from different eras.

1 Introduction

Debating who is the best player (or team) is one of the most discussed topics in any competitive sport and it can stir heated arguments between fans. Objectively quantifying player achievements is not straightforward, even when personal preferences are set aside, since multiple criteria can be used to compare players and the sports themselves evolve throughout the years. Nevertheless, competitive sports require a system that is able to rank players (or teams) according to their performance.

Most existing ranking systems focus on some set of numerical features, with different weights and time spans used depending on the sport under consideration [11].

D. Aparício (✉) · P. Ribeiro · F. Silva
CRACS & INESC-TEC, DCC-FCUP, Universidade Do Porto, Porto, Portugal
e-mail: daparicio@dcc.fc.up.pt

P. Ribeiro
e-mail: pribeiro@dcc.fc.up.pt

F. Silva
e-mail: fds@dcc.fc.up.pt

H. Cherifi et al. (eds.), *Complex Networks VII*, Studies in Computational Intelligence 644, DOI 10.1007/978-3-319-30569-1_12

Professional tennis in particular is governed by the Association of Tennis Professionals (ATP) which ranks players based on their results in official ATP tournaments. The ATP ranking is updated on a weekly basis and aggregates the results from the previous 52 weeks. Points are awarded to players according to the round of the tournament that they reach and the ranking of the tournament itself. Recently, with the emergence of network science, node centrality metrics have been applied to sports datasets in order to derive rankings [5, 7, 9, 10]. The vast majority of these ranking methods are adaptations of the PageRank algorithm [2]. In this work we take a different perspective by instead considering the role of small subgraphs. Subgraph-based metrics have been used to evaluate node importance in other fields such as biology [12]. Our goal is to provide a ranking system that truly captures the dynamics of the network. For that purpose, we devise a ranking mechanism that considers not only the subgraphs themselves but also the position (or orbit) of the players in the subgraphs. Orbit information allows us to discover indirect dominance while at the same time weighting both inward and outward edges. This method contrasts with PageRank which essentially considers only one of the two possible edge directions, giving importance to wins and almost disregarding losses, or vice-versa.

Our approach was tested on one of the most popular individual sports: men's professional tennis. Our results show that, even without any kind of prior knowledge, the methodology put forward is able to produce consistent and meaningful results using only the topology of the dominance network.

2 Network Description

In order to construct the dominance network we first collected the names of all tennis players that have been ranked in the Top-100 of the ATP year-end rankings from 1974 until 2015 and then extracted their match information from Tennis Abstract.[1] Going beyond the Top-100 introduces noise in the data and is not necessary for our purposes since players below the Top-100 only enter a few major tournaments. A total of 856 tennis players have been in the Top-100 throughout the years and they have played about 140,000 matches between themselves. The amount of matches played annually on each surface is presented in Fig. 1 as well as the total number (dotted line). This number increased significantly in the 1990s but has dropped in recent years mostly due to changes in the ranking system that encourage players to only participate in the most prestigious tournaments and also thanks to an increased awareness of the sport's physical demands. Nowadays, most tennis tournaments are contested on either clay or hard courts, with only a handful of matches played on grass each year. Carpet was a popular surface until the mid-1990s but it was discontinued from the ATP Tour in the late 2000s. The surface characteristics affect the pace of the game, favouring different playing styles. Usually, grass is the *fastest* surface to play on, followed by carpet, hard and finally clay.

[1] www.tennisabstract.com.

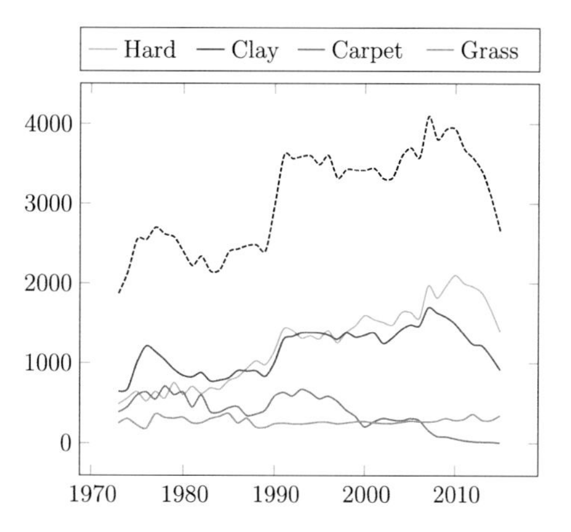

Fig. 1 Matches played by year on each surface

Following data extraction, the information is processed in order to construct 6-tuples of the form ($Player_1$, $Player_2$, *Surface*, *Year*, *Matches*, *WinPercentage*) for each pair of players. Besides creating a tuple for each surface, an additional 6-tuple is necessary to account for overall head-to-head. Using this data, a *dominance network* is created where nodes are players and the orientation of the *colored directed edges* between two players depends on their head-to-head on a given surface. Consider tuple $(p_i, p_j, s, t, m_{ij}^{s,t}, w_{ij}^{s,t})$ and parameters δ and ϕ: a colored directed edge (p_i, p_j) is created if player p_i won at least $\delta\%$ of the matches against p_j on surface s in a given year t (Eq. 1) and they played a minimum ϕ matches in surface s during their careers (Eq. 2). Our networks were built with $\delta = \frac{2}{3}$, meaning that one player only *dominates* another if he has defeated him in more than 66 % of the matches. A minimum of 3 matches ($\phi = 3$) is required to establish a dominance relation between two players on grass courts, and 5 for the other surfaces. An overall dominance relation that disregards playing surface also requires at least 5 matches.

$$w_{ij}^{s,t} \geq \delta\% \tag{1}$$

$$\left(\sum_{t=1974}^{2015} m_{ij}^{s,t} \right) \geq \phi \tag{2}$$

An *aggregated* (or *career*) dominance network is assembled by calculating dominances using the career win-percentage, instead of yearly results. The resulting network has 585 vertices and 5,301 directed edges with 5 possible labels (or *colors*): hard, clay, grass, carpet or overall. The number of *overall* edges is not simply the sum of the edges from all surfaces since an overall dominance is established by playing a minimum ϕ matches on any surface (for instance, one player can dominate another in overall matches without having ϕ encounters with him in any particular surface). Notice that only 585 of the original 856 players are represented in the network since the others did not play the required ϕ matches against any other Top-100

Table 1 Global network statistics of the dominance networks, discriminated by surface

Surface	$\|V\|$	$\|E\|$	$\frac{\|E\|}{\|V\|}$	$\frac{\|E_{\Rightarrow}\|}{\|E\|}$
Hard	301	868	2.88	0.64
Clay	289	793	2.74	0.65
Grass	140	173	1.24	0.90
Carpet	97	188	1.94	0.72
Overall	585	3279	5.61	0.68

player, and consequently have no edges. Requiring the win-percentage to be above a certain threshold δ for a dominance relation to be established results in the creation of bidirectional (or *reciprocal*) edges, meaning that two players met in at least ϕ matches but neither one dominates the other. The *dominant* (unidirectional) edges and *non-dominant* (bidirectional) edges are henceforth represented as $E_{\Rightarrow}$ and $E_{\Leftrightarrow}$, respectively. Table 1 summarizes the networks' global statistics. In regard to individual players, Jimmy Connors dominates the most other players (63), followed by Roger Federer (60) and Ivan Lendl (59). On hard courts Roger Federer leads with 46 out-edges, Guillermo Vilas on clay with 37, John McEnroe on carpet with 23 and Roger Federer on grass with 17 out-edges.

Two visual representations of the network are presented in Fig. 2. The giant component of the aggregate network is shown in Fig. 2a. Each edge color matches a surface: blue for hard, brown for clay, green for grass, pink for carpet and black for overall dominances. Node size depends on the number of out-edges; Roger Federer corresponds to the largest node since he has the most out-edges (132). Figure 2b shows the relations between all 25 players that have been ranked as the ATP Top-1 player. Edges are only relative to overall dominance and the line thickness reflects how unbalanced the relation is. It is interesting to notice that Jimmy Connors, one of the players with most out-edges, does not dominate any Top-1 player. The fact that he faced the others when they were closer to their prime than himself might be the main reason for this. It seems reasonable to expect younger players, which are at their peak, to dominate players declining in form. However, that is generally only the case for players of the same level: very good young players tend to dominate very good older players, but average young players do not usually win against very good older players. Furthermore, considering the players' full history allows us to capture the various stages of their careers. Comparing players from different eras might seem unfair if one inspects only individual relations but what really makes a player *dominant* is the global aspect of his career and the head-to-head results that he had against players from his own era, players from the era preceding his and players from the subsequent era. Therefore, it is difficult to infer that a player p_i from one era dominates another player p_j from a different era, however it is possible to say that p_i is *generally more dominant* than p_j, and those are the relations that we intend to capture using our ranking mechanism.

Figure 3 shows that the networks' in- and out-degrees follow a power law. Results are only presented for overall matches but the surface networks are also scale-free.

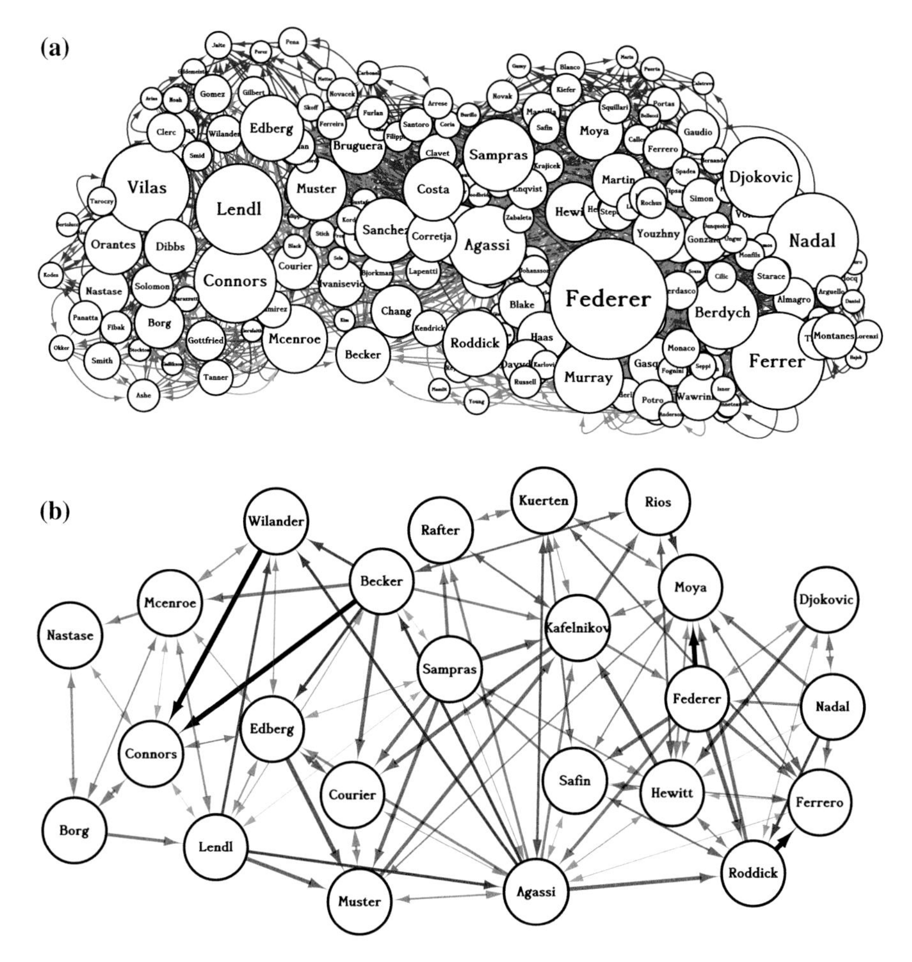

Fig. 2 Player dominance networks: in **a** *blue* edges are drawn for dominances in hard courts, *brown* for clay, *green* for grass and *pink* for carpet. The nodes' size increases proportionally with their out-degree. **b** shows the relations between all ATP Top-1 players, disregarding surface.

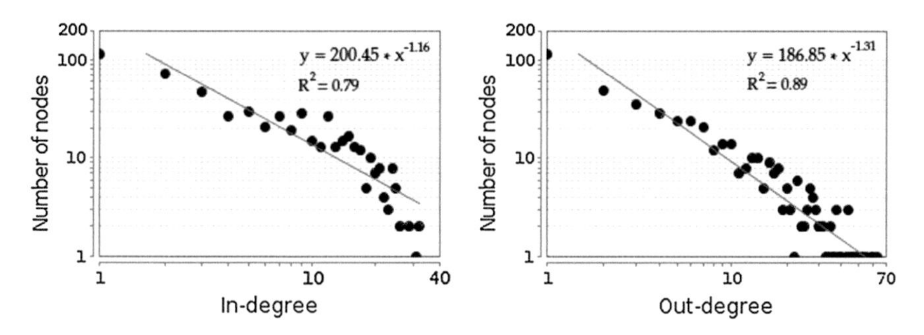

Fig. 3 In- and out-degree distribution of tennis players dominance networks (overall matches)

3 Subgraph-Based Ranking System

3.1 Related Work

The discussion of who is the best tennis player of all-time is open for debate and multiple criteria can be used. Ranking players simply by the number of matches that they won unduly favours players that had very long careers, such as Jimmy Connors, and ranking players by their win-ratio excessively benefits those that, like Björn Borg, retired at the peak of their careers. Furthermore, these possibilities do not take into account the intricate relations between the players. Grand Slam tournament victories (or *grand slams* for short) are often used to compare tennis players; however, before the 1990s several top-ranked players willingly skipped some of the annual Grand Slam tournaments since it was not yet the norm to evaluate players by their number of grand slams.

The work by Radicchi et al. [9] proposes a PageRank-like ranking system for male tennis players. Dingle et al [3] also used Radicchi's ranking system to produce a more up-to-date ranking of both male and female tennis players. The network that Raddicchi et al. built is different from our own since (a) their edges are *weighted* (w_{ij}: number of times that p_i beats p_j) while ours are *simple directed edges* reflecting win-percentages, (b) they used match information from 1969 until 2010 whilst our networks are relative to matches from 1974 to 2015 and (c) they only considered matches played on either Grand Slam tournaments or ATP Masters 1000 whereas we use information from all official ATP tournaments. Traditional PageRank does not decrease the node's rank with respect to its out-edges (in this case, meaning *loses against*) and is therefore not suitable to determine player dominance relations. The prestige score presented in [9] lowers the w_{ij} according to p_j's out-degree (the number of times p_j loses against someone); therefore, dominating a *dominated* node gives less prestige than dominating a more *dominant* player. However, the prestige score is not decreased according to p_i's out-edges, which may result in *dominated* players having a high score as long as they dominate a few *dominant* players. Our scoring system increases the players' score in respect to the players that they dominate and, likewise, decreases their score when they are themselves dominated. Another approach was followed by Motegi and Masuda [7] where they use a dynamic win-loss score that takes into account temporal information and fluctuations in the ranking. They not only consider direct wins and losses but also indirect ones, namely those corresponding to directed paths of size 2. Our work differs because we use subgraphs of size 4, which encapsulate more information than paths of size 2. Furthermore, we consider global dominance relations to obtain an earned ranking, while their work focuses on obtaining a temporal snapshot for a particular point in time and use it for prediction purposes.

3.2 Methodology

A simple way to assess node dominance is to compare its out-degree (*dominant*) with its in-degree (*dominated*). However, tennis players face a limited set of opponents due to their ranking (higher ranked players seldom play against lower ranked players) and career span (players from different periods never face each other). Moreover, requiring at least ϕ matches to be played for a relation to be established further decreases the amount of direct relations, resulting in very sparse networks $\left(\frac{|E|}{|N|^2} \leq 0.05\right)$. Therefore, comparing players only by degree is not sufficient.

Another option is to consider richer structural units: *subgraphs*. Actually, the degree of a vertex $v \in V(G)$ can be regarded as a 2-node subgraph where v occupies one of its two possible positions. In this work, instead of looking only at the directed degree (or subgraphs $\{v \rightarrow u\}$, $\{u \rightarrow v\}$ and $\{v \leftrightarrow u\}$), we analyse slightly larger subgraphs and observe at which position vertex v appears in each occurrence. As illustrated in Fig. 4, this allows not only for direct *dominances* ($a \rightarrow b$) or *equivalences* ($a \leftrightarrow b$) to be captured but also for indirect dominances ($a \rightarrow b \leftrightarrow c$, therefore $a \dashrightarrow c$) and super dominances ($a \rightarrow b \rightarrow c$, therefore $a \twoheadrightarrow c$) due to graph transitivity. This is particularly useful in the tennis players network since, as discussed previously, players have a very limited number of edges (direct dominances). Another advantage lies in the fact that it enables dominance relations to be established between players of different eras by following the path of the subgraph, such as $\{Federer \rightarrow Agassi \rightarrow Becker \rightarrow Connors\}$, which leads to the conclusion that $\{Federer \twoheadrightarrow Connors\}$. However, there are many other possible paths from Federer to Connors and in some of them Connors may actually indirectly dominate Federer. Therefore, all paths from one player to another must be enumerated in order to assess indirect dominances.

Graphlets [8] are subgraphs that take the node position of the subgraph (or *orbit*) into account. Graphlet usage is often restricted to analyzing only the set of 30 undirected graphs of up to five nodes due to computational limitations. Using undirected subgraphs would not produce meaningful results in dominance networks since edge direction is crucial. An extension of graphlets to directed networks was proposed in [1]. Graphlets can be used, for instance, to compare the topology of networks

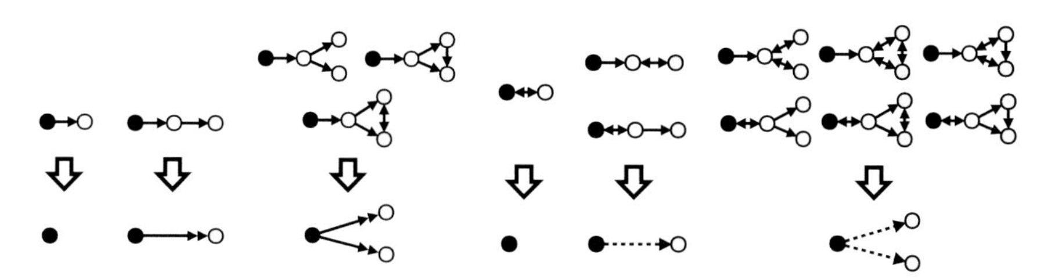

Fig. 4 Graph transitivity translated to *direct dominance* ($a \rightarrow b$), *super dominance* ($a \twoheadrightarrow b$) and *indirect dominance* ($a \dashrightarrow b$)

[4, 8] or nodes [6]. The key idea is to compute how many times a given node appears in an orbit and repeat that process for all possible orbits. Two nodes are more or less alike depending on how similar their orbit frequencies are. For instance, two nodes present at the center of multiple stars are more similar to each other than to another node that appears more frequently at the stars' periphery. Usually, graphlet computation is not concerned with specific types of subgraphs (such as chains, stars or cliques), but instead with all possible subgraphs of a given size. The results presented here are relative to all 199 possible directed subgraphs with 4 nodes.

In a first step, our subgraph-based ranking system receives as input a set of graphlets and assigns scores to their orbits. Then, during subgraph enumeration, the player's score is increased or decreased according to the orbits that he appears in. Orbit scores are calculated using the transitivity closure of the subgraph, as shown in Fig. 5, where d_{ij} is the path length between node n_i and node n_j. Notice that different nodes of a subgraph may be in the same orbit, and will always have the same score (see orbit e from subgraph G_B for instance). Looking at G_B we identify orbit f as *dominant* since it has 3 out-edges and no in-edges, while orbit e has no out-edges and 2 in-edges, representing a *dominated* orbit. Orbits a, b, c and d from G_A constitute a 4-node chain where the orbits at its start are more dominant than the ones at its end since they indirectly dominate more orbits. Orbits h, i and j of G_C form a cycle and are therefore equivalent. However, orbit j dominates k directly while h and i dominate k indirectly. Also, orbit i dominates orbit k more directly than orbit h does. These considerations are taken into account by our scoring mechanism.

Orbit scores are calculated as shown in Eq. 3. The main idea is to subtract the negative points $\left(\sum_{j=0}^{|\mathcal{S}_o|} \beta^{k-d(o_j,o)}\right)$ from the positive ones $\left(\sum_{i=0}^{|\mathcal{I}_o|} \beta^{k-d(o,o_i)}\right)$. Set $\mathcal{I}_o$ is formed by the orbits *inferior* to the orbit being computed while $\mathcal{S}_o$ is the set of orbits *superior* to it. The *distance* between o_i and o_j is given by $d(o_i, o_j)$ and it can be at most $k-1$, where k is the size of the subgraph. Basically, *direct dominant connections* give more points than indirected ones and, conversely, *direct dominated connections* take more points away. Parameter β controls the relative importance of the *directedness*, i.e. a small β (closer to 1) means that direct and indirect dominances

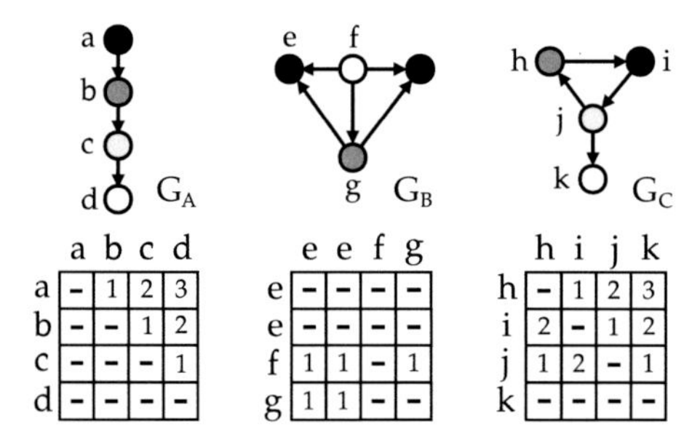

	a	b	c	d
a	-	1	2	3
b	-	-	1	2
c	-	-	-	1
d	-	-	-	-

	e	e	f	g
e	-	-	-	-
e	-	-	-	-
f	1	1	-	1
g	1	1	-	-

	h	i	j	k
h	-	1	2	3
i	2	-	1	2
j	1	2	-	1
k	-	-	-	-

Fig. 5 Graph transitivity of 3 subgraphs. Nodes with the same shade are in the same orbit. Orbit scores are assessed using the transitivity matrix: row values are *positive* points while column values are *negative*. Higher cell values mean that the connection is less direct

give roughly the same points while an high β means that direct dominances are more important. Notice that if β is too big the score becomes almost equivalent to the degree. A parameter $\lambda \in [0, 1]$ is also inserted to control the influence of *dominating* versus *being dominated*. Using $\lambda \approx 1$ means that a player is mostly evaluated by how many players he dominates (out-edges) while the amount of times that he himself is dominated (in-edges) does not have a big impact in the rankings, and vice-versa when $\lambda \approx 0$, i.e. the player is ranked higher if he is dominated by few players. These considerations produce a flexible scoring mechanism with just two parameters. The score of a player p_i is obtained by summing his occurrences in all orbits and multiplying them by their score, as shown in Eq. 4. Finally, players are ordered from the lowest to the highest score to produce the ranking.

$$S(o) = \left(\lambda \times \sum_{i=0}^{|\mathcal{I}_o|} \beta^{k-d(o,o_i)}\right) - \left((1-\lambda) \times \sum_{j=0}^{|\mathcal{S}_o|} \beta^{k-d(o_j,o)}\right) \tag{3}$$

$$S(p_i) = \sum_{o=0}^{|\mathcal{O}|} Fr(p_i, o) \times S(o) \tag{4}$$

4 Results

Table 2 presents the 15 players with the highest scores depending on λ. In the middle column λ is $\frac{1}{2}$, meaning that *dominating* and *not being dominated* is equally important for the players' scores, and this value is used for comparison. When $\lambda < \frac{1}{2}$ the ranking mechanism gives more importance to *not being dominated* that to *dominating* other players. Players such as Björn Borg and Gustavo Kuerten benefit from this parameter choice whereas Guillermo Vilas is penalized. If ones keeps decreasing λ, Rafael Nadal eventually tops the ranking because very few players have a positive win-loss ratio against him. However, making λ too small results in a meaningless ranking since players that have few out-edges unrealistically climb in the rankings as long as they have very few (or none) in-edges. By contrast, when $\lambda > \frac{1}{2}$, players such as Carlos Moya and Guillermo Vilas climb in the rankings while Björn Borg and Novak Djokovic drop some positions. Having $\lambda \approx 1$ still produces meaningful results since the ranking eventually stabilizes and ranks very highly players that dominate many others. Nevertheless, it does not seem fitting to completely disregard the *dominated* edges of the players when building a dominance based ranking system. In the remaining results λ is set to $\frac{1}{3}$, hence giving a slight edge to players that are not dominated by many others while still producing meaningful results.

Table 3 shows the career rankings with $\beta = 1.5$. To illustrate the effect of β take graph G_A from Fig. 5 as an example: if $\beta = 1$, $S(a) = 1^{(4-1)} + 1^{(4-2)} + 1^{(4-3)} = 3$, $S(b) = 1$, $S(c) = -1$ and $S(d) = -3$; if $\beta = 2$, $S(a) = 2^{(4-1)} + 2^{(4-2)} + 2^{(4-3)} = 14$, $S(b) = 4$, $S(c) = -4$ and $S(d) = -14$. In pratice, this means that orbits a and b,

Table 2 Ranking obtained by varying λ: the relative weight between dominating (out-edges) and being dominated (in-edges)

Rank	Player		Player		Player	Player		Player	⇕
1	I. Lendl	1▲	I. Lendl	1▲	R. Federer	R. Federer		R. Federer	
2	R. Federer	1▼	R. Federer	1▼	I. Lendl	I. Lendl		I. Lendl	
3	J. Connors		J. Connors		J. Connors	J. Connors		J. Connors	
4	R. Nadal	1▲	A. Agassi		A. Agassi	A. Agassi		A. Agassi	
5	N. Djokovic	1▲	R. Nadal		R. Nadal	R. Nadal		R. Nadal	
6	B. Becker		N. Djokovic	1▲	B. Becker	B. Becker		B. Becker	
7	A. Agassi	3▼	B. Becker	1▼	N. Djokovic	S. Edberg	1▲	G. Vilas	2▲
8	B. Borg	5▲	S. Edberg		S. Edberg	N. Djokovic	1▼	S. Edberg	
9	S. Edberg	1▼	J. McEnroe	1▲	G. Vilas	G. Vilas		N. Djokovic	2▼
10	P. Sampras	2▲	G. Vilas	1▼	J. McEnroe	J. McEnroe		J. McEnroe	
11	J. McEnroe	1▼	L. Hewitt		L. Hewitt	L. Hewitt		L. Hewitt	
12	A. Murray	4▲	P. Sampras		P. Sampras	P. Sampras		Y. Kafelnikov	2▲
13	L. Hewitt	2▼	B. Borg		B. Borg	Y. Kafelnikov	1▲	P. Sampras	1▼
14	G. Kuerten	7▲	A. Murray	2▲	Y. Kafelnikov	C. Moya	3▲	C. Moya	3▲
15	G. Vilas	6▼	A. Roddick		A. Roddick	B. Borg	2▼	D. Ferrer	2▲
	$\lambda = \frac{1}{6}$		$\lambda = \frac{1}{3}$		$\lambda = \frac{1}{2}$	$\lambda = \frac{2}{3}$		$\lambda = \frac{5}{6}$	

for instance, are much more alike when $\beta = 1$ than when $\beta = 2$. A low β (≈ 1) does not distinguish direct from indirect relations while a high β (≈ 2) penalizes indirect ones too heavily, therefore an intermediate value for β (1.5) was chosen.

Roger Federer is the most dominant player since 1974 according to our ranking system, followed by Jimmy Connors and Ivan Lendl. Evaluating if the results are correct is not straightforward and highly subjective. Nonetheless, one of the most commonly used criteria to judge the quality of a tennis player is the number of grand slams that he won during his career. From Table 3a it can be observed that winning grand slams is correlated with a higher position in our ranking. From the Top-25 players only David Ferrer, Tim Henman and Robin Soderling failed to win any grand slams. Table 3b shows the Top-10 by surface and also the number of grand slams contested on that surface that they won. Roger Federer is the most dominant player both on grass and hard courts, Guillermo Vilas is the best player on clay and McEnroe is ranked first in carpet courts. Again, the number of grand slam victories is correlated with the ranking. We point out that no grand slam tournament was ever contested on carpet. Table 3c gives a more in-depth look at all 25 players that have been the Top-1 player in the ATP rankings from 1974 until 2015. A dash (–) means that the player does not have a single connection on that particular surface, i.e. he did not play the minimum ϕ matches against anyone. The position of the player is presented in bold-face only if our system ranks him among the Top-25 of that particular surface. As can be observed, most (76 %) ATP Top-1 players are also

Table 3 Ranking of tennis players with $\lambda = \frac{1}{3}$ and $\beta = 2$: **a** Top-25 players, **b** Top-10 players by surface and **c** our rankings for all players ranked as Top-1 by the ATP

(a)

Rank	Player
1	R. Federer^{17}
2	J. Connors^{8}
3	I. Lendl^{8}
4	A. Agassi^{8}
5	R. Nadal^{14}
6	J. McEnroe^{7}
7	G. Vilas^{4}
8	N. Djokovic^{10}
9	B. Becker^{6}
10	P. Sampras^{14}
11	S. Edberg^{6}
12	A. Roddick^{1}
13	A. Murray^{2}
14	L. Hewitt^{2}
15	B. Borg^{11}
16	T. Muster^{1}
17	C. Moya^{1}
18	I. Năstase^{2}
19	D. Ferrer*
20	G. Kuerten^{3}
21	Y. Kafelnikov^{2}
22	A. Ashe^{3}
23	JC. Ferrero^{1}
24	T. Henman*
25	R. Soderling*

(b)

Rank	Player (Hard)	Player (Clay)
1	R. Federer^{9}	G. Vilas^{2}
2	N. Djokovic^{7}	R. Nadal^{9}
3	A. Agassi^{6}	T. Muster^{1}
4	A. Murray^{1}	S. Bruguera^{2}
5	A. Roddick^{1}	G. Kuerten^{3}
6	R. Nadal^{3}	M. Orantes^{1}
7	P. Sampras^{7}	B. Borg^{6}
8	L. Hewitt^{1}	M. Wilander^{3}
9	T. Berdych	I. Nătase^{1}
10	I. Lendl^{5}	I. Lendl^{3}

Rank	Player (Grass)	Player (Carpet)
1	R. Federer^{7}	J. McEnroe
2	J. Connors^{4}	B. Becker
3	Edmondson^{1}	I. Lendl
4	J. McEnroe^{3}	J. Connors
5	R. Tanner^{1}	G. Ivanisevic
6	B. Becker^{3}	P. Sampras
7	S. Edberg^{4}	B. Borg
8	N. Djokovic^{3}	A. Ashe
9	P. Cash^{1}	K. Rosewall
10	P. Sampras^{7}	B. Walts

(c)

Player	Overall	Hard	Clay	Grass	Carpet
I. Nătase	**18**	26	**9**	–	**18**
J. Newcombe	38	–	–	128	38
J. Connors	**2**	**11**	27	**2**	**4**
B. Borg	**15**	31	**7**	**12**	**7**
J. McEnroe	**6**	27	297	**4**	**1**
I. Lendl	**3**	**10**	**10**	90	**3**
M. Wilander	27	234	**8**	96	78
S. Edberg	**11**	**12**	118	**7**	70
B. Becker	**9**	192	55	**6**	**2**
J. Courier	41	**13**	37	32	73
P. Sampras	**10**	**7**	66	**11**	**6**
A. Agassi	**4**	**3**	63	28	41
T. Muster	**16**	178	**3**	–	–
M. Rios	33	252	**20**	–	–
C. Moya	**17**	190	149	–	–
Y. Kafelnikov	**21**	269	321	**22**	49
P. Rafter	381	243	193	**20**	–
M. Safin	46	298	31	167	–
G. Kuerten	**20**	**21**	**5**	–	–
L. Hewitt	**14**	**8**	177	496	–
JC. Ferrero	**23**	231	**16**	–	–
A. Roddick	**12**	**5**	76	63	–
R. Federer	**1**	**1**	**14**	**1**	–
R. Nadal	**5**	**6**	**2**	118	–
N. Djokovic	**8**	**2**	35	**8**	–

ranked as one of the Top-25 most dominant players by our system. The exceptions are John Newcombe, Mats Wilander, Jim Courier, Marcelo Rios, Patrick Rafter and Marat Safin. Patrick Rafter is a notable outlier since he is ranked at the bottom half of the table (381th out of 585 players). Notice however that he was only ranked as the ATP Top-1 for one week. Our ranking also detects surface specialists (such as Wilander, Muster, Rios, Kuerten and Ferrero on clay, Courier, Agassi and Hewitt on hard courts, and Newcombe and Rafter on grass), all-round players (such as Năstase, Connors and Federer) and players with an Achilles-heel on a specific surface (such as Sampras and Djokovic on clay, Borg on hard, and Lendl and Nadal on grass). We should note that, for instance, Rafael Nadal has a very low score on grass despite having a $\approx$79 % win-loss ratio in that surface and winning two grand slams on grass. His very low score comes primarily from the fact that he is dominated by Roger Federer on that surface and, because Federer is a hub-like node in grass, Nadal ends up appearing in many different subgraphs with Federer and the other players that Federer dominates. Since Nadal occupies a negative orbit in those subgraphs his score is continuously decreased. This negative effect is primarily felt on small and sparse networks such as the grass network where even a single connection has a very high impact. A possible solution to reduce the influence of hubs would be to ensure that each player only decreases the score of another player once.

5 Conclusion

The first contribution of this work is the distribution[2] of a network summarizing the complete match history between all male Top-100 ATP players since 1974. The data is discriminated by year as well as playing surface. The constructed dominance network models the relations between players: if a player wins against another one more than ϕ times and wins at least $\delta\%$ of the matches, a directed connection is drawn between them. An exploratory analysis was performed in order to verify that these choices are adequate and produce a meaningful representation. It was also observed that, like many real-world networks, both its in- and out-degree distributions follow a power-law, meaning that there are few very dominant players, few very dominated players and many average players.

We present a ranking system based on the subgraph topology of the dominance network that offers a different view than past approaches based on the PageRank algorithm. A complete subgraph enumeration is performed in the original network in order to compute the ranking. During the enumeration process, the position that the player appears in the subgraph is stored and his score is updated: if the player appears in a dominant orbit his ranking is increased, while if he appears in a dominated orbit his ranking is decreased. The ranking system does not require any meta-information about the network such as the tournament or the round that the players faced each other to produce meaningful results, however it could easily be extended to support it by adjusting the edge weights. The system is also flexible since it is possible to control (i) λ the importance of *being dominant* versus *being dominated* and (ii) β the importance of *direct* versus *indirect* dominances.

We assess which values of λ and β are better-suited for this particular tennis network and present rankings for the best overall players since 1974 and the most dominant players by surface. Our ranking system produces results that agree with the ATP ranking while at same time offering a different perspective since wins are not discriminated by tournaments (which some are more valuable than others) nor rounds (where a win in a later round gives more ATP points) and the intricate relations between players are also captured. This approach gives a better idea of actual player dominance which is valuable when trying to assess who are the best tennis players. Using our ranking system it was possible, for instance, to (i) observe that player performance is heavily influenced by the playing surface and (ii) discover which former ATP World Top-1 players were actually dominant players and which ones were not. We also performed a yearly ranking not included here for space concerns

[2]http://www.dcc.fc.up.pt/~daparicio/networks.

where we (i) observed that the most dominant players are usually the ones that reach more tournament finals, semi-finals and quarter-finals but they are not necessarily the ones that win more tournaments due to the unbalanced nature of ATP ranking system, (ii) identified which seasons were *most dominated* by a single (or a few) player(s), (iii) pinpointed tennis transition-eras (1987–1989 and 1999–2003) and (iv) noticed that it is rare for a player be very dominant both on fast (hard or grass) and slow courts (clay).

Acknowledgments This work is partially funded by FCT (Portuguese Foundation for Science and Technology) within project UID/EEA/50014/2013. David Aparício is supported by a FCT/MAP-i PhD research grant (PD/BD/105801/2014).

References

1. Aparício, D., Ribeiro, P., Silva, F.: Network comparison using directed graphlets (2015). arXiv:1511.01964
2. Brin, S., Page, L.: The anatomy of a large-scale hypertextual web search engine. In: Seventh International World-Wide Web Conference (WWW 1998) (1998)
3. Dingle, N., Knottenbelt, W., Spanias, D.: On the (page) ranking of professional tennis players. In: Computer Performance Engineering, pp. 237–247. Springer (2013)
4. Hayes, W., Sun, K., Pržulj, N.: Graphlet-based measures are suitable for biological network comparison. Bioinformatics **29**(4), 483–491 (2013)
5. London, A., Németh, J., Németh, T.: Time-dependent network algorithm for ranking in sports. Acta Cybernetica **21**(3), 495–506 (2014)
6. Milenković, T., Ng, W.L., Hayes, W., Pržulj, N.: Optimal network alignment with graphlet degree vectors. Cancer Inform. **9**, 121 (2010)
7. Motegi, S., Masuda, N.: A network-based dynamical ranking system for competitive sports. Sci. Rep. **2**(904) (2012)
8. Pržulj, N.: Biological network comparison using graphlet degree distribution. Bioinformatics **23**, 177–183 (2007)
9. Radicchi, F., Perc, M.: Who is the best player ever? a complex network analysis of the history of professional tennis. PloS ONE **6**(2), e17249 (2011)
10. Shan, Z., Li, S., Dai, Y.: Gamerank: ranking and analyzing baseball network. In: Social Informatics. pp. 244–251. IEEE Computer Society (2012)
11. Stefani, R.: Survey of the major world sports rating systems. J. Appl. Stat. **24**(6), 635–646 (1997)
12. Wang, P., L, J., Yu, X.: Identification of important nodes in directed biological networks: a network motif approach. PLoS ONE **9**(8), e106132 (2014)

Returners and Explorers Dichotomy in Web Browsing Behavior—A Human Mobility Approach

Hugo S. Barbosa, Fernando B. de Lima Neto, Alexandre Evsukoff and Ronaldo Menezes

Abstract A better understanding of the fundamental mechanisms underlying complex human dynamics is of major interest in contemporary social research. Over the last few years, researchers have made huge strides towards this understanding, thanks especially to the increasing availability of datasets containing digital traces of many human activities. In this work, we investigate Web browsing trajectories using a human mobility approach based on approximately four years of browsing history data. Our findings strongly suggest that return visitation patterns in browsing behaviors and in human mobility exhibit very similar scaling properties. Moreover, we classify Web users as returners and explorers based on their on-line activities, and show that at a population level, the distribution of both profiles agrees with empirical observations in human mobility. Finally, we create a network representation of the most popular websites from the aggregated browsing trajectories and uncover many functional clusters related with different users' activities.

1 Introduction

Uncovering fundamental mechanisms governing human dynamics is one of the ultimate goals of social investigation and research. Recently, the introduction of the complex systems apparatus into the social research has helped the uncovering of many universal regularities in human behaviors. For instance, the scaling proper-

H.S. Barbosa (✉) · R. Menezes
BioComplex Laboratory, Florida Institute of Technology, Melbourne, FL, USA
e-mail: hbarbosa@biocomplexlab.org

R. Menezes
e-mail: rmenezes@cs.fit.edu

F.B. de Lima Neto
Polytechnic School, University of Pernambuco, Recife, PE, Brazil
e-mail: fbln@ecomp.poli.br

A. Evsukoff
COPPE, Federal University of Rio de Janeiro, Rio de Janeiro, RJ, Brazil
e-mail: alexandre.evsukoff@coc.ufrj.br

H. Cherifi et al. (eds.), *Complex Networks VII*, Studies in Computational Intelligence 644, DOI 10.1007/978-3-319-30569-1_13

ties of: structural features of social networks [1, 2], travel distances and visitation frequencies in human mobility [3–6], and time intervals between consecutive events of human activities [3, 6, 7], to name but a few. More surprisingly, many of apparently unrelated behaviors were empirically found to follow common underlying processes. For instance, cumulative advantage–the principle in which popular options are more likely to be chosen rather than unpopular ones–was suggested to play an important role in shaping social networks [8–11] and human trajectories [5], whereas activity bursts–long periods of inactivity between short periods of intensive activity–were observed in distinct modern human behaviors, such as email communications, and Web browsing [12, 13].

Despite the large body of literature suggesting the existence of general mechanisms underlying behaviors of apparently distinct nature (e.g. human displacements and social interactions), the roots of such universal mechanisms are not widely agreed, let alone understood.

The difficulty of searching for universal mechanisms for human dynamics stem from the differences among certain human activities. For instance, if at one hand, human trajectories are shaped by environmental factors (e.g. physical or spatial constraints) [14, 15], Web browsing, on the other hand, cannot be said to be *physically* constrained or spatially bounded, in the sense that the amount of time necessary to *move* from one location to another is irrelevant. In this work we trace a parallel between human mobility and Web browsing behaviors. The motivation is that if such different behaviors are indeed regulated by common fundamental mechanisms, a better understanding of one can provide important insights on the other.

The contribution of this work is threefold. First, we applied methods from the human mobility analysis framework to Web browsing history data, unveiling that visitation patterns in both activities share similar scaling properties, suggesting that both behaviors might be generated by a common fundamental process. Second, we employed a graph-based approach to generate *abstract mobility networks*, high-level representations of Web browsing *trajectories*, and created a spatial distribution of the most visited Web sites in our data. Finally, we focused our attention in profiling Web users as *returners* and *explorers* as proposed by Pappalardo et al. [16].

2 Materials and Methods

2.1 Datasets

Our findings are based on the empirical analyses of 44 months of Web browsing history data generated by 521 anonymous users between September 2010 and May 2014. For comparison reasons, we also analyzed 6 months of mobile phone traces of 30,000 users and 2.5 years of a location-based social networking check-ins produced by approximately 13,000 users. The datasets can be summarized as follows:

- **Web browsing history:** The dataset consists of more than 5M anonymized Web browsing history entries corresponding to visits to 187,680 hosts by 524 users between September 21, 2010 to May 24, 2014.[1]
- **Mobile phone traces:** 6 month of CDR data produced by 30,000 users in one of the largest metropolitan areas in Brazil.
- **BrightKite:** approximately 2.5 years of geo-tagged check-ins produced by more than 13,000 BrightKite users, in three metropolitan areas in the US, namely Los Angeles, San Francisco and New York.[2]

Although the Web browsing dataset provides information at the URL level, we decided to use the host information since we are interested in the websites being accessed rather than the unique pages. Also, only requests that originated actively by the user were considered. Thus, we analyzed only three types of requests[3]:

- LINK: When a user followed a link and got a new toplevel window;
- TYPED: A user typed the page's URL in the URL bar, selected it from URL bar autocomplete results or clicked on it from a history query;
- BOOKMARK: When a user followed a bookmark;

Additionally, since all the location information in the dataset is hashed, we generated a lookup table to retrieve the actual domains being accessed.

2.2 *The Web Browsing Mobility Network*

The traces of Web browsing mobility were extracted from the browsing history data where *locations* are represented by the *host* of the pages. Paths correspond to the sequence in which the locations were visited. The weight $w_{i,j}$ of an edge is the number of *jumps* from site i to site j, representing a *proximity* between the two *locations*. The greater the weight, the closer the two locations must be in the abstract mobility space. For each user in the dataset we extract a visitation sequence $S = [s_i]$, $i \in 1 \ldots n$, from which we produced the Web browsing mobility network. Figure 1 illustrates the process of creating this network.

Because Web browsing trajectories have no spatial dimension, we used the Fruchterman-Reingold graph layout algorithm [19] to generate what we called the *Abstract Mobility Space (AMS)*, defined as: *a 2D space where locations are arranged based on how frequent they are visited in sequence (i.e. based on the weight $w_{i,j}$).* At some extent, this approach is similar to the work of Dragulescu with visualization

[1]The browsing history data is provided by the Web History Repository project http://webhistoryproject.blogspot.com/.

[2]Brightkite was a location-based social networking service launched in 2007 and closed in 2011 [17, 18].

[3]For more details on the structure of the browsing history, visit http://forensicswiki.org/wiki/Mozilla_Firefox_3_History_File_Format.

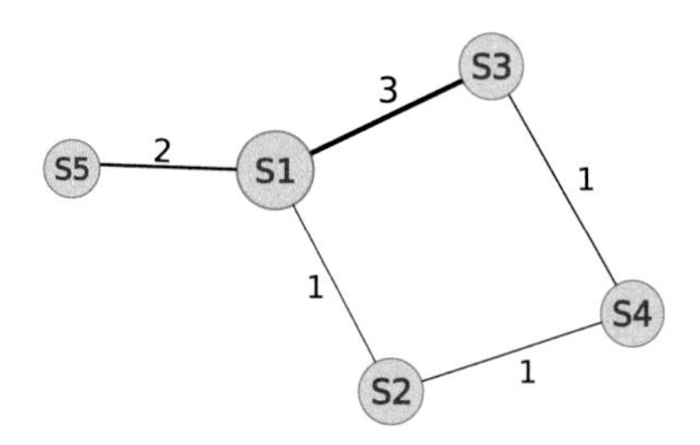

Fig. 1 Web browsing mobility network generated from the subtrajectory $S = [s_1, s_3, s_4, s_2, s_1, s_3, s_1, s_5, s_1]$

of on-line communities,[4] in the sense that both works use a force-directed algorithm to create a *geographical abstraction* of a network. It is worthy noting that the AMS is generated from the aggregated data across all users in the datasets.

Notice that the AMS is, by definition, a visual metaphor of the mobility networks, with no intrinsic scale or orientation. Hence, its only important spatial property is that if two *locations* i and j are frequently visited consecutively (i.e. the edge weight $w_{i,j}$ is high), their corresponding vertices will be separated by a distance $d \propto w_{i,j}^{-1}$. Notice that the AMS can be generated from any mobility network, not exclusively from Web browsing mobility. In fact, we also generated the AMSs for actual mobility networks extracted from the mobile phone and BrightKite datasets, using the same logic depicted in Fig. 1.

It is needless to say that the graph-layout algorithm of choice is by no means intended to model, reproduce or mimic any actual geographical or physical property of the graph (even though it is inspired in a mechanical system of springs [19]). Rather, its purpose is to generate aesthetically-pleasing pictures of graphs, such that their topological features are preserved and emphasized [19].

Nevertheless, for the purpose of this work they can be helpful in the sense that they can reproduce the indented properties in the graph layout. One limitation however is the fact that the graph layout is bounded by the plotting area. Therefore, very long edges are absent from the graph, which places a rigid upper limit to the jump lengths as measured from the AMS.

However, the soundness of the AMS approach relies heavily on the topological similarities between the traveling and browsing mobility networks. Significant discrepancies between them would weaken any evidence of a universal common process underlying both behaviors. Hence, we compared the degree distribution of the mobility networks.

We can see that the Browsing and BrightKite Mobility networks both exhibit a power-law degree distribution

$$p(k) \sim (k + k_0)^{-\alpha},$$

with comparable fitting parameters (see Table 1). In Fig. 2, the solid line corresponds to a power law with $\alpha = 1.90 \pm 0.008$. The phone data on the other hand, totally

[4]Ekisto: an interactive visualization of online communities http://ekisto.sq.ro/.

Table 1 Estimated parameters of the power-law fits for the mobility networks degree distribution with parameter estimation error σ_α

Data	k_0	α	σ_α
Web mobility	13.0	1.907	0.008
BrightKite–Los Angeles	24.0	2.15	0.04
BrightKite–New York	26.0	2.30	0.06
BrightKite–San Francisco	29.0	2.19	0.04

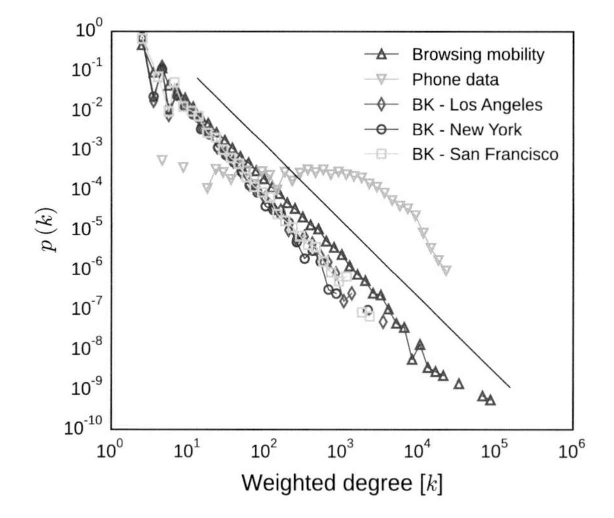

Fig. 2 Weighted degree distribution of mobility networks. As one can see, the Browsing and BrightKite Mobility networks both exhibit a power-law degree distribution, with comparable fitting parameters (see Table 1). The *solid line* corresponds to a power law with $\alpha = 1.90$ and is plotted as a guide to the eye. The phone data on the other hand, totally deviates from the power-law shape, being better approximated by a log-normal distribution with parameters $k_0 = 185$, $\mu \approx 7.62$ and $\sigma \approx 1.042$

deviates from the power-law shape, being better approximated by a log-normal distribution with parameters $k_0 = 185$, $\mu \approx 7.62$ and $\sigma \approx 1.042$.

The deviation in the phone data can be explained by the communication infrastructure. Cell phone towers are distributed according to factors such as population density, expected traffic load and technical specifications of the equipments. Since the weighted degree distribution in the mobility network is directly related with its traffic load, it is clear that the presence of a power-law degree distribution would represent a poor infrastructure design, with many underused antennas.

3 Results

Before approaching the *explorer* and *returners* dichotomy, the validity of our approach depends heavily on level of agreement between browsing visitation and human mobility patterns, especially the visitation frequencies distribution. The reason is because very different functional forms would conflict with the hypothesis of a common underlying process shared by distinct behaviors.

Maybe one of the most widely-reported features of human returning behavior is the heavy-tailed visitation frequencies distribution [5, 20, 21], i.e. most of the visits of an individual are to very few highly-visited locations (e.g. home and work); visitation frequencies are better approximated by a power-law distribution.

For the Web browsing behavior, our analyses have shown that the frequency distribution is indeed very similar to what we observed in human mobility data (see Fig. 3). As one can see, the parameters of the power-law fittings do not differ much across the different datasets (see Table 2). Moreover, the scaling parameter observed in the browsing mobility data ($\alpha \approx 1.85 \pm 0.05$) is very close to what was previously observed in human trajectories as reported by Song et al. ($\alpha \approx 1.83 \pm 0.07$) [5].

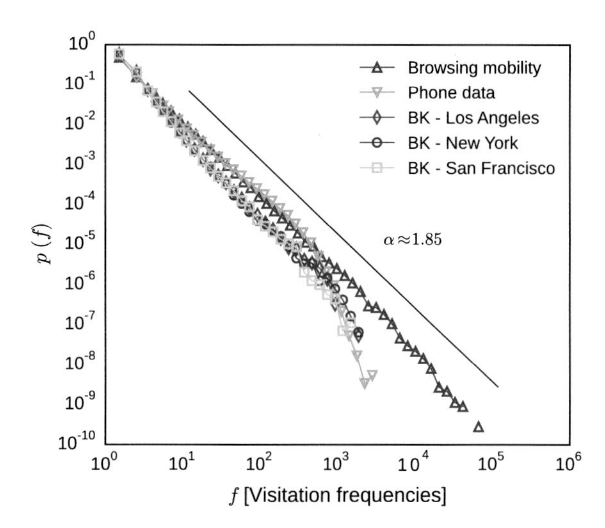

Fig. 3 Distribution of visitation frequencies. We can see that all distributions are approximated by a power law. The *solid line* correspond to a power-law fit with $\alpha \approx 1.85$ shown as a guide to the eye. See Table 2 for the best fit parameters

Table 2 Estimated parameters of the power-law fits for the visitation frequencies distributions with parameter estimation error σ_α

Data	f_0	α	σ_α
Web mobility	12	1.85	0.005
Phone data	2	1.74	0.001
BrightKite–Los Angeles	10	1.98	0.02
BrightKite–New York	11	1.98	0.02
BrightKite–San Francisco	8	2.03	0.02

Such similarities at functional forms and scaling parameters suggests that both mechanisms indeed might share a common underlying mechanism.

Notice, however, that the heavy tail of the browsing visitation frequencies spans up to two orders of magnitude in comparison to the mobility data. Such discrepancy can offer interesting insights regarding the origins of the upper limit observed in the human mobility data. One possible explanation is that because human mobility is strongly influenced by the physical constraints of the system such as displacement distances and times, our mobility is bounded to these spatial limitations. Web browsing, on the other hand, does not face physical constraints, since the amount of time necessary to reach any point is negligible for all practical purposes. If indeed browsing and traveling are both driven by a common underlying process, the browsing patterns could represent an upper limit of such process.

3.1 Characterizing Explorers and Returners in Web Browsing Navigation Trajectories

In a recent work, Pappalardo et al. [16] investigated the impact of recurring movements to the overall mobility, from which they could identify distinct profiles of individuals: *returners* and *explorers*. Returners are characterized by their radii of gyration being dominated by recurrent visits between a few preferred locations. Explorers, on the other hand, tend to travel between a larger number of different locations.

For such, the authors analyzed the characteristic distances traveled by a person when visiting the most frequented locations in comparison with their overall characteristic distance. The characteristic travel distance of an individual can be estimated by the *radius of gyration*, r_g, defined as

$$r_g = \sqrt{\frac{1}{N} \sum_{i \in L} n_i (\mathbf{r}_i - \mathbf{r}_{cm})^2},$$

where:

- L is the set of visited locations;
- n_i is the number of visits to a location i;
- $N = \sum_{i \in L} n_i$ is the total number of visits to location i;
- $\mathbf{r}_i$ is the vector of the geographical coordinates of location i;
- $\mathbf{r}_{cm}$ is the center of mass of the trajectory, defined in terms of the visitation frequency of the locations.

In order to characterize the movements to frequently visited locations the authors defined a k-radius of gyration $r_g^{(k)}$ as an estimate of characteristic trip lengths computed over the kth most-visited locations. The k-radius of gyration can be formalized as

$$r_g^{(k)} = \sqrt{\frac{1}{N_k} \sum_{i=1}^{k} \left(\mathbf{r}_i - \mathbf{r}_{cm}^{(k)} \right)^2},$$

where k is the frequency rank of a location, $N_k = \sum_{j=1}^{k} n^{(j)}$ where $n^{(j)}$ represents the total number of visits to the jth most-visited location.

To assess the capability of the AMS approach in capturing and reproducing plausible structural properties of the mobility networks, we compared the true radii of gyration r_g extracted from the mobile phone data with the AMS radius of gyration, ν_g, defined as

$$\nu_g = \sqrt{\frac{1}{N} \sum_{i \in L} n_i (\boldsymbol{\rho}_i - \boldsymbol{\rho}_{cm})^2}$$

where in this case, $\boldsymbol{\rho}_i$ is the 2D coordinates vector of vertex i within the AMS and $\boldsymbol{\rho}_c m$ is center of mass of the corresponding trajectory.

Additionally, we define the $\nu_g^{(k)}$ equivalent to the k-radius of gyration

$$\nu_g^{(k)} = \sqrt{\frac{1}{N_k} \sum_{i=1}^{k} \left(\boldsymbol{\rho}_i - \boldsymbol{\rho}_{cm}^{(k)} \right)^2},$$

where $\boldsymbol{\rho}_i$ is the 2D coordinates of the vertices corresponding to kth most-visited location within AMS whereas $\boldsymbol{\rho}_{cm}^{(k)}$ is the center of mass of the movements between the top k locations.

From the definitions above, we measured the Pearson correlation between r_g and ν_g as measured from the mobile phone data. Our objective is to verify if people with high r_g also have a high ν_g and vice-versa, which would add more supporting evidence that the AMS is a plausible approximation for the spatial dimension of human trajectories. Also, we measured the correlations for the corresponding $r_g^{(k)}$ and $\nu_g^{(k)}$ for $k = 2, 4, 8$.

As one can see, indeed the AMS was able to capture and reproduce reasonable spatial properties of the mobility patterns. The moderate correlation between the actual users' radii of gyration and those extracted from the AMS (boldface values in Table 3) confirms that a graph layout produced by a force-directed algorithm is able to capture some of the spatial features of human trajectories. It is noteworthy the fact that our analyzes have shown that although r_g and ν_g are correlated, their distributions do not have the same functional form.

Therefore, more analyses were necessary to assess the validity of our approach regardless the differences in the radii of gyration distributions. Then, we applied the method recently proposed by Pappalardo et al. [16] in order to profile users as *returners* and *explorers*, based on their characteristic travel distances. Additionally, the authors also proposed the idea of k-returners as those individuals whose characteristic traveled distance is dominated by their top k most-visited locations locations.

Table 3 Pearson correlation between the actual radii of gyration (r_g) and the radii of gyration extracted from the AMS (ν_g)

	r_g	$r_g^{(2)}$	$r_g^{(4)}$	$r_g^{(8)}$	ν_g	$\nu_g^{(2)}$	$\nu_g^{(4)}$	$\nu_g^{(8)}$
r_g	1.00	0.71	0.89	0.96	**0.51**	0.32	0.42	0.46
$r_g^{(2)}$	0.71	1.00	0.82	0.76	0.36	**0.46**	0.40	0.38
$r_g^{(4)}$	0.89	0.82	1.00	0.95	0.44	0.38	**0.46**	0.45
$r_g^{(8)}$	0.96	0.76	0.95	1.00	0.46	0.35	0.44	**0.46**
ν_g	**0.51**	0.36	0.44	0.46	1.00	0.70	0.90	0.96
$\nu_g^{(2)}$	0.32	**0.46**	0.38	0.35	0.70	1.00	0.82	0.76
$\nu_g^{(4)}$	0.42	0.40	**0.46**	0.44	0.90	0.82	1.00	0.95
$\nu_g^{(8)}$	0.46	0.38	0.45	**0.46**	0.96	0.76	0.95	1.00

Table 4 Proportion of k-returners in the Mobile Phone data

	$k = 2$ (%)	$k = 4$ (%)	$k = 8$
$\nu_g^{(k)}$	53.6	83.0	96.8 %
$r_g^{(k)}$	46.9	73.88	90.32

For instance, a 2-returner is an individual whose traveled distances are mainly determined by the visits to two locations (e.g. home and work). More precisely, the authors defined a k-returner as one whose $r_g^{(k)} > r_g/2$.

Hence, we use this approach to classify the Web browsing users in the dataset as k-returner or k-explorer. Once again we performed the same analyses to the mobile phone datasets for both r_g and ν_g to serve as reference. Table 4 shows that the proportion of k returners for $\nu_g^{(k)}$ and $r_g^{(k)}$. For the Web browsing data, the proportion of k-returners is 56.3, 93.9 and 99.8 % for $k = 2, 4, 8$ respectively; not far from what was observed in the mobility data.

Another way to look at the distribution of k-returners and k-explorers in the population is by analyzing the ratio $S_k = r_g^{(k)}/r_g$ for each user in the datasets. This ratio characterizes to what extent a user is a returner or explorer. Values close to one corresponds to returners while values close to 0 correspond to explorers. Also, we measured the S_k ratios to the respective $\nu_g^{(k)}$ and ν_g. Here we are not interested in the actual S_k distribution but rather to compare the distributions generated from r_g and ν_g. The closer these distributions are, the stronger is the support for the validity of our method. As seen in Fig. 4 (middle and bottom rows), the distribution of ratios when measured from r_g and ν_g are indeed very similar. Moreover, when we look at the distribution of ratio values for the browsing data, we can see that they have similar shapes as observed from the actual mobility data, especially for $k \geq 4$.

We used a hierarchical clustering algorithm [22] to generate a representation of more than the 400 most-visited sites (see Fig. 5). We can observe that in some cases, sites with similar contents were grouped together. For instance, cluster (**b**) can be interpreted as the *Sharing* region, whose sites are predominantly torrent search

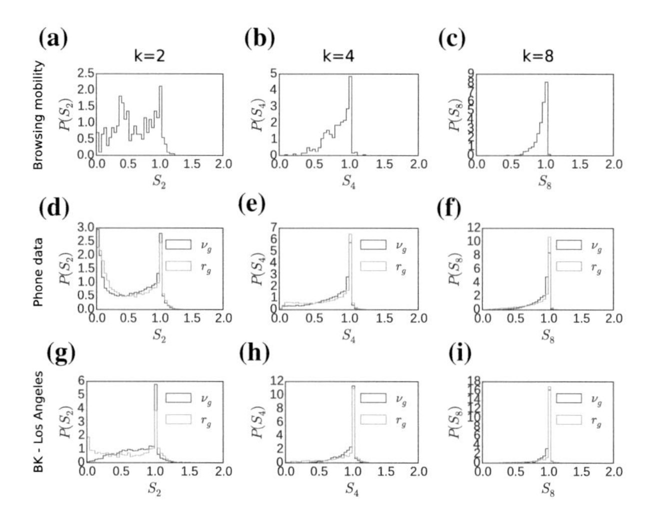

Fig. 4 The ratio between recurrent and overall mobility as estimated by the S_k ratios for $k = 2, 4, 8$. **a–c** Browsing mobility data. We can notice that phone traces (**d–f**) and BrightKite (Los Angeles) data (**g–i**). We omitted from the plot the San Francisco and New York BrightKite data since they had the same form as in the LA data

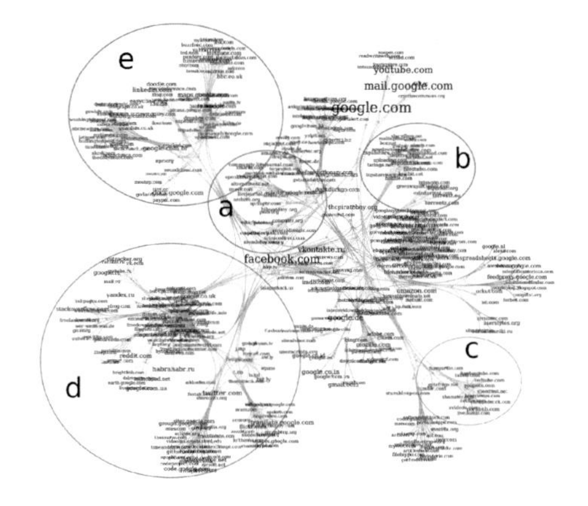

Fig. 5 A visualization of the top 423 hosts generated based on the browsing behaviors and visitation frequencies (high resolution available on-line http://my.fit.edu/~hbarbosafilh2011/webmap.pdf). Label sizes represent the PageRank of the vertices in the mobility network whereas node colors encode the clusters the vertices belong

engines and file hosting services. Also, adult sites were all clustered together (**c**). In other cases, the clusters represent websites related with particular set of activities. The cluster (**a**) can be seen as the *Science* region, whose members are mostly scientific publications and bibliography management systems whereas (**d**) represents a broad *Development* region whose sites range from source code repositories (e.g. Github and Google Code) to Q&A sites (e.g. StackOverflow). Moreover, (**e**) seems to be also related with academic-related activities, since travel sites (e.g. Expedia and TripAdvisor) and professional social networks (e.g. Linkedin and Xing) appear together with the conference management system EasyChair.

4 Discussion and Conclusions

In this work, we investigated patterns of Web browsing behaviors from a human mobility perspective. Our results strongly suggest that both behaviors are driven by common fundamental processes, especially when it comes to return visitation patterns, here shown to exhibit similar scaling properties.

Also, our approach made it possible to classify Web users as returners and explorers based on their on-line activities and their corresponding Web browsing mobility networks. From the aggregation of these mobility networks we created a hierarchical representation of the most popular websites uncovering many functional clusters of semantically-related sites, whose proximity could be traced to different users' activities.

Moreover, our findings were consistently in agreement with the human mobility literature, adding a small but relevant piece of evidence in favor of a universal law of human dynamics, possibly rooted the cognitive processes of memory formation and decision-making, although an ultimate answer is still far from be known. In other words, this work provides important insights to help us in this journey for the understanding of the inner-most mechanisms underlying the complex nature of human behaviors.

Acknowledgments The authors acknowledge the partial support from National Science Foundation (NSF) grant No. 1152306. Any opinions, findings, and conclusions or recommendations expressed in this material are those of the authors and do not necessarily reflect the views of the NSF.

References

1. Clauset, A., Shalizi, C.R., Newman, M.E.J.: Power-law distributions in empirical data. SIAM Rev. **51**(4), 661–703 (2009)
2. Albert, R., Barabási, A.-L.: Statistical mechanics of complex networks. Rev. Mod. Phys. **74**(1), 47–97 (2002). Jan
3. Brockmann, D., Hufnagel, L., Geisel, T.: The scaling laws of human travel. Nature **439**(7075), 462–465 (2006)
4. González, M.C., Hidalgo, C.A., Barabási, A.-L.: Understanding individual human mobility patterns. Nature **453**(7196), 479–482 (2008)
5. Song, C., Koren, T., Wang, P., Barabási, A.-L.: Modelling the scaling properties of human mobility. Nat. Phys. **6**(10), 818–823 (2010)
6. Szell, M., Sinatra, R., Petri, G., Thurner, S., Latora, V.: Understanding mobility in a social petri dish. Sci. Rep. **2**, 1–6 (2012)
7. Brockmann, D.: Statistical mechanics: the physics of where to go. Nat. Phys. **6**, 720–721 (2010)
8. Barabási, A.-L., Albert, R., Jeong, H.: Mean-field theory for scale-free random networks. Phys. A: Stat. Mech. Appl., 1–19 (1999)
9. Krapivsky, P., Redner, S.: Network growth by copying. Phys. Rev. E **71**(3), 036118 (2005). Mar
10. Dorogovtsev, S.N., Mendes, J.F., Samukhin, A.N.: Structure of growing networks with preferential linking. Phys. Rev. Lett. **85**(21), 4633–4636 (2000)

11. Simkin, M.V., Roychowdhury, V.P.: Do copied citations create renowned papers? Ann. Improb. Res. **11**(1), 24–27 (2005). Jan
12. Barabási, A.-L.: The origin of bursts and heavy tails in human dynamics. Nature (2005)
13. Vázquez, A., Oliveira, J.G., Dezsö, Z., Goh, K.I., Kondor, I., Barabási, A.-L.: Modeling bursts and heavy tails in human dynamics. Phys. Rev. E **73**(3), 036127 (2006)
14. Thiemann, C., Theis, F., Grady, D., Brune, R., Brockmann, D.: The structure of borders in a small world. PloS One **5**(11), e15422 (2010). Jan
15. Zhao, K., Musolesi, M., Hui, P., Rao, W., Tarkoma, S.: Explaining the power-law distribution of human mobility through transportation modality decomposition. Sci. Rep. **5**, 9136 (2015)
16. Pappalardo, L., Simini, F., Rinzivillo, S., Pedreschi, D., Giannotti, F., Barabási, A.-L.: Returners and explorers dichotomy in human mobility. Nat. Commun. **6**, 8166 (2015)
17. Grabowicz, P.A., Ramasco, J.J., Gonçalves, B., Eguíluz, V.M.: Entangling mobility and interactions in social media. PloS One, 1–16 (2014)
18. Cho, E., Myers, S.A., Leskovec, J.: Friendship and mobility. In: Proceedings of the 17th ACM SIGKDD International Conference on Knowledge Discovery and Data Mining—KDD '11. ACM Press, New York, New York, USA (2011), p. 1082
19. Fruchterman, T.M.J., Reingold, E.M.: Graph Drawing by Force-directed Placement. Softw.-Pract. Exp. **21**(November), 1129–1164 (1991)
20. Hasan, S., Schneider, C.M., Ukkusuri, S.V., González, M.C.: Spatiotemporal patterns of urban human mobility. J. Stat. Phys. **151**, 304–318 (2012)
21. Krumme, C., Llorente, A., Cebrian, M., Pentland, A.S., Moro, E.: The predictability of consumer visitation patterns. Sci. Rep. **3**, 1645 (2013)
22. Peixoto, T.P.: Hierarchical block structures and high-resolution model selection in large networks. Phys. Rev. X **4**(1), 011047 (2014)

Part VI
Dynamics and Spreading Phenomena on Networks

Modeling Memetics Using Edge Diversity

Yayati Gupta, Akrati Saxena, Debarati Das and S.R.S. Iyengar

Abstract The study of meme propagation and the prediction of meme trajectory are emerging areas of interest in the field of complex networks research. In addition to the properties of the meme itself, the structural properties of the underlying network decides the speed and the trajectory of the propagating meme. In this paper, we provide an artificial framework for studying the meme propagation patterns. Firstly, the framework includes a synthetic network which simulates a real world network and acts as a testbed for meme simulation. Secondly, we propose a meme spreading model based on the diversity of edges in the network. Through the experiments conducted, we show that the generated synthetic network combined with the proposed spreading model is able to simulate a real world meme spread. Our proposed model is validated by the propagation of the Higgs boson meme on Twitter as well as many real world social networks.

1 Introduction

"We ape, we mimic, we mock, we act" is a law universal to all human beings. Imagine a lady in an elevator, heading to the fifth floor of her office. Suddenly, one by one, every person in the elevator turns back, what does she do now? According to Elevator Groupthink psycology experiment [1], most of us would turn back in such

Y. Gupta (✉) · A. Saxena · S.R.S. Iyengar
Indian Institute of Technology Ropar, Rupnagar, India
e-mail: yayati.gupta@iitrpr.ac.in

A. Saxena
e-mail: akrati.saxena@iitrpr.ac.in

S.R.S. Iyengar
e-mail: sudarshan@iitrpr.ac.in

D. Das
PES Institute of Technology, Bangalore, India
e-mail: debarati.d1994@gmail.com

H. Cherifi et al. (eds.), *Complex Networks VII*, Studies in Computational Intelligence 644, DOI 10.1007/978-3-319-30569-1_14

a situation. Usually, most of us become followers of the crowd when faced with our sense of conformity. If ants follow each other with the help of the pheromone trail, humans too involuntarily imitate and follow each others' behaviours and ideas. Behaviours like obesity, smoking and altruism are also seen to spread through social networks [2]. Today, Online Social Networks (OSNs) like Facebook and Twitter provide a platform to fulfill people's penchant for information sharing, arguing and mudslinging. Used by approximately 1.4 billion people worldwide [3], Facebook's "Read, Like and Share" tradition has today become a way of living. Understanding these spreading phenomena can help us in diverse ways such as accelerating the spread of useful information i.e. health related advices or disaster management related announcements as well as for viral marketing of products and memes. Predicting the trajectory of a meme's propagation in a network can also prevent the spread of malicious rumors and misinformation. Social networks play an instrumental role in the spread of influence in today's world. Hence, contagion prediction models are an extensively studied field in complex networks research. Such models evolve frequently with time, aiming to depict real world information propagation more accurately. Initially, meme propagation models were inspired from compartmental epidemiological models [4]. These models [5] were too simplistic and did not consider the role of edges in the spreading of information. Later on, the advent of independent cascade [6] and linear threshold models [7] proved seminal and these became the standardised models for meme propagation. However, most of these models did not take into consideration the network structure and the calculation of parameters for these models also remained a challenge.

Consider an anecdote about a small child Bob who went to visit the theme park, Six Flags Magic Mountain in California, with his parents. Bob got lost in the Fright Fest, which is the biggest and most terrifying maze at the theme park known for its complex spider-web like structure. Confused by the many turns the maze took at every step, poor Bob could not find his way out of the Fright Fest. When Bob did not return, his worried parents contacted the park authorities for help. These authorities having complete knowledge about the structure of the maze and the possible paths that could be traversed by the players, could easily locate Bob. Similarly, real world networks also have a complex yet distinct structure and if one could understand this structure and estimate the paths that can be taken by the meme in its trajectory, could she also not behave like the park authorities in the above analogy? In connection with the above anecdote, the knowledge of a network's structure is important for understanding meme propagation. It is known that the real world social networks have a very well defined structure. We employ this well known structure for the simulation of a meme. The major contributions of the paper are:

1. Generation of an artificial synthetic network that mimics a real world social network in terms of network structure.
2. We propose a spreading model for meme propagation based on the structure of the network. This model is based on the difference in spreading probabilities of different edges which is recognised from the network itself.

The proposed synthetic network and spreading model give a synthetic simulation environment which serves as a test-bed to study meme propagation patterns. Further, it gives a way of organising the edges in an hierarchy based on varying probabilities of information transmission across these edges. We validate the proposed spreading model against the real world spreading of the Higgs boson meme on Twitter. If one could extract the structure of offline social networks, our framework can be used for understanding a wide range of phenomena on offline networks as well in addition to online networks. In addition to controlling information flow on OSNs, we can decrease the increasing behavioral spreading of obesity and depression in the world and promote altruism and positive movements. Inspired from the diversity of edges in a social network, the paper lays light on a novel aspect of looking at information propagation.

The rest of the paper is organised as follows: Sect. 2 describes the related work. Section 3 explains the synthetic networks in addition to describing the real world networks used for simulation. The network structure based spreading model is proposed in Sect. 4. Section 6 is devoted to results and discussion. Section 7 concludes the paper alongwith the future work.

2 Related Work

An enormous amount of work has been done to study the information propagation pattern on an online social network [8, 9]. Initially, memes in a social network were considered analogous to a virus in a biological network [4]. As a result, most information spreading models were inspired from compartmental epidemiological models like SIS and SIR models [5] introduced in 1989. However these models assumed a homogenous mixing of people constituting the population and did not take into account interactions between the individuals. Later, independent cascade (IC) [6] and linear threshold (LT) [7] models were investigated which are now used as the standard models for information propagation [10]. However, these models did not consider factors like network structure and model simulation parameters. There were some studies that predicted the parameters associated with the information propagation models [11], but these are largely based on the utilisation of the past data, obtaining which is a difficult process. Studying considering the impact of network structure on a meme's propagation provide a relative view of the meme spread. For example, the spread of epidemics is faster on scale free networks as compared to the random networks due to the presence of hubs [12]. Zhang et al. presents a stochastic model for the information propagation phenomenon [13]. Studying the information propagation may help the scientists in a number of ways like halting the spread of misinformation [14] and accelerating useful information [15] through a network.

Meme Virality prediction is an active research area in social network analysis [16, 17] and meme propagation models can be used extensively in fields like Viral Marketing. Viral Marketing can be done by targeting a set of nodes in a network as done by Kleinberg et al. in their paper on influence maximisation [18, 19]. Influential spreaders play a significant role in information propagation as shown by Kitsak et al. in their work [20]. Meme virality can not only depend on network structure and nodes in a network, it seems to be intuitive that meme content also has a role to play in the meme becoming viral [16, 17]. Though most studies consider nodes in their study of meme virality, we consider the property of edges in the spread of information. An edge connecting a vulnerable node to an influential node may have more impact as compared to a vulnerable node to another. Our study takes the diversity of edges into consideration and then probes into the meme pattern that can be formed.

3 Generation of Networks for Meme Simulation: SCCP Networks

It has been observed that most of the social networks are scale free and can be generated by the preferential attachment model. Further, these networks have communities because of the phenomenon of homophily that leads to the formation of dense clusters in the network. We also consider one more meso scale characteristic in the formation of network- core-periphery structure. It has been shown that the scale free networks possess an implicit core-periphery structure. Considering these 3 characteristics, we have tried to simulate real world networks via SCCP networks which show properties like **S**cale-free structure, presence of **C**ommunities and **C**ore-**P**eriphery structure. We introduce a modification to the algorithm [21] employed by Wu et al. to generate these synthetic networks. The modified algorithm can be found in detail in [22]. There is an implicit core-periphery structure in the generated network. It follows from the work done by Della et al. [23][1] which proves the existence of core-periphery structure in scale free networks.The modified algorithm generates synthetic scale free networks having a varying number and sizes of communities, which is a feature prevalent in real world networks.

4 Proposed Spreading Model

Meme propagation on a real world network follows the pattern of a complex contagion. Unlike a simple contagion, the spreading pattern of a complex contagion depends on factors like homophily and social reinforcement.[2] A simple contagion is

[1]The work shows that scale free networks possess a core-periphery structure. They define cp-centralisation value which is a measure of the degree to which a network contains a core-periphery structure. According to this study, the average cp-centralisation value for 1000 instances of scale free networks with 100 nodes and average degree 4 is 0.668.

[2]Homophily is the name given to the tendency of similar people becoming friends with each other. This leads to more number of ties between like minded people and hence leads to the formation

like an infectious disease which spreads with equal probability across all the edges, while a complex contagion spreads with different probabilities depending on the factors like social reinforcement and homophily [17]. In addition, user influence also plays a prominent role in meme propagation. We take into consideration all these factors in modeling the diffusion of a meme permeating through the ties in the network.

Our model is based on two key ideas:

1. *Diversity in Tie Strength*: "Birds of the same feather flock together". We are more engaged and connected with the people in our own community as compared to people from other communities [24]. Hence, the probability associated with the edges connecting people of the same community should be higher than the edges connecting people of different communities. This observation gains motivation from the theory of weak ties [25].
2. *The social status of nodes*: The social influence of a person in a network plays a big role in acceptance of information propagated by that person. A person's social status also decides if that person is vulnerable to adopting information. Simply stated, *lower the status, higher the vulnerability* and vice versa. *Higher the status, more the influence* and vice versa.

Because of the presence of core-periphery structure in SCCP networks, there are two kinds of nodes in a SCCP network: core nodes and periphery nodes (periphery nodes are further divided into many communities). Initially, all the nodes are uninfected and a node turns infected as soon as it adopts a meme. We call an infected node u, the sender and an uninfected neighbour of u say v, the receiver of an infection. The probability of infection transmission across an edge depends on the types of both nodes—the sender and the receiver. In our model, the probabilities of infection across edges are divided into five categories:

$$P_{cc}, P_{cp}, P_{pc}, P_{pp_0}, P_{pp_1}$$

Here, 'P' represents probability. The type of edge is represented by the subscript. The subscript's first alphabet denotes the type of sender node and second alphabet denotes the type of receiver node. 'c' represents core, 'p' represents periphery. 0 in the subscript denotes same community membership of sender and the receiver node, while 1 represents sender and receiver belonging to different communities. We worked towards predicting the most plausible order for these edge probabilities, which is initially proposed to be as: $P_{cc} > P_{cp} > P_{pp_0} > P_{pp_1} > P_{pc}$.

Our model can be considered as an extension of the simple cascade model, with a slight change in the definition in every iteration, each infected node tries infecting its uninfected neighbours in accordance with the above probability hierarchy.

(Footnote 2 continued)
of communities in the network. Social reinforcement is the phenomenon by which multiple exposures of an information to a person leads to him adopting it. Social reinforcement and homophily tend to block the information inside one community.

5 Datasets

In addition to the SCCP networks, we have used a number of other real world networks to simulate meme's spread. These datasets are listed in Table 1. For the simulation of our spreading model, we have considered the two most widely used online social networks- Facebook and Twitter having approximately 1371 and 271 million users. For comparing our complete framework, we use the Higgs boson meme propagation information on Twitter (dataset1). The dataset 1 gives a complete picture of a meme spreading on an online social network along with the information "who infected whom at every step . All these considered datasets are the examples of SCCP kind of networks. We have picked the datasets for the most popular social networking sites. Furthermore, the datatset of Twitter in addition to the network, also gives the retweet information of the Higgs boson meme which is used to validate the spreading model. A detailed explanation of these datasets is given below.

We detect communities in datasets 1(a), 2 and 3 using fast greedy modularity optimization algorithm. This algorithm is given by Newman et. al. [29] and is used to detect community structure for very large graphs. We also find out the core-periphery structure for all the above listed datasets using k-shell decomposition algorithm. We assign a coreness value to each node equal to the shell value assigned to it by the algorithm. Then, we pick top 10 % of the nodes having highest coreness values and call them the core nodes. The remainder of the nodes are termed periphery nodes.

Table 1 Datasets used for experiments

Dataset	Specification
Dataset 1(a):	This dataset is an induced directed unweighted subgraph on Twitter users who were involved in any of the activities (reply, retweet, or mention) regarding the Higgs boson meme.[a] [26] It is an undirected unweighted graph containing 456631 nodes and 14855875 edges
Dataset 1(b):	This is a directed weighted graph between the Twitter users who were involved in retweeting [27] of the Higgs Boson meme. There is an edge from B to A if A retweets B. This graph contains 425008 nodes and 733647 edges. In datasets 1(a) and 1(b), the tweets posted in Twitter about this discovery between 1st and 7th July 2012 are considered
Dataset 2:	This dataset is an undirected unweighted induced subgraph on Facebook with 4039 nodes and 88234 edges [28]
Dataset 3:	This dataset is an induced undirected unweighted subgraph on Twitter with 81306 nodes and 1768149 edges [28]
Dataset 4(a):	These datasets have been derived from the algorithm proposed in the previous section. This is a SCCP network on 65800 nodes, 591750 edges and 11 communities
Dataset 4(b):	This is a SCCP network on 4000 nodes, 170314 edges and 11 communities
Dataset 5	This is an Erdos-Renyi graph on 4000 nodes and 34650 edges

[a]Higgs boson is one of the most elementary elusive particle in modern physics. A meme in Twitter is considered to be a Higgs Boson meme if it contains at least one of these keywords or tags: lhc, cern, boson, higgs

6 Experiments and Results

6.1 *Spreading Model Validation*

Our model was validated using datasets 1(a) and 1(b), where 1(a) gives us the information about the structure of a social network and 1(b) is the cascading pattern of a meme over 1(a). Let the dataset 1(a) be represented by $G(V, E)$. Based on the structure of G, we partition its nodes in two subsets C and P. C is the set of core nodes and P is the set of periphery nodes such that $C \cup P = V$ and $C \cap P = \emptyset$. We also associate a variable δ_{ij} with each edge E_{ij}. $\delta_{ij} = 1$ if nodes i and j belong to the same community, else 0. We divide the edges in the retweet network (dataset 1(b)) in four categories based on the types of users an edge is connecting. These categories are as follows:-

1. $E_{cc} = \{E_{ij} \in E : (i \in C) \wedge (j \in C)\}$
2. $E_{cp} = \{E_{ij} \in E : (i \in C) \wedge (j \in P)\}$
3. $E_{pc} = \{E_{ij} \in E : (i \in P) \wedge (j \in C)\}$
4. $E_{pp} = \{E_{ij} \in E : (i \in P) \wedge (j \in P)\}$
   - $E_{pp_0} = \{E_{ij} \in E : (i \in P) \wedge (j \in P) \wedge \delta_{ij} = 1\}$
   - $E_{pp_1} = \{E_{ij} \in E : (i \in P) \wedge (j \in P) \wedge \delta_{ij} = 0\}$

The types of nodes for 1(b) are extracted from its main graph 1(a).

In retweet networks, the weight of an edge from A to B specifies the amount of information flowing from A to B (number of times B retweeted a message from A). Therefore, more the weight, higher the probability of information transmission across that edge. We calculate the following weights from the above graphs:

Let $W(E_{ij})$ be the weight of an edge from node i to node j and N_{xy} represent the type of edges E_{xy} where x and y are the types of nodes hence having the possible values p and c. Then, we calculate W_{xy} ,the sum of weights of all the edges from a node of type x to a node of type y.

1. $W_{cc} = \sum(W(E_{ij}))/N_{cc}$ such that $E_{ij} \in E_{cc}$
2. $W_{cp} = \sum(W(E_{ij}))/N_{cp}$ such that $E_{ij} \in E_{cp}$
3. $W_{pc} = \sum(W(E_{ij}))/N_{pc}$ such that $E_{ij} \in E_{pc}$
4. $W_{pp} = \sum(W(E_{ij}))/N_{pp}$ | such that $E_{ij} \in E_{pp}$
   - $W_{pp_0} = \sum(W(E_{ij}))/N_{pp_0}$ such that $E_{ij} \in E_{pp_0}$
   - $W_{pp_1} = \sum(W(E_{ij}))/N_{pp_1}$ such that $E_{ij} \in E_{pp_1}$

The weights obtained show that the observed order is the same as we have proposed earlier thereby validating the ordering we proposed i.e. $W_{cc} > W_{cp} > W_{pp_0} > W_{pp_1} > W_{pc}$.

6.2 Simulation Results

We introspect on the extent as well as rate of infection of the network, while propagating a meme on it. We simulate EBH as well as uniform spreading model on a number of datasets and report the results. For the simulation of our proposed model, we use the following probabilities: E_{pc} : 0.00001, E_{pp_0} : 0.0003, E_{pp_0} : 0.0001, E_{cc} : 0.006, and E_{cp} : 0.004. For the simulation of uniform spreading model, every edge is considered to have an equal probability of infection i.e. $E_{ij} = 0.0002$, where i and j are the endpoints of an edge. We have chosen these probabilities such that we can visualise the spreading pattern of a meme to the best possible extent. For all the figures in this section, X axis represents the number of iterations and Y axis represents the cumulative number of nodes infected up to that iteration. The results of this paper are structured in three parts.

6.2.1 Meme Spreading Patterns on Different Networks Using the EBH and Uniform Spreading Models

Figure 1(3) shows the actual spreading pattern of the Higgs boson meme which indicates that in the real world, a meme does not have a constant growth rate. The rate remains constant upto some point, after which the popularity of a meme shoots up steeply and then slowly fades, giving rise to a sigmoid curve which is characterised by the equation: $F(x) = 1/(1 + e^{-kx})$ Fig. 1(1, 2) shows the simulation of our proposed spreading model on the SCCP network and two real world networks of Facebook and Twitter and Higgs boson Twitter network respectively. It can be seen that in both these cases, the curve for the spreading pattern is seen to be sigmoidal just like Fig. 1(2). Figure 1(5, 6) shows the difference in the spreading patterns when the simulation is done through an uniform spreading model and our model respectively. It can be seen that the simulation through an uniform spreading model is also a sigmoid function but has a lesser value of parameter x. Figure 1(4) shows the simulation of the proposed spreading model on 3 different kinds of networks. Despite simulating the EBH spreading model on all the three graphs, the value of x is observed to be lower only in the case of random networks.[3] Thus we can say that the sharp S shaped infection pattern is observed only for the SCCP kind of networks. *These graphs show that the presence of both- a SCCP kind of network as well as EBH spreading model are required to mimic a real world meme propagation.*

[3] In the case of random network, even though the declared 10 % core nodes have a high probability of infecting their neighbours, the connections between the core nodes are not dense enough to result in an overshoot in the number of infected nodes.

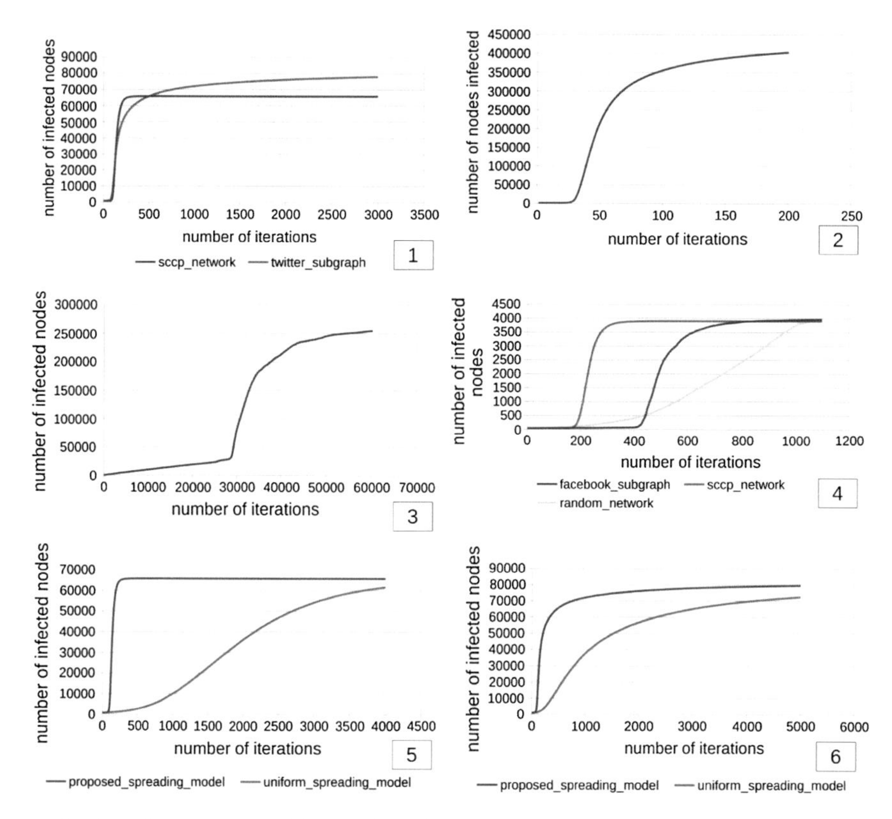

Fig. 1 Spreading patterns on different kinds of networks and its comparison to real world data—*1*: Spreading patterns on datasets 3 and 4(a) *2*: Spreading pattern on dataset 1(a) *3*: Actual spreading pattern of Higgs boson meme(dataset 1(a) and 1(b)) *4*: Comparison between the proposed spreading model on datasets 2, 4(a), and 5 *5*: Spreading patterns for dataset 4(a) *6*: Proposed and uniform spreading models on dataset 3

6.2.2 Explanation of the Plateau Structure Observed in the Meme Pattern

Figure 2(3) shows the pattern of infection of core nodes and periphery nodes for the actual Higgs boson meme. As in the previous case, all iterations are considered to be of equal length (10 timestamps). We observe the cumulative number of core nodes and periphery nodes infected in every iteration. When we started infection from periphery nodes (Fig. 2(1)), the plateau structure of the curve continues till a core node is infected and then the infection shoots up suddenly. Figure 2(2) shows the plot when the infection is started only from the core nodes. We can see that in this case, infection shoots up immediately without the plateau structure. *This solidifies the observation that the number of periphery nodes infected increases sharply as soon as a sufficient fraction of the core nodes gets infected.*

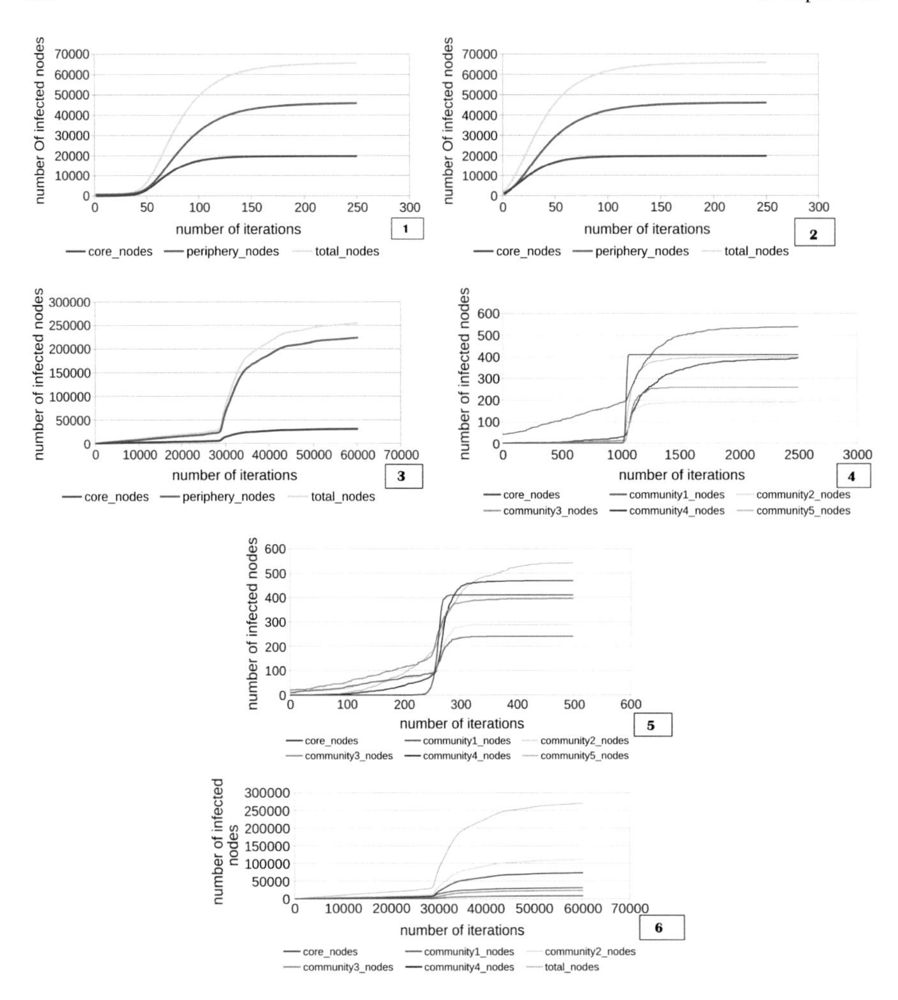

Fig. 2 Spreading patterns starting from different types of seed nodes and its comparison to real world data—*1*: Proposed spreading model on dataset 4(a) where spreading starts from periphery nodes *2*: Proposed spreading model on dataset 4(a) where spreading starts from core nodes *3*: Actual spreading pattern for the Higgs boson meme(from dataset 1(a) and 1(b)) *4*: Spreading patterns starting from single community *5*: Spreading patterns starting from multiple communities *6*: Actual spreading pattern for the Higgs boson meme(from dataset 1(a) and 1(b))

6.2.3 Effect of Communities and Core Nodes on Meme Virality

In Fig. 2(4), we start the infection from a single community and show that the infection spreads in multiple communities only when the meme infects the core sufficiently and gets viral. Figure 2(5) shows the spreading pattern when the infection starts from multiple communities. But the meme becomes viral only after the infection of core nodes. So, *whether the infection starts from single community or multiple communities, the infection of core nodes is sufficient to predict its virality*. Figure 2 shows the actual spreading pattern of Higgs boson meme.

7 Conclusion and Future Work

A lot of researchers are working towards proposing the models that can predict the pattern of meme spread in a real world network today. A number of models have been proposed for this ranging from simple epidemiological models to the standard models like Linear Threshold and Independent Cascade. Most of these models do not give an approach to identify the parameters required to simulate them. Moreover, they are proposed for all kind of networks though they can be improved upon and specialised for a particular kind of network. Hence, improving these models to better simulate a meme propagation is possible. It is shown that, together, SCCP and EBH models effectively simulate real world meme propagation. The sigmoid curve with a sharp slope is shown to be the characteristic pattern of an internet meme. Furthermore, the importance of core nodes in marking the virality of a meme is emphasised. It is also shown that infecting multiple communities also require the infection of core nodes. The study is validated with the Higgs boson meme spreading on Twitter in addition to various other real world networks. This study opens a new direction of considering edge diversity in meme propagation models. One can extend our problem to predict the exact values of the probabilities influencing the meme propagation. This can greatly help in prediction of a future cascade pattern. If such cascades could be predetermined then we could exert a control on our otherwise ever changing social networks. Not only could preventive checkpoints be placed in the network but also useful information could be accelerated through the network by using the predicted meme trajectory.

Acknowledgments S.R.S. Iyengar was partially supported by the ISIRD grant(Ref. No. IITRPR/ Acad./359) from IIT Ropar. Further, we express our gratitude to the Indian Academy of Sciences,Bangalore for providing us with partial funding to carry out this research.

References

1. Clissold, B.D.: Candid camera and the origins of reality tv. In: Understanding Reality Television, pp. 33–53 (2004)
2. Christakis, N.A., Fowler, J.H.: The spread of obesity in a large social network over 32 years. New Engl. J. Med. **357**(4), 370–379 (2007)
3. S.N. Statistics.: Statistic Brain (July 9, 2014)
4. Daley, D.J., Kendall, D.G.: Epidemics and rumours. Nature **204**, 1118 (1964)
5. Hethcote, H.W.: Three basic epidemiological models. In: Applied Mathematical Ecology. Springer, Berlin (1989), pp. 119–144

6. Goldenberg, J., Libai, B., Muller, E.: Talk of the network: a complex systems look at the underlying process of word-of-mouth. Mark. Lett. **12**(3), 211–223 (2001)
7. Granovetter, M.: Threshold models of collective behavior. Am. J. Sociol., 1420–1443 (1978)
8. Bakshy, E., Rosenn, I., Marlow, C., Adamic, L.: The role of social networks in information diffusion. In: Proceedings of the 21st International Conference on World Wide Web. ACM (2012), pp. 519–528
9. Lerman, K., Ghosh, R.: Information contagion: an empirical study of the spread of news on digg and twitter social networks. ICWSM **10**, 90–97 (2010)
10. Kleinberg, J.: Cascading behavior in networks: algorithmic and economic issues. Algorithmic Game Theory **24**, 613–632 (2007)
11. Saito, K., Nakano, R., Kimura, M.: Prediction of information diffusion probabilities for independent cascade model. In: Knowledge-Based Intelligent Information and Engineering Systems. Springer, New York (2008), pp. 67–75
12. Pastor-Satorras, R., Vespignani, A.: Epidemic spreading in scale-free networks. Phys. Rev. Lett. **86**(14), 3200 (2001)
13. Zhang, Z.Y.-C.L.Y., Fei, H.-F.C.H.X.: The research of information dissemination model on online social network. Acta Phys. Sinica **5**, 10 (2011)
14. Budak, C., Agrawal, D., El Abbadi, A.: Limiting the spread of misinformation in social networks. In: Proceedings of the 20th International Conference on World Wide Web. ACM (2011), pp. 665–674
15. Scanfeld, D., Scanfeld, V., Larson, E.L.: Dissemination of health information through social networks: twitter and antibiotics. Am. J. infect. Control **38**(3), 182–188 (2010)
16. Guerini, M., Strapparava, C., Özbal, G.: Exploring text virality in social networks. In: ICWSM (2011)
17. Weng, L., Menczer, F., Ahn, Y.-Y.: Virality prediction and community structure in social networks. Sci. Rep. **3** (2013)
18. Kempe, D., Kleinberg, J., Tardos, É.: Influential nodes in a diffusion model for social networks. In: Automata, Languages and Programming. Springer, Heidelberg (2005), pp. 1127–1138
19. Kempe, D., Kleinberg, J., Tardos, É.: Maximizing the spread of influence through a social network. In: Proceedings of the Ninth ACM SIGKDD International Conference on Knowledge Discovery and Data Mining. ACM (2003), pp. 137–146
20. Kitsak, M., Gallos, L.K., Havlin, S., Liljeros, F., Muchnik, L., Stanley, H.E., Makse, H.A.: Identification of influential spreaders in complex networks. Nature Phys. **6**(11), 888–893 (2010)
21. Wu, J.-J., Gao, Z.-Y., Sun, H.-J.: Cascade and breakdown in scale-free networks with community structure. Phys. Rev. E **74**(6), 066111 (2006)
22. Saxena, A., Iyengar, S., Gupta, Y.: Understanding spreading patterns on social networks based on network topology (2015). arXiv:1505.00457
23. Della Rossa, F., Dercole, F., Piccardi, C.: Profiling core-periphery network structure by random walkers. Sci. Rep. **3** (2013)
24. McPherson, M., Smith-Lovin, L., Cook, J.M.: Birds of a feather: homophily in social networks. Annu. Rev. Sociol., 415–444 (2001)
25. Granovetter, M.S.: The strength of weak ties. Am. J. Sociol., 1360–1380 (1973)
26. De Domenico, M., Lima, A., Mougel, P., Musolesi, M.: The anatomy of a scientific rumor. Sci. Rep. **3** (2013)
27. Yang, Z., Guo, J., Cai, K., Tang, J., Li, J., Zhang, L., Su, Z.: Understanding retweeting behaviors in social networks. In: Proceedings of the 19th ACM International Conference on Information and Knowledge Management. ACM (2010), pp. 1633–1636
28. Leskovec, J., Mcauley, J.J.: Learning to discover social circles in ego networks. In: Advances in Neural Information Processing Systems (2012), pp. 539–547
29. Clauset, A., Newman, M.E., Moore, C.: Finding community structure in very large networks. Phys. Rev. E **70**(6), 066111 (2004)

Growing Networks Through Random Walks Without Restarts

Bernardo Amorim, Daniel Figueiredo, Giulio Iacobelli and Giovanni Neglia

Abstract Network growth and evolution is a fundamental theme that has puzzled scientists for the past decades. A number of models have been proposed to capture important properties of real networks. In an attempt to better describe reality, more recent growth models embody local rules of attachment, however they still require a primitive to randomly select an existing network node and then some kind of global knowledge about the network (at least the set of nodes and how to reach them). We propose a purely local network growth model that makes no use of global sampling across the nodes. The model is based on a continuously moving random walk that after s steps connects a new node to its current location, but never restarts. Through extensive simulations and theoretical arguments, we analyze the behavior of the model finding a fundamental dependency on the parity of s, where networks with either exponential or a conditional power law degree distribution can emerge. As s increases parity dependency diminishes and the model recovers the degree distribution of Barabási-Albert preferential attachment model. The proposed purely local model indicates that networks can grow to exhibit interesting properties even in the absence of any global rule, such as global node sampling.

B. Amorim · D. Figueiredo · G. Iacobelli (✉)
Department of Computer and System Engineering (PESC),
Federal University of Rio de Janeiro (UFRJ), Rio de Janeiro, Brazil
e-mail: giulio@land.ufrj.br

B. Amorim
e-mail: bamorim@land.ufrj.br

D. Figueiredo
e-mail: daniel@land.ufrj.br

G. Neglia
MAESTRO Team, INRIA, Sophia-Antipolis, France
e-mail: giovanni.neglia@inria.fr

H. Cherifi et al. (eds.), *Complex Networks VII*, Studies in Computational Intelligence 644, DOI 10.1007/978-3-319-30569-1_15

1 Introduction

The growth and evolution of networks is a fundamental problem in Network Science specially in the light that networks are constantly changing over time. Explaining how and why different real networks grow and evolve the way they do has kept researchers busy for the past decades. Not surprising, various mathematical models for network growth and evolution have been proposed in the literature, either ad-hoc models tailored to specific domains, or general models aiming to capture general principles. A celebrated general network growth model is the Barabási-Albert (BA) model [1] which embodies the principle of *preferential attachment* found in various real networks.

A recognized drawback of most proposed network growth and evolution models is the assumption of global information about the network [2, 4, 5, 8]. For example, the BA model has a primitive to randomly select a node from the existing network according to the degree distribution. To relax such assumption, models that attach new nodes and edges to the existing network using local attachment rules, such as the Random Walk Model [9, 10], have been proposed. Clearly, random walks require knowledge of the current node degree and its neighbors, a much more localized information. Moreover, it seems more reasonable that new nodes connect to nearby nodes (through some local process) rather than selecting new neighbors from the entire population (through some global process). However, the Random Walk Model studied in [9, 10] and others [5, 8] still require a primitive to randomly select a node from the network (for the purpose of restarting the walker, for example) and are thus not purely local, because they need to know the number and the identity of all network nodes as well as a way to reach them. Such models have local attachment rules, but global "entry point" selection. More recently, models that have no global primitives have started to be explored [6, 7]. A drawback of these other models is that they rely on an initial network already containing all nodes such as a lattice or a regular tree, that is then modified according to local rules, and thus are technically not growth models.

In this work we propose and explore a network growth model that is purely local, requiring no global selection over the nodes or any initial network. The model works as follows:

0. Start a network with a single node with a self-loop and place a random walk on this node.
1. Let the random walk take exactly s steps.
2. Connect a new node to the node where the walker resides.
3. Stop if the number of nodes in the network is n, otherwise go to Step 1.

Intuitively, the random walk moves around continuously and after every s steps a new node is added and connected to its current location. The new node immediately becomes part of the network and the walker sees no difference between it and any other node. Note that the model has two parameters s and n and grows an undirected tree (apart from the self-loop at the initial node) since every new node starts with

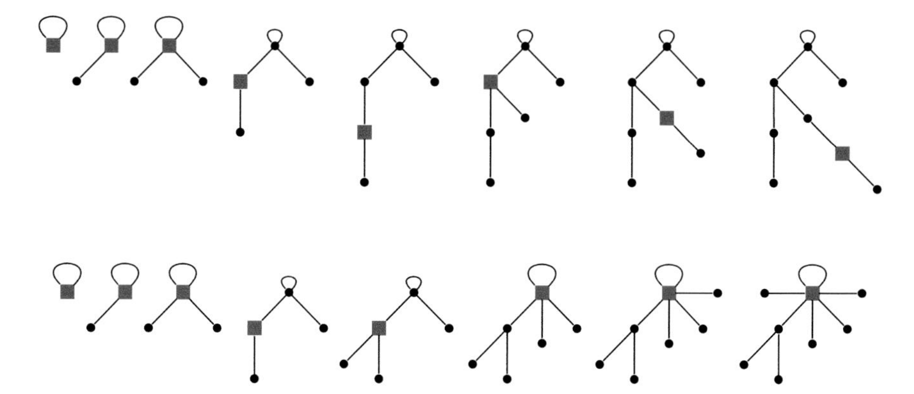

Fig. 1 Examples of sample path for network growth for $s = 1$ (*top*) and $s = 2$ (*bottom*). The *red square* denotes the walker position. The *snapshots* represent the growing network just after the new node is connected

degree one. Moreover, the random walk is uniform on the neighbors and is never restarted, thus its name NRRW (No Restart Random Walk) model. Figure 1 illustrates a sample path of network growth with $s = 1$ and $s = 2$. Can such purely local model give rise to interesting network structures such as networks that exhibit a power-law degree distribution?

Interestingly, we uncover various non-trivial features of this model such as the fundamental dependency on the parity and magnitude of s and its relationship to the degree distribution. If s is odd and small we find that networks generated by the model tend to have very short-tailed degree distribution and very long distances. On the contrary, if s is even and small, networks exhibit a special kind of power law degree distribution (to be formalized later) and very short distances! As s increases the effect of parity decreases and networks exhibit a heavy-tailed degree distribution. Interestingly, with s large enough, the observed degree distribution follows a power law with exponent identical to the network generated by the BA model, recovering the effect of preferential attachment. We also rigorously prove that for $s = 1$ the random walk is transient and the degree of every node is bounded from above by a geometric distribution. Other interesting features will be highlighted in what follows.

The model here proposed is very related to the Random Walk Model [9] which also allows a random walk to take s steps before connecting a new node. The key difference is that in [9], after a new node is attached to the network, the random walk is restarted uniformly at random across all existing nodes in the network. Our random walk never restarts, and is therefore a purely local model. Interestingly, the authors of [9] show (through simulations and approximations) that their model is closely related to the BA model and yields a power law degree distribution independently of s. However, recently this finding has been questioned and for $s = 1$ it was mathematically proven that this is not the case [3]. Our model and findings contributes to this debate and possibly sheds light on how both results could be reconciled (more on this on Sect. 6).

The remainder of this paper is organized as follows. Section 2 discusses the model and its intuitive behavior, as well as the connection with prior works. Section 3 presents the evolution of node degree induced by the model. Section 4 analyzes the depth of the tree generated by the model. Section 5 presents our theoretical findings for the case $s = 1$, showing the transient nature of the model in this case. Finally, we summarize our findings and present a brief discussion in Sect. 6.

2 Network Growth Model

As presented in Sect. 1, NRRW (No Restart Random Walk) model can be interpreted as a simple random walk that attaches a new node to its current location every s steps. Similar proposed random walk models for growing networks assume that the random walk *restarts* either after connecting a new node or adding some number of edges to the new node [9, 10]. A restart consists of randomly selecting a node from the existing network (usually uniformly) and placing the random walk on that node. Despite the similarities, the lack of restarts makes NRRW fundamentally different from models with restart. In particular, the restart significantly reduces the correlation between consecutive node additions since it is very unlikely that the random walk will visit the previous new node when walking to add a new node. Intuitively, the random walk loses memory at every restart. Moreover, restarts have the drawback of assuming knowledge of all network nodes and random access to any such node, and is thus not a purely local growth model.

What is intuitively the behavior of NRRW? In a sense, when s is large the random walk will have little memory between node additions. However, this behavior is different from restarts since the random walk will not find itself on a node chosen uniformly at random but on a node chosen randomly proportional to its degree.[1] Thus, when s is large the NRRW seems similar to the BA model since new nodes connect to random nodes chosen proportionally to their degree. However, since s is fixed and the network grows, will NRRW indeed exhibit a behavior similar to BA model when $s \ll n$ and then s will finally become small in comparison to the network size?

What about small values for s? Intuitively, the random walk will frequently stumble over the newly created nodes. Interestingly, this local behavior depends fundamentally on the parity of s. If $s = 1$ then the random walk can always walk to the newly created node and add a new node to it. Such behavior is just not possible if $s = 2$ and the walker is not on the root. This qualitative difference is not limited to $s = 1$ and $s = 2$. When s is odd the walker can always land to the most recently added node after s steps and then add a new node. For s even, this is impossible unless the walker does not traverse the loop at the initial node.

[1] Recall that the steady state distribution of a random walk on a fixed network is given by $d_i / \sum_j d_j$, where d_i is the degree of node i.

The above observation justifies why in the NRRW model we consider a single node with a self-loop as a starting point. If this was not the case, for any s even the random walk would only add nodes to the original node, trivially constructing a star since it can never step on a newly created node. The loop allows a change of "parity" with respect to the levels of the tree where new nodes can be added. In fact if the random walk is at level k of the tree and s is even, the random walk can only add new edges at the levels $k + 2h$ for $h = -\lfloor \frac{k}{2} \rfloor, \ldots, -1, 0, 1, 2, \ldots$ until it does not traverse the loop. For s odd this is not necessary as the walker can step on a newly created node to be able to add to nodes in any level of the tree without returning to the root. Thus, yet another fundamental difference between s even and odd.

Will these differences between small and large s and even and odd s manifest themselves in structural properties of the trees generated by the model? In particular, will the degree distribution fundamentally depend on s? In what follows we explore the degrees and other properties of the trees generated by the model showing in fact, that s plays a key role.

2.1 *Simulations*

In order to study the model we designed and implemented an efficient simulator (in C++) for the NRRW model which has as parameters s, n and r, with r denoting the number of independent runs. For each run, we start with a single node with a self loop, move the random walk s steps, connect a new node to its current location, and repeat. We collect statistics for the various properties merging the results across the r simulation runs, such as degree distribution (fraction of nodes with degree k across all runs). The worst case time complexity of a simulation run is $O(sn \log n)$ but the amortized time complexity is $O(sn)$, as we use a growing vector to represent the neighbors of a node that doubles its capacity when needed. Thus, a walker step requires $\Theta(1)$ time and a node addition takes $O(1)$ amortized time.

3 Degree Behavior

In this section we study the degree distribution of NRRW through extensive simulations illustrating its behavior and dependencies. Figure 2 shows the Complementary Cumulative Distribution Function (CCDF) of nodes' for various values for s. Surprisingly, when s is small (between 1 and 8) the respective degree distributions are fundamentally different, exhibiting a kind of power law for s even and an exponential tail for s odd. Note that when $s = 1$ we do not observe nodes with degree larger than 40 while for $s = 2$ a non-negligible fraction of nodes have degree greater than 10^5. We also observe opposite trends in the degree distribution as s increases. For s odd, increasing s yields a distribution with heavier tails, while for s even increasing s yields a distribution with a lighter tails. As s increases into a medium range (between

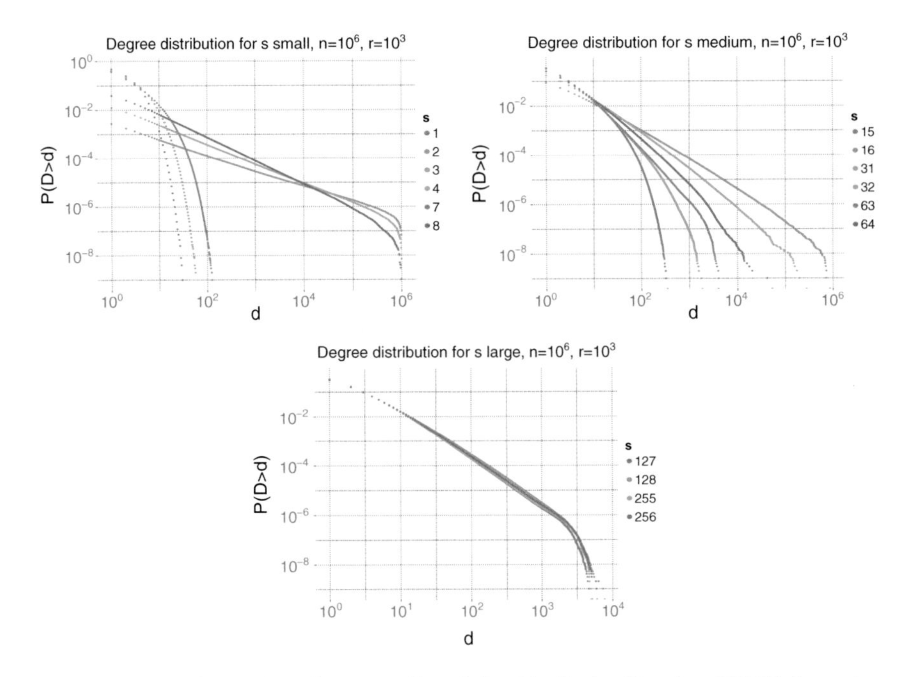

Fig. 2 Empirical degree Complementary Cumulative Distribution Function (CCDF) for various values of s in log – log scale ($n = 10^6$, $r = 10^3$)

15 and 64) the trends continue and the two distributions approach each other. For even larger s (between 127 and 256) the degree distributions become very close, being almost indistinguishable. Interestingly, with large s the degree follows a power law distribution, suggesting that the effect of even s value dominates the dynamics. Moreover, for large s the CCDF exhibits a power law with exponent approximately -2 as it is also the case for the BA model which is based on linear preferential attachment.[2] This supports our initial intuition that when s is large, the random walk samples nodes (adding a new node and connecting to it) with probability proportional to their degree, behaving similarly to the BA model.

Figure 3 shows the degree distribution for $s = 2$ but over different values for n. Interestingly, note that independent of n the degree distribution exhibits the same power law exponent. However, as n increases the fraction of nodes greater than k becomes smaller for any fixed $k > 0$ (with the exception of the cut-off regime which occurs when k is near n). This implies that the fraction of nodes with $k = 1$, the minimum degree, is increasing with n. This is clear by observing $d = 1$ (leftmost point in x-axis) and noting that the fraction of nodes with degree greater than 1 is decreasing

[2]Recall that if D follows a power law distribution, then $P(D = k) \sim k^{-\alpha}$ where $\alpha > 1$ is the power law exponent, and it follows that $P(D \geq k) \sim k^{-(\alpha-1)}$. Thus, the CCDF has an exponent that is one unit less than the PDF.

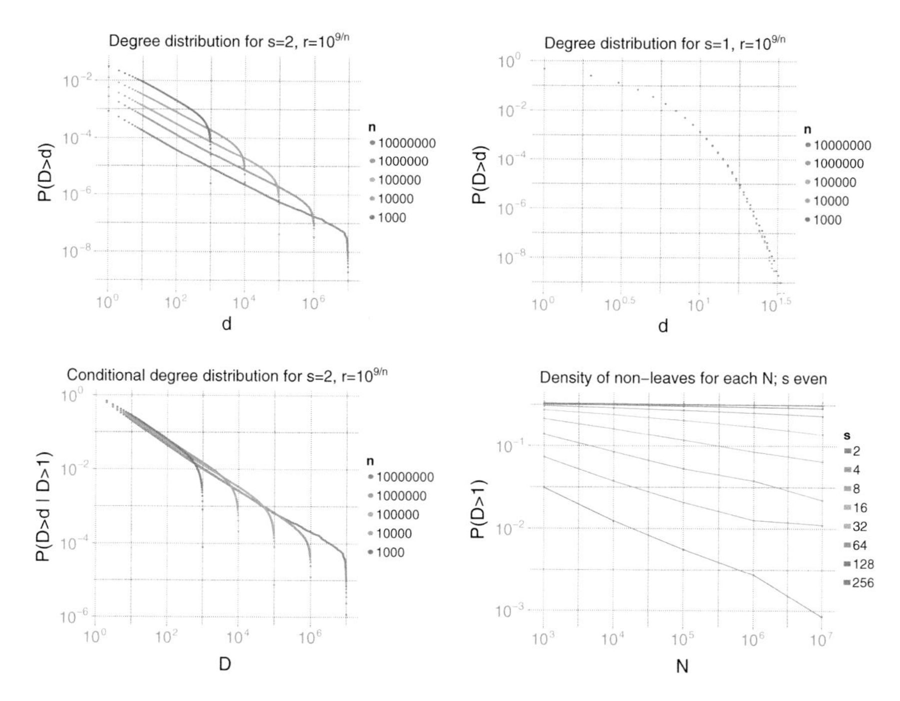

Fig. 3 Empirical degree CCDF for various values of n in log – log scale (*top plots*). Empirical degree CCDF conditioned on the degree being greater than 1; fraction of nodes with degree greater than 1 (*bottom plots*)

with n. Note that such behavior does not occur for $s = 1$ which maintains its degree distribution as n increases (the dots for different n values are barely distinguishable in plot).

If for $s = 2$ the fraction of nodes with degree 1 increases and converges to 1 as n goes to infinity, then we cannot claim that the degree distribution follows a power law. However, we can consider the degree distribution of the nodes that do not have degree 1. In particular, the conditional degree distribution, conditioned on $D > 1$, is shown in Fig. 3. Note that the conditional degree distribution does not show dependance on n and moreover seems to follow a power law. This finding is quite interesting since the fraction of nodes with degree 1 can converge to 1 (as $n \to \infty$) while the remainder of nodes can still follow a power law. This may shed new light on the contrasting results in [3, 9]. We return to this discussion in Sect. 6.

Figure 3 also shows the fraction of nodes with degree greater than 1, $P(D > 1)$, as a function of n for different even s values (for s odd, it does not depend significantly on n as shown in the top right part of Fig. 3 for $s = 1$). Note that for small s the fraction goes to zero reasonably fast (and thus, the fraction of nodes with degree 1 goes to one). As s increases the rate at which $P(D > 1)$ decreases also decreases. Note that for large s (128 or 256) this decrease is barely noticeable, despite being present. Interestingly, as s odd increases, $P(D > 1)$ decreases but without showing any dependency on n. When $s = 257$, $P(D > 1)$ approaches the value shown in Fig. 3 for $s = 256$ (result not shown due to space constraints).

As shown, the NRRW model has a very particular behavior with respect to the degree distribution. In Sect. 6 we provide a further discussion with a few conjectures for its asymptotic behavior with n.

4 Level Behavior

We now investigate the level of the nodes on the trees generated by the NRRW model.[3] As we have shown above, the model dynamics has a fundamental dependence on the parity of s, specially when s is small. Indeed, this dependence also manifests itself on the level of the nodes. Figure 4 shows the level distribution (fraction of nodes at level larger than ℓ) for a few small values of s separated into odd and even, respectively. The level distribution for s even decreases very fast. Note that although $n = 10^6$, when $s = 2$, 90 % of nodes are at level 4 or less and no node is at level 7 or higher. As s even increases the level distribution decreases relatively slower, with 90 % of nodes found at level greater than 4 when $s = 8$. Still, no node is found at level greater than 10. The behavior is completely different for s odd, and the level distribution seems to be uniform (straight line on a linear-linear plot). For $s = 1$ the distribution has the heaviest tail with about 4 nodes per level, giving rise to $2.5 \cdot 10^5$ different levels. For $s = 7$ there are about 40 nodes per level, giving rise to $2.5 \cdot 10^4$ different levels. Interestingly, as s even increases the level distribution becomes heavier while as s odd increases the level distribution becomes lighter. Figure 4 also shows the level distribution for large s. Indeed, as s increases the level distributions for s even and s odd become more similar and the dependency on the parity diminishes. This behavior is similar to what observed for the degree distribution, illustrated in Sect. 3.

Note that from the level distribution we can infer the kind of trees that NRRW generates. When s is small and even, the trees generated are "fat and short", with most nodes near the root and a few with very large degrees. When s is small and odd, the trees are "thin and long" with few nodes spread across many levels and no node with large degree. As s increases, the two kind of trees move in each other's direction, becoming more and more similar.

[3]Recall that the level of node on a tree is given by its distance to the root, and thus the root is at level zero.

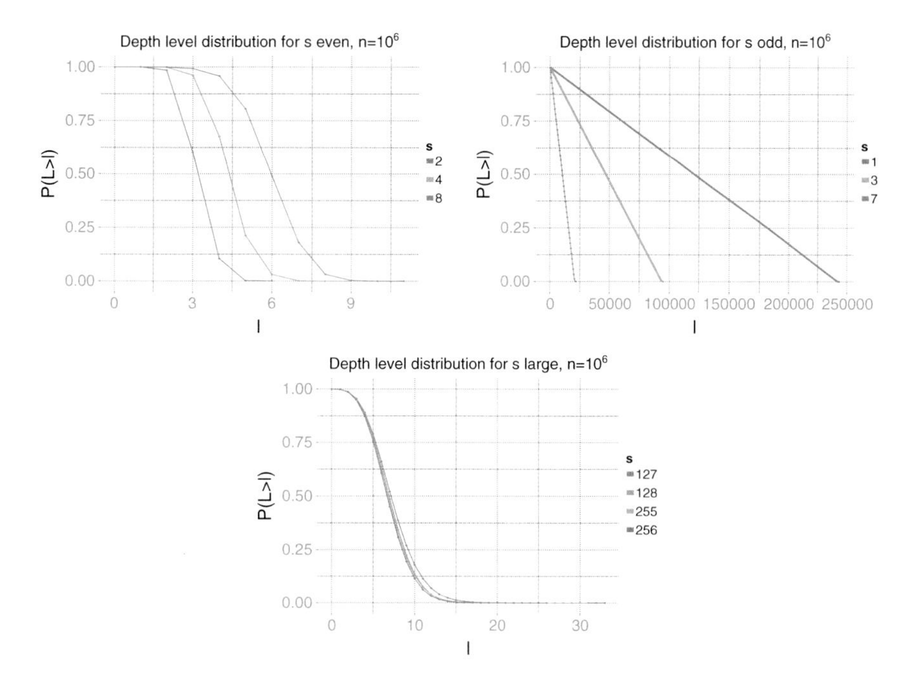

Fig. 4 Empirical CCDF of the node level for different s values ($n = 10^6$, $r = 10^3$)

5 Theoretical Findings for $s = 1$

The numerical simulations with s odd and in particular with $s = 1$ suggest that trees grow in depth as the number of nodes increases. In particular, the growth in depth seems linear on the number of nodes. This is an indication that the random walk is continuously pushing the tree to lower depths just never to return to its origins. In a nutshell, the random walk is transient and visits each node in the tree only a relative small number of times, with high probability. The following Theorem rigorously formalizes this intuition.

Theorem 1 *In the NRRW model with* $s = 1$, *the number of visits to a node is stochastically dominated by* 1 *plus a geometric random variable with support on* $\mathbb{Z}_{>0}$.

Proof We consider here that the initial network consists of a single node with no self-loop. This simplifies the notation and does not compromise the main argument.

Let r denote the initial node of the growing network hereafter referred to as the root of the tree (at any step the growing network is a tree) where the walker resides at time zero. Note that r is the only node at level zero. Let X_n be the level (i.e. the distance from the root) of the node visited by the NRRW at step n. We call the process $\{X_n, n \in \mathbb{Z}_{\geq 0}\}$ the *level process*. Note that the random walk visits r the same number of times that the level process visits level zero. At step $n > 0$ the NRRW is in a node v_n with at least two edges: the one the NRRW has arrived from and the new one added as a consequence of the NRRW's arrival. Let $d_n \geq 2$ denote the degree of node v_n. If $v_n \neq r$, the NRRW jumps from v_n to a node with larger level with probability $\frac{d_n-1}{d_n} \geq \frac{1}{2}$ and with the complementary probability $\frac{1}{d_n} \leq \frac{1}{2}$ to a node with smaller level. If $v_n = r$, then the level can obviously only increase. Note that, due to the fact that degrees keep changing because of the arrival of new edges, the level process is non-homogeneous (both in time and in space).

We now study the evolution of the level process every two steps, i.e. we consider the process $Y_n \triangleq X_{2n}$. Given that the network is a tree and $X_0 = 0$, the two-step level process can be seen as a non-homogeneous reflecting 'lazy' random walk on $2\mathbb{Z}_{\geq 0} = \{0, 2, 4, \ldots\}$. We denote by $p_{k,h}(n)$ the probability that the level at step $n+1$ is h conditioned on the fact that it is k at step n. Although the notation hides it, we observe that the probabilities $p_{k,h}(n)$ depend on the whole history of the NRRW until step n. The reason to consider the two-step level process is that we can get bounds on the transition probabilities $p_{k,h}(n)$ that allow a simple comparison with a (biased) homogeneous random walk. The bounds derived above for X_n lead immediately to conclude that $p_{k,k+2}(n) \geq \frac{1}{2}\frac{1}{2} = \frac{1}{4}$ for any level $k \geq 0$ and $p_{k,k-2}(n) \leq \frac{1}{2}\frac{1}{2} = \frac{1}{4}$ for $k \geq 2$, but we can get a tighter bound for $p_{k,k-2}(n)$. If the NRRW is at level k, all the nodes on the path between its current position and the root r have degree at least 2. If it then moves to node v at level $k-1$, a new edge is attached to v, whose degree is now at least 3. The probability to move from v further closer to the root to a node with level $k-2$, is then at most $\frac{1}{3}$. It follows then that $p_{k,k-2}(n) \leq \frac{1}{2}\frac{1}{3} = \frac{1}{6}$ for $k \geq 2$.

We consider now a homogeneous biased lazy random walk $(Y_n^*)_{n\geq 0}$ on $2\mathbb{Z}_{\geq 0}$ starting from 0 with transition probabilities $p^*_{k,k+2} = \frac{1}{4}$ for all $k \in 2\mathbb{Z}_{\geq 0}$ and $p^*_{k,k-2} = \frac{1}{6}$ for $k \in 2\mathbb{Z}_{\geq 0}$ and $k \neq 0$. We show that if $(Y_n^*)_{n\geq 0}$ also starts in 0 ($Y_0^* = 0$), it is stochastically dominated by $(Y_n)_{n\geq 0}$. We prove it by coupling the two processes as follows. Let $(\omega_n)_{n\geq 0}$ be a sequence of independent uniform random variables over $[0, 1]$. We use them to generate sample paths for both processes $(Y_n)_{n\geq 0}$ and $(Y_n^*)_{n\geq 0}$ as follows:

$$Y_{n+1} = \begin{cases} Y_n - 2, & \text{if } \omega_n \in [0, p_{k,k-2}(n)) \\ Y_n + 2, & \text{if } \omega_n \in [1 - p_{k,k+2}(n), 1] \\ Y_n & \text{otherwise} \end{cases} \quad Y^*_{n+1} = \begin{cases} Y_n^* - 2, & \text{if } \omega_n \in [0, p^*_{k,k-2}) \\ Y_n^* + 2, & \text{if } \omega_n \in [1 - p^*_{k,k+2}, 1] \\ Y_n^* & \text{otherwise} \end{cases}$$

where $p_{k,k-2}(n)$ and $p^*_{k,k-2}$ are 0 if $k = 2$. We start observing that if Y_n and Y_n^* have the same value k, then every time Y_n increases also Y_n^* increases because $p^*_{k,k+2} = \frac{1}{4} \leq p_{k,k-2}(n)$. On the contrary if Y_n^* decreases (as it can happen only for $k \geq 2$), then Y_n may decrease or not because $p_{k,k-2}(n) \leq \frac{1}{6} = p^*_{k,k-2}$. It follows that if Y_n and Y_n^* are at the same level, then $Y^*_{n+1} \leq Y_{n+1}$.

We are going to prove by induction on n that $Y^*_{n+1} \leq Y_{n+1}$ for every n. With a slight abuse of terminology we say that Y_n increases (resp. decreases) if $Y_{n+1} > Y_n$ (resp. $Y_{n+1} < Y_n$). We start observing that indeed $Y_0^* \leq Y_0$, because both processes start in 0. Let us assume that $Y_n^* = h \leq k = Y_n$. For all values of h, every time Y_n^* increases also Y_n increases because $p^*_{k,k+2} = \frac{1}{4} \leq p_{k,k+2}(n)$ and then $Y^*_{n+1} = h+1 \leq k+1 = Y_{n+1}$. If $h \geq 2$, then $p^*_{h,h-2} = \frac{1}{6} \geq p_{k,k-2}(n)$ and if Y_n decreases then Y_n^* must also decrease ($Y_{n+1}^* = h - 1 \leq k - 1 = Y_{n+1}$). It follows that for $h \geq 2$ then $Y_{n+1}^* \leq Y_{n+1}$. The only case when Y_n may decrease without Y_n^* decreasing is when $h = 0$ and $k \neq 0$, but in this case $Y_{n+1}^* = 0$ and $Y_{n+1} \geq 0$. This proves that $Y_{n+1}^* \leq Y_{n+1}$ for every n.

Given that $Y_n^* \leq Y_n$ and both processes start at level zero, the number of visits of $(Y_n)_{n\geq 0}$ to level zero is bounded by the number of visits of $(Y_n^*)_{n\geq 0}$ to level zero. The homogeneous biased lazy random walk $(Y_n^*)_{n\geq 0}$ is transient since $p_{k,k+2}^* = 1/4 > p_{k,k-2}^* = 1/6$. Thus, the probability of the first return time to level 0 is $f_0 < 1$. By the strong Markov property, the number of visits to level 0 is geometrically distributed on the set $\mathbb{Z}_{>0}$ with parameter equal to $1 - f_0$. Since a visit to level zero in $(X_n)_{n\geq 0}$ (one level process) implies a visit to level zero in $(Y_n)_{n\geq 0}$ (two level process), then it follows that the number of visits of $(X_n)_{n\geq 0}$ to level zero is bounded by a geometric random variable and then even more so by 1 plus the same geometric random variable.

Now let us consider any node v in the growing network. If the NRWW never visits v, then the degree of b is 1 and the thesis follows immediately. Otherwise, let consider the first time the NRWW visits v to be time $t = 0$ and let consider v to be the root of the current tree. We can retrace the same reasoning and conclude that the number of visits to v for $t > 0$ is bounded by a geometric random variable on $\mathbb{Z}_{>0}$ with parameter equal to $1 - f_0$. Then the total number of visits to v is bounded by 1 plus such random variable. This concludes the proof. □

Corollary 1 *In the NRRW model with $s = 1$, the degree distribution of any node is bounded by a geometric distribution.*

This follows since the degree of every node equals the number of visits of the random walk to the node plus 1 (the plus 1 accounts for the fact that any node joining the network, although not yet visited by the walker, has degree 1).

6 Discussion and Conclusion

As we have shown, the NRRW model exhibits interesting features that fundamentally impact the networks it generates. For $s = 1$ the random walk is transient and node degree is bounded by a geometric distribution (Theorem 1). For $s = 2$, the fraction of nodes with degree 1 seems to converge to 1 as $n \to \infty$. However, the conditional degree distribution seems to follow a power law. Can such results be made mathematically rigorous? Other interesting questions also emerge from our analysis of the NRRW model. In particular, our numerical simulations seem to indicate that for any s even, the fraction of nodes with degree 1 will converge to 1 as $n \to \infty$. On the other hand, our simulations also indicate that this is not the case for any s odd. So will there be a fundamental difference between a fixed but arbitrarily large even and odd s? It is hard to imagine that $s = 2^{10}$ and $s = 2^{10} - 1$ would have fundamentally different behavior, since in both cases the random walk moves quite a lot before adding a new node. Of course, any fixed s will be small as $n \to \infty$. Thus, we make the following conjecture:

Conjecture 1 For any fixed s even, the fraction of nodes with degree one converges to 1, as $n \to \infty$. For any fixed s odd, the fraction of nodes with degree one converges to a number strictly less than 1, as $n \to \infty$.

If true, such conjecture would imply that the degree distributions are also never identical, for any fixed s even or odd. However, our numerical results indicate that the conditional degree distribution (conditioned on degree being greater than one) for s even, seems to converge to a power law as $n \to \infty$. On the other hand, for $s = 1$ we have proved that the random walk is transient and degree distribution is bounded by a geometric distribution (Theorem 1). Can fixed odd s values really generate power laws? If this is the case, then there would be a phase transition on s, from inducing a network with degree distribution with an exponential tail ($s = 1$) to a power law tail. Despite the numerical results indicating the heavy tail degrees for $s = \{127, 255\}$, we make the following conjecture:

Conjecture 2 For any fixed s even, the conditional degree distribution is bounded from below by a power law, as $n \to \infty$. For any fixed s odd, degree distribution is bounded from above by an exponential, as $n \to \infty$.

Such conjectures consider that n diverges. In practice n must be finite when generating a network with NRRW model. Thus, for a fixed n, the differences induced by an even or odd s may diminish as s increases. In particular, the degree distribution generated by even and odd s values may become arbitrarily close as s increases, as we have observed in numerical simulations for a fixed n (Fig. 2).

Last, we return to the recent dispute if the Random Walk model with restarts generates a power law degree distribution, independently of s [3, 9]. It has been mathematically proved that when $s = 1$ the fraction of nodes with degree one converges to 1, as $n \to \infty$ [3]. At the same time, simulation results suggest that the degree distribution follows a kind of power law [9]. We can attempt to reconcile such findings by leveraging our own findings on NRRW model. When $s = 1$ the new node is connected to a given existing node u if i) a neighbor of u is selected at the restart and then ii) the random walk moves to u. A node whose neighbors are all leaves would be selected with a probability proportional to its degree. Now it has been shown that when $n \to \infty$ the fraction of nodes that are leaves converges to 1, then most of the neighbors of a non-leaf node are leaves and this node is essentially selected proportionally to its degree, similarly to the BA model embodying preferential attachment. Thus, the conditional degree distribution, leaving out degree 1 nodes, will follow a power law distribution with the same exponent as in the BA model. In some sense, this reconciles the findings of the two prior works [3, 9].

To conclude, as exemplified above, a fundamental understanding of NRRW model adds to our understanding of purely local network growth models. In particular, besides requiring a less strict assumption to operate, models that do not rely on any global primitive can also generate networks with rich and diverse structural properties.

Acknowledgments Research conducted within the context of the THANES Associate Team, jointly supported by Inria (France) and FAPERJ (Brazil). This work has also been partially funded through research grants from CNPq and CAPES (Brazil).

References

1. Barabási, A.L., Albert, R.: Emergence of scaling in random networks. Science **286**(5439), 509–512 (1999)
2. Blanchard, P., Krueger, T., Ruschhaupt, A.: Small world graphs by iterated local edge formation. Phys. Rev. E **71**, 046139 (2005)
3. Cannings, C., Jordan, J.: Random walk attachment graphs. Electron. Commun. Probab. **18**, 1–5 (2013)
4. Colman, E.R., Rodgers, G.J.: Local rewiring rules for evolving complex networks. Phys. A: Stat. Mech. Appl. **416**, 80–89 (2014)
5. Gabel, A., Redner, S.: Sublinear but never superlinear preferential attachment by local network growth. J. Stat. Mech.: Theory Exp. **2013**(02), P02043 (2013)
6. Ikeda, N.: Network formation determined by the diffusion process of random walkers. J. Phys. A: Math. Theor. **41**(23), 235005 (2008)
7. Ikeda, N.: Network formed by movements of random walkers on a bethe lattice. In: Journal of Physics: Conference Series, vol. 490, p. 012189. IOP Publishing (2014)
8. Li, M., Gao, L., Fan, Y., Wu, J., Di, Z.: Emergence of global preferential attachment from local interaction. New J. Phys. **12**(4), 043029 (2010)
9. Saramäki, J., Kaski, K.: Scale-free networks generated by random walkers. Phys. A: Stat. Mech. Appl. **341**, 80–86 (2004)
10. Vázquez, A.: Growing network with local rules: preferential attachment, clustering hierarchy, and degree correlations. Phys. Rev. E **67**(5), 056104 (2003)

Pseudo-Cores: The Terminus of an Intelligent Viral Meme's Trajectory

Yayati Gupta, Debarati Das and S.R.S. Iyengar

Abstract Most memes die soon after they have been released, but only few go viral and spread worldwide. Identifying the secret recipe for the success of such viral memes is a very interesting ongoing research question. While many researchers have attributed the success of a meme to its content and place of origin, we propose taking into consideration the underlying network structure that the meme propagates on. In this paper, we induce artificial virality in a meme by intelligently directing its trajectory in the network. This induction is based upon the spreading power of core nodes in a core-periphery structure. This paper puts forward two greedy hill climbing approaches to determine the path from a node in the periphery shell (where the memes generally originate) to the core of the network. We also unearth specialized shells—*Pseudo-Core*, which emulate the behavior of the core in terms of spreading power. We consider two sets for the target nodes, one being core and the other being any of the pseudo-cores. We show that our algorithms perform better than random and degree based approaches and have a worst case time complexity of O(n). The paper highlights the importance of core-periphery structure in a network and the role of pseudo-cores in making a meme go viral.

1 Introduction

Behind every viral meme, there is a story. Korean pop star Psy launched a quirky music video "Gangnam Style" in 2012 which received almost 1 billion views on Youtube. In India, the song "Why this Kolaveri Di" by Dhanush became an instant social rage in 2011. It earned more than 10,500,000 Youtube views by November

Y. Gupta (✉) · S.R.S. Iyengar
Indian Institute of Technology Ropar, Rupnagar, India
e-mail: yayati.gupta@iitrpr.ac.in

S.R.S. Iyengar
e-mail: sudarshan@iitrpr.ac.in

D. Das
PES Institute of Technology, Bangalore, India
e-mail: debarati.d1994@gmail.com

H. Cherifi et al. (eds.), *Complex Networks VII*, Studies in Computational Intelligence 644, DOI 10.1007/978-3-319-30569-1_16

2011 and investigating the reasons behind the virality of these videos became a pressing research question [1]. The unique dance moves, upbeat tone and the visual nature of the Gangnam style were considered instrumental in making the video viral [2]. Similarly "Why this Kolaveri di" gained a cult status because of its distinct "Tanglish[1]" style. These memes required only a few seconds to absorb the attention of the viewer because of their novel characteristics.

Though at first glance it does seem that novelty is the only factor responsible for making a meme viral. But then, on 8 January 2010, a Californian resident, Paul "Bear" Vasquez uploaded a video of a double rainbow. It was a video exclaiming about the double rainbow he had witnessed in the front yard of his home. What was surprising is that this seemingly mundane video got close to 23 million views in 2010. If only the unique intrinsic characteristics of a meme are responsible in making it viral, what made this not so innovative "Double Rainbow" video go viral?

Kevin Alloca, Youtube's trends manager elegantly addressed this question in his TED talk—"Why Videos go Viral ?" [3]. He highlights the importance of tastemakers in introducing a video to the crowd and make it viral. Tastemakers are the people on Youtube having a large number of followers. Kevin showed that both the videos Double Rainbow" and "Friday" became popular after being referred by the tastemakers Jimmy Kimmel and Michael Nelson respectively. A broader study by Kitsak et al. showed that the influence of a node can be better marked by the coreness [4]. Coreness of a node takes into account both the degree and the influence of its neighbours in account. Kitsak et al. identify the core nodes in a network and term these nodes as superspreaders.

If we could hit the core of a network with the meme traversing over the network, it will quickly go viral. Hence, we probe upon the question: *"How can we intelligently target links in the network to quickly reach the superspreaders in a network?"*.

1.1 Superspreaders : The Destination for a Viral Meme

In our proposal, we have visualized the social network as having a meso scale characteristic called as core-periphery structure. The notion of a core-periphery structure was defined by Borgatti and Everett in their seminal paper [5] in 2000.[2] It has been proved that most networks existing in the nature possess core-periphery structure [6].[3] This result has far reaching consequence on meme virality.

High status people enjoy higher privileges, hence becoming well connected amongst themselves as well as with the rest of the network. This makes them easily

[1] portmanteau word of Tamil and English.

[2] The core was defined as a set of nodes densely connected to each other having a large number of connections to the periphery nodes. On the other hand, the periphery nodes although connected to the core nodes are largely disconnected amongst themselves.

[3] Most of the networks in real world are scale free [7]. Works done by Della et al. [8] and Liu et al. [6] prove that scale-free networks usually possess a core-periphery structure. Therefore by transition, it becomes evident that a social network is a scale free network as well as a core-periphery structure.

accessible. Moreover, Kitsak et al. [4] have shown that the influential spreaders in a network are those which lie at its core. These two properties of high reachability and maximum spreading power validate the core nodes in a network as superspreaders.

Taking all the aforementioned facts in account, we propose algorithms that can help a user intelligently guide a meme towards the core of a network. So, if the information originates in the periphery of a network, our problem can be reduced to a path finding algorithm, given that the source is a node in the periphery of the network and destination is the core. The main contributions of the paper are as follows:

1. Evaluation and employment of different properties of the shells provided by the k-shell decomposition algorithm to take an intelligent walk towards the destination in the network. The key idea of the paper is to utilise the presence of multiple shells in a network to effectively reach its core starting from the periphery.
2. Unearthing specialized shells which mimic the core in terms of spreading power and help a meme go viral. We call these shells as *"Pseudo-Cores"*.

These revelations of our experiments may impact the fields of information propagation as well as epidemiology. It can help in the formulation of intelligent pathways for information propagation and placement of preventive checkpoints to halt infection spread.

2 Presence of Pseudo-Cores in a Network

In this section, we define the pseudo-cores and describe their importance in making a meme go viral. We have done it with the help of certain experiments which are performed on the networks [6, 9] mentioned in Table 1. We use k-shell decomposition[4] algorithm to unravel the pseudo-core shells. Reference [11] is one of the most popular algorithms to decompose a network into multiple shells of influence. In this decomposed network, the innermost shell, or the core, is the most dense subgraph having the highest closeness centrality. As we move outward towards other shells, density and closeness both decrease and the outermost shells are called the periphery shells. Core-periphery structures have been studied greatly ever since they were introduced in 2000. However the research attention was focused mostly on the core, and scientists have not yet tried to harness the power of the intermediary shells in the network. In this paper we investigate the properties of all the shells in a network and then observe how the network structure in these shells can be efficiently used to guide the meme in a correct direction and make it viral.

[4]This algorithm is explained in detail in [10].

Table 1 Datasets used for experiments

Dataset	Description
Facebook	Facebook is the most popular Social Networking Site today. This dataset consists of anonymized friendship relations from Facebook [12]. The network contains 4,039 nodes and 88,234 edges
Google plus	Google plus is a social layer for Google Services [12]. The network contains 107,614 nodes and 13,673,453 edges
Slashdot	Slashdot is a website where the users can submit and evaluate the news stories on science and technology. It is famous for its specific user community [13]. This dataset contains 82,168 nodes and 948,464 edges
Flickr	This is an image and video hosting site. It is mainly used for sharing and embedding personal photographs [6]. This dataset has 80513 nodes and 5899882 edges
Livemocha	Livemocha is the world's largest online language learning community [6]. This dataset has 104438 nodes and 2196188 edges
DBLP	The DBLP computer science bibliography is a collaboration network. It provides a detailed list of research papers in computer science [14]. The network contains 317,080 nodes and 1,049,866 edges
Buzznet	Buzznet is social media network used for sharing photos, journals, and videos. It has 101168 nodes and 4284534 edges

2.1 Cascading Power of Different Shells in a Network:

Cascading power of a shell: We know that a shell is a group of nodes. For finding the cascading power of a shell, we infect the network starting from some of the random nodes in the particular shell. The network is infected using independent cascade model. At every iteration, the most recently infected nodes infect their adjacent nodes with certain probability. The process stops as soon as no new node is infected in an iteration. It is to be noted that every node gets at most one chance of infection. We note the number of nodes infected at the end of this process. Corresponding to one shell, we repeat this process 100 times starting from different seed nodes. The average number of nodes infected in 100 iteration is termed as the cascading power of a shell.

The plots in Fig. 1 represent the meme cascade size produced if the cascade starts from some of the nodes of a particular shell. One ideally expects the cascade to accelerate when core shell is encountered, but it is observed that the acceleration point is reached much before the point where the meme reaches the core. This intriguing fact led us to investigate the existence of a "pseudo-core shell" or a shell which provides something akin to an escape velocity for the meme to become viral i.e. once the meme reaches this shell, it goes viral and there is no longer a need for it to target some higher shell node. This hypothesis may have many large scale implications. For example-Imagine a political analyst trying to find which person to infect in a political network, she would no longer have to infect the most influential politicians or relatively insulated core nodes. Infecting someone relatively less influential(if this person lies in a pseudo-core shell) would cause the same effect.

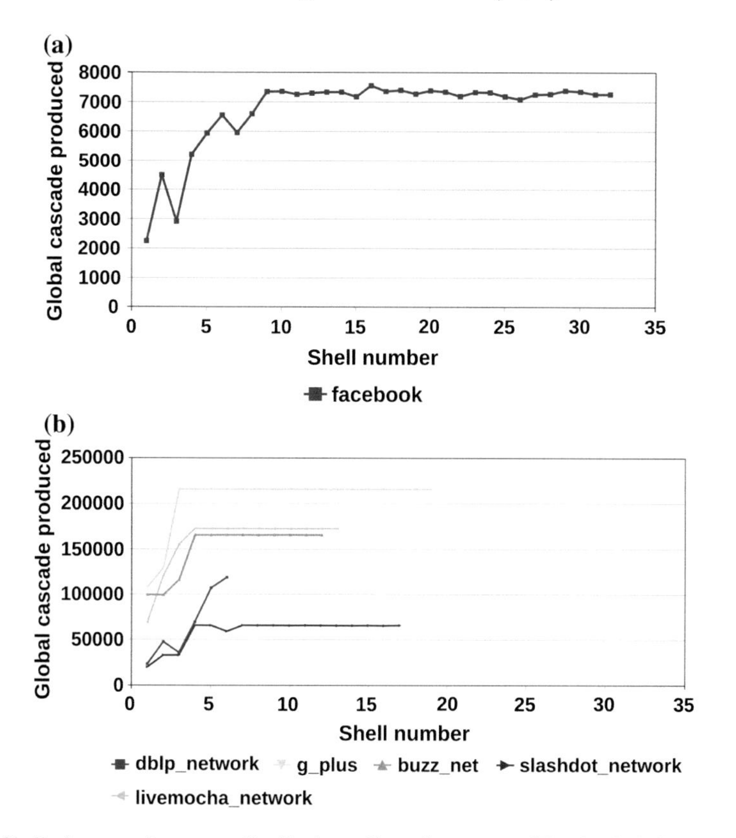

Fig. 1 Shell wise cascading power distribution. **a** Cascading power of Facebook shells. **b** Cascading power of other real world network's shells

We propose and compare a set of path finding algorithms in the next section and further observe if changing the destination to pseudo-core could impact the time taken by the algorithm.

3 Algorithms

We describe algorithms in this section to find a path from the periphery of a network to the destination, which we initially assume as core. Later we apply the same algorithm with pseudo-cores as the destination and report the improved results. We describe two already implemented algorithms in Table 2.

Below, we propose two hill climbing algorithms based on the shell numbers of the nodes. After the network has been decomposed into shells, the entire system can be visualized as a circular maze made up of concentric circles(shells). The goal of

Table 2 Existing algorithms

Random walk algorithm [15]	Degree based hill climbing [16]
This algorithm involves a node inspecting its neighbours at every step and selecting one of them randomly. If the chosen neighbour is a core node, the algorithm terminates, else the selection of the random neighbours continues. Random walk algorithm (without repetition of nodes) has a time complexity of $O(n)$	This algorithm uses a hill climbing approach based on the degree of the nodes in the network. At every step, a node looks at its neighbours and chooses the unexplored node having the highest degree. If the chosen node is a core node, the algorithm terminates, else the process continues. As hill climbing algorithms have a complexity of $O(n)$ where n is the number of nodes and finding degree of all nodes takes $O(m)$ time, degree based hill Climbing has a time complexity of $\max[O(n), O(m)] \sim O(n)$ in sparse graphs

the algorithm is to intelligently move from the outermost shell to the innermost shell. There are inter-shell edges that help a user in taking such a walk across shells, while the intra-shell edges help the user to traverse a shell.

3.1 Algorithm 1—Shell Based Hill Climbing Approach (SH):

Let $G(V, E)$ represent the graph where $V(G)$ is the set of vertices and $E(G)$ is the set of edges. Let the number of vertices and edges in G be n and m respectively. $shell(u)$ represents the shell number of a node u as calculated by the k-shell decomposition algorithm. $start$ is the periphery node from where the meme starts spreading. $N_G(u)$ represents the set of neighbours of node u in the graph G. The proposed SH approach has a complexity of $max[O(m + n), O(n)] \sim O(n)$ in sparse graphs.

3.2 Algorithm 2—Intershell Hill Climbing with Intrashell Degree Based Approach(SA):

Algorithm 2 is a modification of Algorithm 1 and utilises the idea that a node with very high degree will cover most of the shell. If this node is chosen, it would greatly reduce the number of steps required to traverse a shell. Let the number of vertices and edges in G be n and m respectively. $N_G(u)$ represents the set of neighbours of node u in the graph G. The proposed SA approach has a complexity of $max[O(m + n), O(n), O(m)] \sim O(n)$ in sparse graphs (Fig. 2).

Algorithm 1 Shell Based Hill Climbing(SH)

procedure FINDNUMSTEPS
  Input:- Graph $G(V, E)$, Starting node $start$
  Output: Number of steps taken by the algorithm to terminate
  Apply k-shell decomposition and calculate $shell(u)\ \forall u \in V(G)$
  $visited[u] \leftarrow `false'\ \forall u \in V(G)$
  $numsteps \leftarrow 0$
  $current \leftarrow start$
  $visited[current] \leftarrow `true'$
  while $current$ is not a core node **do**
    $v_1 \leftarrow argmax_{u \in N_G(current) \wedge visited[u]=`false'} shell(u)$
    if $shell(v_1) \leq shell(current)$ **then**
      $v_2 \leftarrow random\ node\ u \in N_G(current) \wedge visited[u] = `false'$
      $current \leftarrow v_2$
    else
      $current \leftarrow v_1$
    end if
    $numsteps \leftarrow numsteps + 1$
  end while
  return $numsteps$
end procedure

Algorithm 2 Improved Shell Based Hill Climbing(SA)

1: **procedure** FINDNUMSTEPS
2:   **Input**:- Graph $G(V, E)$, Starting node $start$
3:   **Output**: Number of steps taken by the algorithm to terminate
4:   Apply k-shell decomposition and calculate $shell(u)\ \forall u \in V(G)$
5:   $visited[u] \leftarrow `false'\ \forall u \in V(G)$
6:   $numsteps \leftarrow 0$
7:   $current \leftarrow start$
8:   $visited[current] \leftarrow `true'$
9:   **while** $current$ is not a core node **do**
10:     $v_1 \leftarrow argmax_{u \in N_G(current) \wedge visited[u]=`false'} shell(u)$
11:     **if** $shell(v_1) \leq shell(current)$ **then**
12:       $v_2 \leftarrow argmax_{u \in N_G(current) \wedge visited[u]=`false'} degree(u)$
13:       $current \leftarrow v_2$
14:     **else**
15:       $current \leftarrow v_1$
16:     **end if**
17:     $numsteps \leftarrow numsteps + 1$
18:   **end while**
19:   return $numsteps$
20: **end procedure**

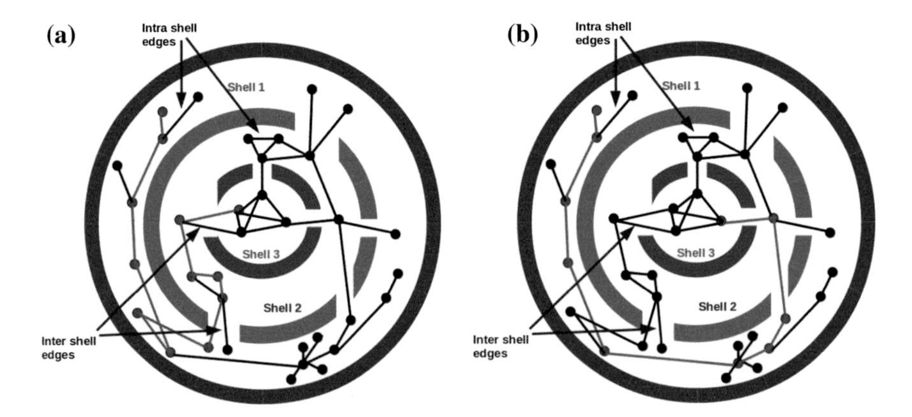

Fig. 2 Proposed Algorithms: The path denoted in the *pink edges* is the path chosen by the corresponding algorithm to move towards the core. **a** Algorithm1—shell based hill climbing (SH). **b** Algorithm2—modified shell based hill climbing (SA)

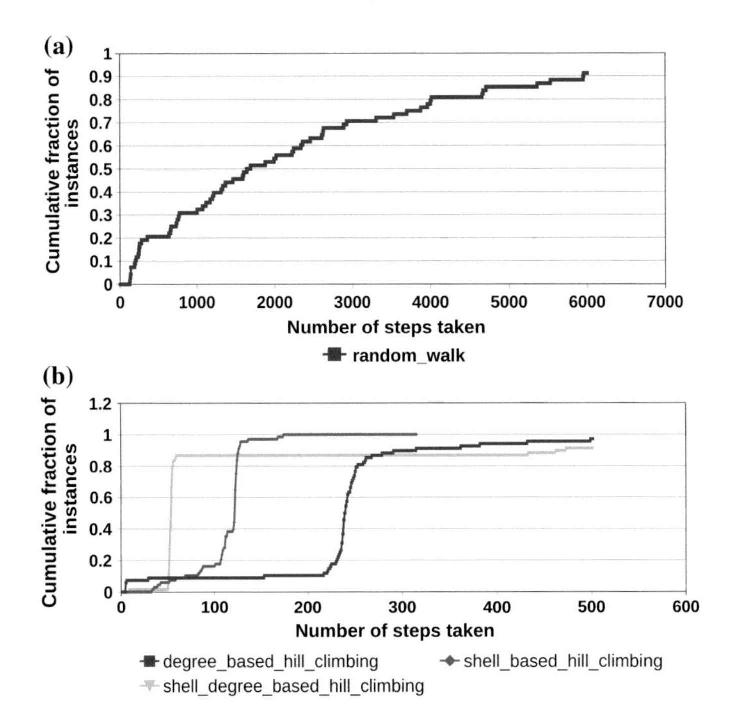

Fig. 3 Comparison of algorithms for facebook. **a** Random walk. **b** Shell based hill climbing algorithms

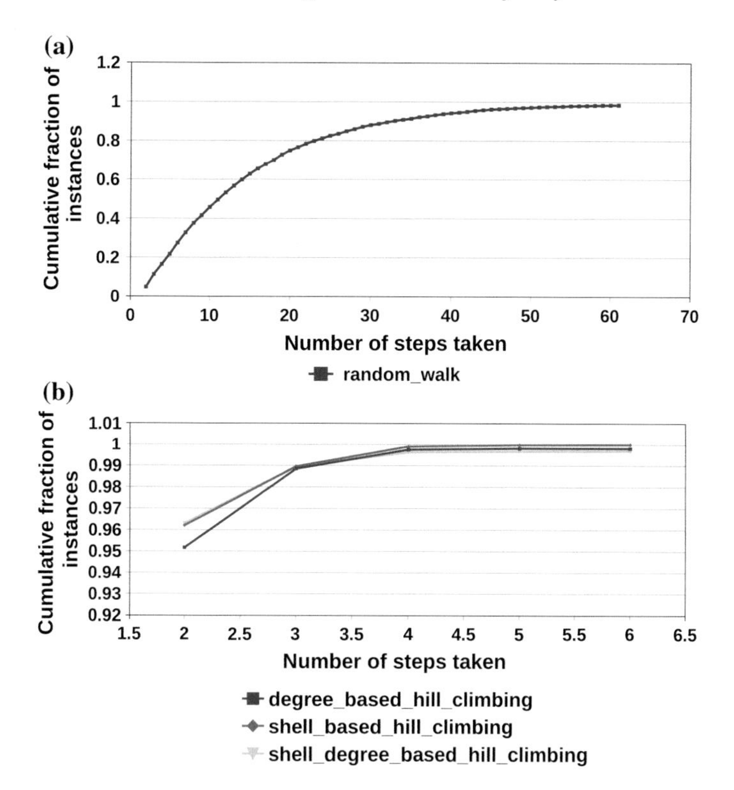

Fig. 4 Comparison of algorithms for google plus. **a** Random walk. **b** Shell based hill climbing algorithms

4 Experimental Results

To evaluate the performance of the algorithms mentioned in the above section, we select periphery nodes from shell 1(periphery) in a network and for each of these nodes, we find the number of steps taken to reach the core. We term each run from a periphery node as an instance of the problem. Therefore, we can say the number of instances is equal to the number of periphery nodes. It is observed that more than 80 % of the walks conclude in a maximum of 15 steps in most of the datasets.[5]

Let R be a random variable depicting the number of steps taken by the algorithm to terminate. Let $P(R = k)$ be the Probability of R being k, where $k \geq 2$. We plot the cumulative probability distribution function of R. X axis indicates all possible values of R while Y axis shows the probability of $R \leq k$.

The plots given below validate that the proposed algorithms cover most of the instances in very less number of steps as compared to the existing path finding

[5]Without loss of generality, we have ignored the trivial case where source nodes are directly connected to the core as the path length in these cases is 1.

algorithms. The highest line in the curve represents the most efficient algorithm. In the case of Facebook network, the proposed algorithms cover 80 % of the instances in less than 100 steps. Degree based hill climbing requires around 200 steps to cover 80 % of the instances. In the case of Google Plus, all the three hill climbing based approaches cover 90 % of the instances in less than 3 steps. The results for the rest of the networks can be found in [10]. In all the cases, the algorithms proposed reach their peak at the earliest proving that they are more optimal with respect to the time taken to reach the destination.The random walk algorithm clearly performs the worst (Figs. 3 and 4).

Next, we modify the destination to be the pseudo-core shells and observe the cumulative frequency distribution of R as given in Fig. 5. In this case also, our proposed algorithms perform better than the other algorithms. Interestingly, the performance of even the random walk algorithm increases drastically when the target is changed to be the pseudo-cores. This indicates that the virality which seems frequent and random in our social as well as biological networks may be because of the presence of pseudo-cores in the network. It is fairly intuitive that it is difficult to target a core node. However, it would be easier to hit the pseudo-core and this could be one of the possible explanations for meme virality in a network. The results for these simulations are shown in Fig. 5.

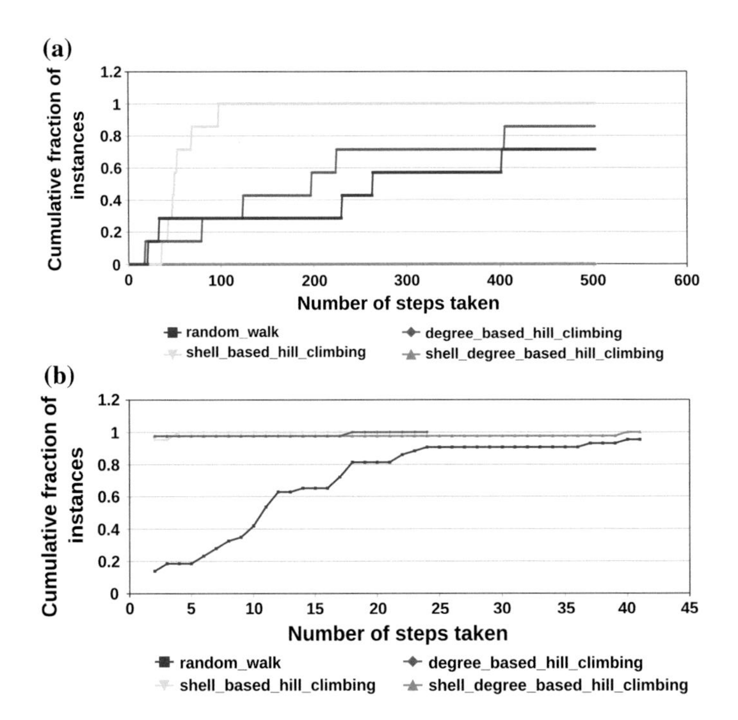

Fig. 5 Comparison of algorithms for infecting pseudo-cores. **a** Facebook. **b** Google plus

5 Related Work

The information derived from the internet is being harnessed in a myriad of applications today. Culotta et al. [17] have used the information potential of a social network to predict epidemics in a population. Social networks act as reservoirs of data which can be used to predict the results of elections [18] as well as patterns in crime [19]. Meme is a term used to describe a unit of information traversing in a network. These memes behave like biological viruses and evolve over time as suggested by Daley et al. in their work [20]. Memetics or the study of memes has a wide range of applications in several research areas like Digital Marketing and Epidemiology. This is not surprising as deciphering patterns in any kind of data or trajectories in information flow in the network can have wide range impacts. If for some reason an information goes viral and impacts a large portion of the network, then the meme holds more potential in the network for analysis as to why it went viral.

Many approaches have been employed to understand the cause of meme virality. Berger and Milkman [21] employed the content of a meme to predict its virality while Weng et al. observed the similarity between a simple contagion and a viral meme [22]. The existence of communities and core-periphery structure [5] are two major discoveries with respect to complex network structure. We applied the studies on complex network structures to understand meme spread and probed upon the question : *"Can we intelligently alter the path of a meme flowing through a network to make it go viral?"*.

Milgrams experiment [23] had a similar aim to find the shortest path from a source person to the target person. For this experiment, breadth first search approach is not suitable as it would lead to flooding of letters in the network. We cannot also assume that a person will advertise a product to each of his/her neighbours. Similarly a DFS might result in several paths which are not optimal. However, It was observed that even though people did not possess an overview of the entire network, they were still able to trace the average 6-hop path between two individuals. This spawned the idea of a decentralised search approach [24]. This greedy heuristic aimed at providing a path from a source to a destination exploring only one new node per iteration which is nearest to the target. We were inspired by this decentralised approach to propose a method to direct a meme in an optimal direction. Our work differentiates from the decentralised algorithms in two ways:

1. Instead of focusing on one target node, we are trying to attack exactly one node in a group of nodes, termed as core.
2. We are proposing the algorithm for real world networks instead of very well defined lattice like structures while most of the decentralised algorithms are proposed for very well defined structures.

In this paper, we propose a hill climbing technique by virtue of which a user needs to focus only on one neighbour who gives him/her more benefit as compared to distributing his/her efforts among all the neighbours. Our work is the first of its kind to the best of our knowledge.

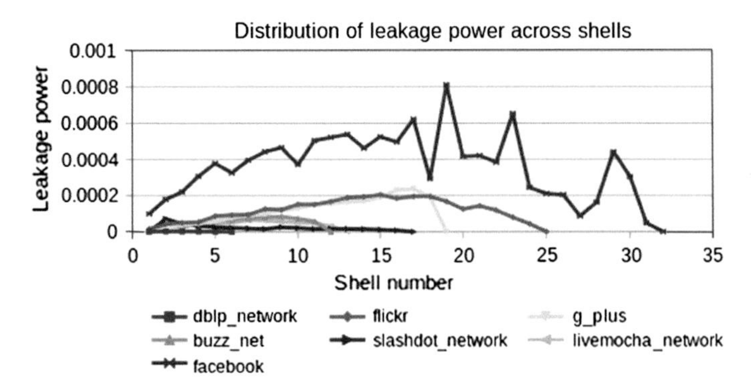

Fig. 6 Distribution of leakage power across shells for various networks

6 Conclusion

The paper unravels the potential of a core-periphery structure in imparting virality to a passable/non-exemplary meme. We empirically observed that the innermost core shell has the greatest tendency to trigger a global cascade in the network, thereby increasing the necessity to infect the core quickly in order to cause virality. We have proposed two shell based hill climbing approaches that help a meme to pave an intelligent path to the core, when it originates in the periphery. One of the most important contributions of the paper is the unveiling of the concept of *"Pseudo Core"* shells that have the same cascading impact on the network as a core shell. Intelligently hitting the pseudo-core shell achieves the same virality as that achieved by core shell. As a result the path taken during the trajectory of a viral meme can be reduced. These revelations introduced by our experiments may prove to have huge impact across several disciplines.

7 Future Work

While performing experiments to analyse network property effects on shells, we defined a shell parameter which we deem *"Leakage Power"*. Leakage power denotes a shell's potential to connect to the higher numbered shells.[6] We plotted leakage power against shell number and observed that the leakage power was not necessarily the highest in the case of core. There were indications of high leakage power in intermediary shells as well. This is shown in Fig. 6. This led us to investigate the ideas of *"teleportation shells"* and *"barricade shells"*. The shells having higher leakage powers act as the teleportation shells and can trigger a meme to take longer jumps on its path to the core. On the other hand, the shells having low leakage powers

[6] The exact calculation for finding the leakage power of a shell can be found in [10].

may tend to block a meme inside it, hence suggesting why some memes are non viral. Based on these observations, the algorithms may be altered to provide better results and most importantly answer the bigger question which is *"How to make a meme go viral ?"*.

Acknowledgments The authors would like to thank the IIT Ropar HPC committee for providing the resources to perform experiments. S.R.S. Iyengar was partially supported by the ISIRD grant (Ref. No. IITRPR/Acad./359) from IIT Ropar. Further, we express our gratitude to the Indian Academy of Sciences, Bangalore for providing us with partial funding to carry out this research.

References

1. Jiang, L., Miao, Y., Yang, Y., Lan, Z., Hauptmann, A.G.: Viral video style: a closer look at viral videos on youtube. In: Proceedings of International Conference on Multimedia Retrieval, p. 193. ACM (2014)
2. H.G.S. went Viral: Gangnam style takes the world by storm
3. Allocca, K.: Why videos go viral. Ted talk (2011)
4. Kitsak, M., Gallos, L.K., Havlin, S., Liljeros, F., Muchnik, L., Stanley, H.E., Makse, H.A.: Identification of influential spreaders in complex networks. Nat. Phys. **6**(11), 888–893 (2010)
5. Borgatti, S.P., Everett, M.G.: Models of core/periphery structures. Soc. Netw. **21**(4), 375–395 (2000)
6. Zafarani, R., Liu, H.: Social computing data repository at ASU (2009). http://socialcomputing.asu.edu
7. Barabási, A.-L., et al.: Scale-free networks: a decade and beyond. science **325**(5939), 412 (2009)
8. Della Rossa, F., Dercole, F., Piccardi, C.: Profiling core-periphery network structure by random walkers. Scientific reports, vol. 3 (2013)
9. Leskovec, J., Krevl, A.: SNAP Datasets: Stanford large network dataset collection, Jun 2014. http://snap.stanford.edu/data
10. Gupta, Y., Das, D., Iyengar, S.: Pseudo-cores: the terminus of an intelligent viral meme's trajectory (2015). arXiv preprint arXiv:1507.07833
11. Batagelj, V., Zaversnik, M.: An o (m) algorithm for cores decomposition of networks (2003). arXiv preprint arXiv:cs/0310049
12. Leskovec, J., Mcauley, J.J.: Learning to discover social circles in ego networks. In: Advances in Neural Information Processing Systems, pp. 539–547 (2012)
13. Leskovec, J., Lang, K.J., Dasgupta, A., Mahoney, M.W.: Community structure in large networks: natural cluster sizes and the absence of large well-defined clusters. Internet Math. **6**(1), 29–123 (2009)
14. Yang, J., Leskovec, J.: Defining and evaluating network communities based on ground-truth. Knowl. Inf. Syst. **42**(1), 181–213 (2015)
15. Noh, J.D., Rieger, H.: Random walks on complex networks. Phys. Rev. Lett. **92**(11), 118701 (2004)
16. Adamic, L.A., Lukose, R.M., Puniyani, A.R., Huberman, B.A.: Search in power-law networks. Phys. Rev. E **64**(4), 046135 (2001)
17. Culotta, A.: Towards detecting influenza epidemics by analyzing twitter messages. In: Proceedings of the First Workshop on Social Media Analytics, pp. 115–122. ACM (2010)
18. Sang, E.T.K., Bos, J.: Predicting the 2011 dutch senate election results with twitter. In: Proceedings of the Workshop on Semantic Analysis in Social Media. Association for Computational Linguistics, pp. 53–60 (2012)

19. Gerber, M.S.: Predicting crime using twitter and kernel density estimation. Decis. Support Syst. **61**, 115–125 (2014)
20. Daley, D.J., Kendall, D.G.: Epidemics and rumours (1964)
21. Berger, J., Milkman, K.L.: What makes online content viral? J. Mark. Res. **49**(2), 192–205 (2012)
22. Weng, L., Menczer, F., Ahn, Y.-Y.: Virality prediction and community structure in social networks. Sci. Rep. **3**, 2522 (2013)
23. Milgram, S.: The small world problem. Psychol. Today **2**(1), 60–67 (1967)
24. Kleinberg, J.: Complex networks and decentralized search algorithms (2006)

Modeling Evolutionary Dynamics of Lurking in Social Networks

Marco A. Javarone, Roberto Interdonato and Andrea Tagarelli

Abstract Lurking is a complex user-behavioral phenomenon that occurs in all large-scale online communities and social networks. It generally refers to the behavior characterizing users that benefit from the information produced by others in the community without actively contributing back to the production of social content. The amount and evolution of lurkers may strongly affect an online social environment, therefore understanding the lurking dynamics and identifying strategies to curb this trend are relevant problems. In this regard, we introduce the *Lurking Game*, i.e., a model for analyzing the transitions from a lurking to a non-lurking (i.e., active) user role, and vice versa, in terms of evolutionary game theory. We evaluate the proposed Lurking Game by arranging agents on complex networks and analyzing the system evolution, seeking relations between the network topology and the final equilibrium of the game. Results suggest that the Lurking Game is suitable to model the lurking dynamics, showing how the adoption of rewarding mechanisms combined with the modeling of hypothetical heterogeneity of users' interests may lead users in an online community towards a cooperative behavior.

1 Introduction

Most members of online communities and social networks do not actively contribute to the shared online space, i.e., they only consume (e.g., read, watch) information without sharing their knowledge or expressing their opinion. These users are

M.A. Javarone
Department of Mathematics and Computer Science, University of Cagliari, Cagliari, Italy
e-mail: marcojavarone@gmail.com

R. Interdonato · A. Tagarelli (✉)
Department of Computer, Modeling, Electronics, and Systems Engineering, University of Calabria, Rende, Italy
e-mail: andrea.tagarelli@unical.it

R. Interdonato
e-mail: rinterdonato@dimes.unical.it

H. Cherifi et al. (eds.), *Complex Networks VII*, Studies in Computational Intelligence 644, DOI 10.1007/978-3-319-30569-1_17

commonly defined as *lurkers*, since they remain quite unnoticed while benefiting from others' information or services. Remarkably, lurkers feel themselves as community members, and should not be trivially regarded as totally inactive users, i.e., registered users who do not use their account to join the online community.

The characterization of lurking in online communities has been a controversial issue from a social science and computer-human interaction perspective [9]. One common perception of lurking is related to the infrequency of active participation to the community life [24], while other definitions refer to legitimate peripheral participation [17], individual information strategy of microlearning [16], and knowledge sharing barriers [5]. In general, in the realm of online social networks (OSNs), neutral or even positive views of the presence of lurkers have normally supplanted negative views. The silent presence of lurkers can indeed be seen as harmless as it reflects a subjective reticence (rather than malicious motivations) to contribute to the community wisdom [24]. Moreover, lurking can be expected or even encouraged because it allows newcomers to learn the netiquette before they might decide to provide a valuable contribution over time. On the other hand, if users are worried that their private information may be revealed or their security may be threatened by posting, they may decide to lurk to protect themselves.

Lurkers hold great potential in terms of *social capital*, because they acquire knowledge from the OSN; further, they might decide to use this knowledge in order to form their own opinions, although these will never or rarely be unveiled to the community. Within this view, it is highly desirable to *delurk* such users, i.e., to apply a mix of strategies aimed at encouraging lurkers to return their acquired social capital, through a more active participation to the community life. As a matter of fact, even though a massive presence of lurkers is typical in a large-scale social environment, too many lurkers would impair the virality of the online community, which instead needs to be sustained over time with fresh ideas and initiatives. Social science and human-computer interaction research studies have addressed the delurking problem mainly focusing on the conceptualization of the strategies to adopt, such as [27]: reward-based external stimuli (e.g., badges [3]), providing encouragement information, improvement of the usability of the online platform, and guidance from elders/master users to help lurkers become familiar with the system as quickly as possible. However, given the variety of influencing factors that drive online participation, developing a computational approach to turn lurkers into active members of an OSN is an emerging yet challenging problem, regardless of the delurking strategy adopted.

Contributions. Our intuition in this work is that the behavioral dynamics underlying the transition from a lurking to non-lurking (i.e., active) user role, and vice versa, can suitably be modeled via an *evolutionary game theory* approach [15, 21–23, 33]. We define the *Lurker Game*, in which active users are regarded as *cooperators* and lurkers as *defectors*. Cooperators contribute to the system by adding information represented by "virtual coins" to a common pool, while defectors do not contribute. The total amount of virtual coins in the common pool increases according to two key aspects: (i) the collective effort of cooperators and (ii) the different impact that information naturally has on each agent, depending on her/his preferences. Our *Lurker*

Game employs a Fermi-like function [29] to model the transition probability from one agent strategy to another. Having considered the importance of rewarding mechanisms [4, 29, 30] towards an ordered phase of cooperation (i.e., delurking), we also introduce a prize structure for promoting cooperation. We evaluate the *Lurker Game* on random graph models that resemble the complexity of real-world OSNs, focusing on the effect that the network topology may have on the final equilibrium of the game. This work represents, to the best of our knowledge, the first attempt for quantitatively understanding lurking and delurking dynamics in OSNs via the evolutionary game theory.

The remainder of the paper is organized as follows. Section 2 introduces the *Lurker Game* on complex networks. Section 3 shows results of numerical simulations, and Sect. 4 provides a discussion on main experimental findings. Related works are discussed in Sect. 5, finally Sect. 6 concludes the paper.

2 The *Lurker Game*

User-generated communications and social content produced in an OSN represent a rich source of knowledge whose value can, in principle, be increased by collective efforts. Within this view, evolutionary games provide a powerful tool to model the dynamics of OSNs [1, 10, 26].

Our aim in this work is the definition of a novel game, named *Lurker Game*, to analyze the dynamics of OSN populations, by focusing on the two main roles played by network members: active contributors and lurkers. The former are regarded as contributors, whereas the latter as defectors. Information generated by contributors is expressed in terms of *virtual coin* (vc), which is assumed to be unitary by default. Note that we adopt the term information with its more general meaning, which includes any type of social content produced in an OSN (i.e., posts, comments, preferences, etc.).

Our *Lurker Game* entails important aspects in cooperator-defector games. The collective effort is represented by a synergy factor r ($r > 0$), which is usually adopted in *public goods games* (PGGs) [29, 30], and used to grant groups of cooperators. However, *Lurker Game* has two main differences from classic PGG. First, due to its nature, the "public good" in our game (i.e., information generated by contributors) is not divided but rather *equally shared* among all users of a group. Second, we observe that information may acquire a different value for each individual (e.g., one may contribute by writing posts on politics but it is not interested in reading about music); to model this heterogeneity of user interests and preferences, we introduce a further parameter, denoted by ν, ranging in $(0, 1]$, such that the common pool of virtual coins, shared in the OSN environment, is diversified by means of ν.

2.1 Basic Dynamics

Given a set of N agents, the dynamics of *Lurker Game* unfolds in discrete time steps and is defined as follows. At each time step, agents have to put into the common pool a virtual coin if they take the role of cooperators, otherwise (i.e., they are lurkers) do nothing. The accumulated amount of virtual coins is increased by r and ν, and then equally shared among all agents. The *payoff equations* in *Lurker Game* are defined as follows:

$$\begin{cases} \pi^c = r\nu\sum_1^{N^c} \upsilon c - \upsilon c \\ \pi^d = r\nu\sum_1^{N^c} \upsilon c \end{cases} \tag{1}$$

with N^c number of cooperators, r synergy factor, and ν representing the heterogeneity of interests of users. Due to its evolutionary nature, *Lurker Game* allows agents to change their strategy [33], i.e., from cooperation to defection and vice versa. In particular, when considering two agents at a time, we adopt a Fermi-like function to implement a transition probability from one strategy to another. Given two agents x and y, this probability is defined as:

$$W(s^x \rightarrow s^y) = \left(1 + \exp\left[\frac{\pi^y - \pi^x}{K}\right]\right)^{-1} \tag{2}$$

where s^x and s^y denote the strategies of the players x and y, respectively, π^x and π^y denote their respective payoff, and K indicates uncertainty in adopting a strategy. By setting $K = 0.5$, we implement a rational and meritocratic approach during the strategy revision phase [29]. Like in the PGG, behaving as defectors is much more convenient than behaving as cooperators and the Nash equilibrium of *Lurker Game* corresponds to defection.

2.2 Mean Field Analysis

We perform a *mean field* analysis [8] of *Lurker Game*, in order to investigate if the Nash equilibrium corresponds to the final ordered phase. Hence, we assume that the population is composed of only one big community and every agent interacts with all the others. Under this assumption, the evolution of a population with N agents is described by the following set of equations [14]:

$$\begin{cases} \frac{d\rho^c(t)}{dt} = p^c \cdot \rho^c(t) \cdot \rho^d(t) - p^d \cdot \rho^d(t) \cdot \rho^c(t) \\ \frac{d\rho^d(t)}{dt} = p^d \cdot \rho^d(t) \cdot \rho^c(t) - p^c \cdot \rho^c(t) \cdot \rho^d(t) \\ \rho^c(t) + \rho^d(t) = 1 \end{cases} \tag{3}$$

with $\rho^c(t)$ and $\rho^d(t)$ densities of cooperators and defectors, $p^c(t)$ probability that cooperators prevail, and $p^d(t)$ probability that defectors prevail. These probabilities are computed according to the payoffs obtained, at each time step, by cooperators and defectors as defined in Eq. 6. Therefore, we have to consider the difference between the payoffs accumulated by the two agents randomly chosen at each time step. If we denote with x a cooperator and with y a defector, the probability p^c corresponds to $W(x \rightarrow y)$, so we consider the difference $\pi^{(d)} - \pi^{(c)}$. While, p^d corresponds to $W(y \rightarrow x)$, then we consider $\pi^c - \pi^d$. Few algebraic steps lead to the following solutions:

$$\begin{cases} \pi^d - \pi^c = rvN\rho^c - rvN\rho^c + 1 = 1 \\ \pi^c - \pi^d = rvN\rho^c - 1 - rvN\rho^c = -1 \end{cases} \tag{4}$$

By substituting results of Eq. 4 in Eq. 2, one obtains $p^c \sim 0.12$ and $p^d \sim 0.88$. Remarkably, the mean field approach to *Lurker Game* leads to dynamics completely independent both from r and v. Given the values computed in Eq. 4, the solution of the system in Eq. 3 confirms the expected result, i.e., defection prevails according to the Nash equilibrium.

2.3 Rewarding Mechanisms

The above result leads us to focus on *rewarding mechanisms* to drive a population towards an ordered phase of cooperation. Therefore, we introduce a variation in the basic formulation of *Lurker Game* by introducing a prize structure for promoting cooperation. This impacts on the payoff equation of cooperators, whereby the set of payoff equations is modified as follows:

$$\begin{cases} \pi^c = rv\sum_1^{N^c} vc - vc + \Phi(\Delta t^c) \\ \pi^d = rv\sum_1^{N^c} vc \end{cases} \tag{5}$$

with $\Phi(\Delta t^c)$ rewarding function that allows cooperators to receive a further amount of virtual coins. This function takes in input Δt^c, i.e., the amount of time each agent behaves as a cooperator. The prize structure S grants cooperative agents at a fixed rate, i.e., every k time steps: $S : \Delta t^c = \{k, k, \ldots, k\}$. This way, each prize consists of an amount of vc equal to that paid by a cooperator over time (between two achieved prizes). We define the *prize function* as follows:

$$\Phi(\Delta t^c) = \begin{cases} \Delta t^c \cdot vc & \text{if } \Delta t^c \in S \\ 0 & \text{if } \Delta t^c \notin S \end{cases} \tag{6}$$

Analogously to the basic dynamics of *Lurker Game*, after every iteration agents undergo a strategy revision phase based on Eq. 2. Algorithm 1 sketches the main steps performed in *Lurker Game*.

Algorithm 1 *Lurker Game*

Require: A population of N agents, where N^c are cooperators and N^d are defectors ($N = N^c + N^d$).
The synergy factor $r > 0$.
The user preference coefficient $\nu \in (0, 1]$.
A network topology G that models the connectivity of the N agents, otherwise agents are fully connected to each other (mean field).
1: **repeat**
2: Compute the payoff of cooperators and defectors, according to Eq. 5
3: Randomly select two agents x and y (with different strategies) s.t. x, y are linked w.r.t. G
4: Agent y takes the strategy of agent x according to Eq. 2
5: **until** all agents have the same behavior (Nash equilibrium)

2.4 Lurker Game on Networks

Since social networks constitute the natural environment to observe the phenomenon of lurking, we also study the *Lurker Game* on complex networks. Following the lead of previous studies on evolutionary games (e.g., [10, 15, 18, 21, 28, 34]), we focus our attention on two relevant models: Barabasi-Albert model [6] (hereinafter BA) and Watts-Strogatz [35] model (hereinafter WS). Since the topological properties of networks generated by both considered models (i.e., BA and WS) are well-known (see, e.g., [7]), all outcomes of the proposed model can be analyzed seeking relations with the considered topology.

Note that when agents are arranged on networks, the dynamics of the game are different from those adopted in the mean field case, which in topological terms, can be viewed as a fully-connected network. Adopting complex networks, only few agents are considered at each iteration. In particular, at each time step two randomly chosen agents play *Lurker Game* with all groups of belonging. Therefore, the accumulated payoffs are computed for each group and the final prize is assigned only to cooperative agents that played the game. Next, as previously discussed, the xth agent tries to enforce its strategy to the yth agent with probability defined in Eq. 2.

Memoryless and memory-aware payoff. We introduce a further aspect of the proposed model, related to the way agents manage their accumulated payoffs. We distinguish between two scenarios of payoff accumulation, namely *memoryless* and *memory-aware*.

The memoryless case entails that every time two agents are selected to play *Lurker Game* with their groups, they reset their accumulated payoff. Therefore, when computing the transition probability of Eq. 2, they consider only the payoff accumulated during the present time step. Instead, the memory-aware case entails agents save their payoff over time. Note that while the memory-aware case is closer to a real scenario (e.g., online users may accumulate several badges over time), the memoryless case avoids noise effects in numerical simulations that can emerge in Eq. 2 for large payoffs.

We investigate both cases, by introducing a *cutoff* in the difference between the payoffs of the two considered agents (i.e., x and y). In doing so, for large payoffs, the Fermi function behaves like a simple rule with only two possible results: 1 and 0, i.e., 1 if the payoff of the xth agent is greater than that of the yth, and 0 otherwise. Thus, the interesting granularity, in terms of transition probabilities, introduced by the Fermi function is lost in the memory-aware case, after few time steps.

It is also relevant to observe that a similar problem may arise when dealing with scale-free networks since, even in the memoryless case, nodes with high degree (i.e., hubs) can accumulate at each iteration a very high payoff. As a result, we expect that simulations performed on scale-free networks in the memoryless case yield outcomes similar to those achieved by the memory-aware case, at least by considering the same topology (i.e., scale-free in both cases).

Identifying critical parameters. Numerical simulations will be primarily devoted to the identification of critical values of k and ν, i.e., the step adopted in the prize structure S and the variety of information (or users' interests) in the social network, respectively. These values together with the final equilibrium achieved in both networks, provide a useful indicator for studying the dynamics of *Lurker Game* and for comparing different network topologies. Remarkably, we are dealing with a disordered system [11, 13, 14], in terms of states (i.e., cooperators and defectors), having only two possible equilibria: one characterized by the prevalence of one species (i.e., cooperators or defectors) and one characterized by a coexistence of both species at equilibrium. The former corresponds to a ferromagnetic phase, whereas the latter to a paramagnetic phase [13]. Thus, both the Nash equilibrium and its opposite case correspond to the ferromagnetic phase. The paramagnetic phase has been observed in games like the PGG, obtained by tuning the synergy factor and without adopting rewarding mechanisms [29].

3 Results

Experimental setting. We evaluated *Lurker Game* by arranging agents on different networks, generated according to the BA and WS models. The former generates scale-free networks, i.e., networks characterized by the presence of nodes with a very high degree, defined hubs. The WS model generates different kinds of networks by tuning a rewiring parameter, β, which ranges within [0, 1]. In particular, $\beta = 0$ yields a regular ring lattice topology, intermediate values of β yield small-world-networks (characterized by relatively low average path lengths and high clustering coefficients), while completely random networks are obtained for high values of β. In this work, we considered the following values: $\beta = \{0.0, 0.3, 0.5, 0.8\}$. Figure 1

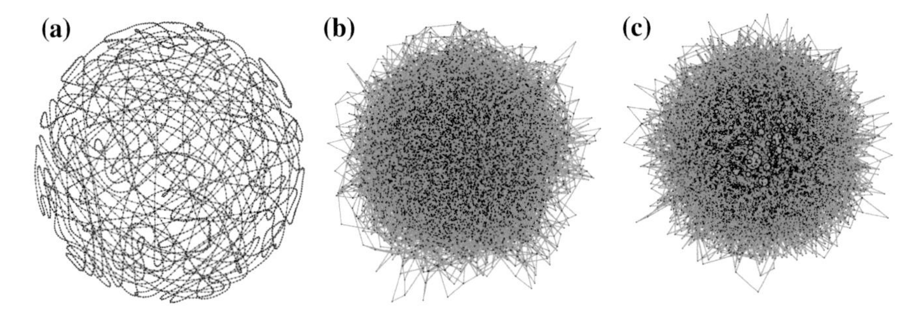

Fig. 1 Evaluation networks: **a** WS with $\beta = 0.0$, **b** WS with $\beta = 0.5$, **c** BA

Table 1 Structural properties of evaluation networks

Network model	Avg. path length	Diameter	Clust. coeff.
WS $\beta = 0.0$	625.38	1250	0.500
WS $\beta = 0.3$	7.89	14	0.165
WS $\beta = 0.5$	6.99	12	0.054
WS $\beta = 0.8$	6.67	11	0.005
BA	4.85	9	0.002

shows a pictorial representation of each kind of networks, whereas Table 1 reports some of their structural properties (achieved with 5000 nodes).

Numerical simulations were performed with $N = 5000$ agents, with an equal initial density of cooperators and defectors (i.e., $\rho^c(0) = \rho^d(0) = 0.5$), and an average degree $\langle k \rangle = 4$. We set the synergy factor r to 2, as we found that this value does not allow cooperators to survive without rewarding mechanisms (see also [29] for a discussion about the critical thresholds of the synergy factor). Parameter ν was instead varied considering values from 0 to 1. It is worth noting that for $\nu = 0.2$ the game, in the memoryless case and without the adoption of rewarding mechanisms, corresponds to the PGG in networks with the same topology. Simulations were carried out for a maximum number of time steps equal to 10^8, then results were averaged over several different runs.

Evolution of the system. We initially analyzed the density of cooperators over time in all networks. We found three main behaviors: cooperators vanish (Fig. 2b) or prevail (Fig. 2a) after a number of time steps, or both cooperators and defectors coexist over time. This finding clearly indicates that a population playing the *Lurker Game* can reach both ordered phases and disordered phases at equilibrium. In particular, since agent strategies can be mapped to spins $\sigma \pm 1$ respectively and, as observed, there are only two possible equilibria, the evolution of the system can be analyzed in terms of ferromagnetic phase transitions [8, 20]. Thus, mapping our model to a spin system allows us to identify the conditions that can lead towards the different kinds of equilibrium. The relevance of identifying a description based on the language of phase transitions, lays in the fact that it opens the way to further

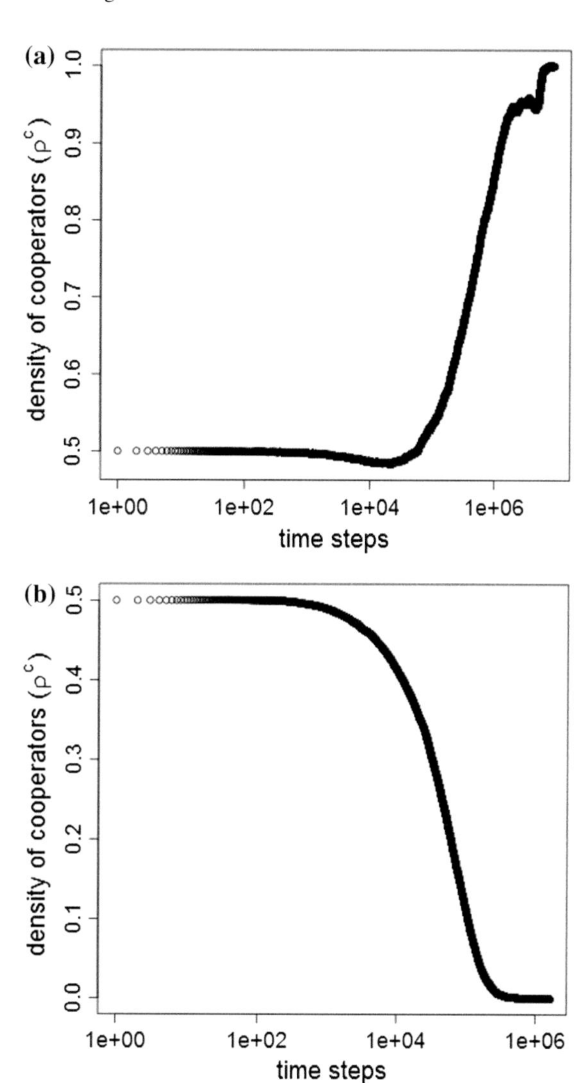

Fig. 2 Possible behaviors of the *Lurker Game* system. Time evolution of density of cooperators: **a** cooperators prevail, **b** defectors prevail. Results correspond to WS model ($\beta = 0.5$) with 5000 agents and $k = 2$, for ν equal to **a** 0.5 and **b** 0.3

analytical investigations [14] that can potentially lead to get new insights on the proposed model.

Critical values of ν. We finally analyzed the role of ν. Results are reported in Table 2 for the memoryless case. For the memory-aware case, results indicate a more complex scenario, which is discussed next.

Table 2 Memoryless agents

Network model	Critical ν	k-range
WS $\beta = 0.0$	0.6	[1...5]
WS $\beta = 0.3$	0.44	[1...5]
WS $\beta = 0.5$	0.42	[1...5]
WS $\beta = 0.8$	0.41	[1...5]
BA	0.22	[1...5]

4 Discussion

Results of our investigations suggest that our *Lurker Game* has a rich behavior, which can be described by considering the main degrees of freedom of the system: ν, k, network topology and the evolution of payoffs over time.

Results on WS Networks. In the memoryless case, for each considered β, we found a well recognized critical ν. In particular, by increasing β, cooperators require a smaller ν to prevail. This suggests that, in general, random topologies support cooperation better than regular ones. It is worth noting that in all cases critical ν showed a certain robustness towards the considered k values, i.e., $k \in [1 \ldots 5]$. In this regard, further investigations will be devoted to better clarify the relation between ν and k, since we hypothesize that for high k values defectors may prevail even for ν values greater than the identified thresholds (see Table 2). On the other hand, results achieved by memory-aware agents indicate that, in general, critical ν are smaller than those found in the memory-less case, e.g., for $\beta = 0.0$ we obtained $\nu \sim 0.4$. However, we found that even for values greater than the minimal threshold of ν, sometimes defectors may prevail. Before trying to mind a hypothesis about this behavior, we have to recall that in the memory-aware case some noise may arise resulting from high payoffs. Moreover, the memory-aware case may easily promote cooperation than its counterpart as groups of cooperative agents tend to increase their payoff unboundedly. Therefore, as a future work, we aim to investigate this aspect of the model. Also note that a mixed phase (i.e., composed of both species) has been found for values close to the critical ν.

Results on scale-free networks. When considering the BA model, a major finding is that cooperators need a smaller ν to prevail than those computed in WS network; specifically, $\nu = 0.22$ and $\nu \sim 0.1$, in the memory-less and in the memory-aware case, respectively. Moreover, scale-free networks in the memory-aware case show an interesting bistable behavior for small values of ν. We suggest again that this may result from noise introduced by the utilization of large payoff in the Fermi function that we faced by adding a numerical cutoff. It is relevant to note that our results are in accord with those reported in [25], as stated above, since scale-free networks have been found to foster cooperation better than other topologies. Also, like for WS networks, critical ν are robust to variations of k in the considered range.

Overall, the proposed *Lurker Game* suggests that the adoption of rewarding mechanisms combined with the modeling of hypothetical heterogeneity of users' interests (ν) may lead a population towards a cooperative behavior. This supports our initial intuition that *Lurker Game* is suitable to model the dynamics of such a complex phenomenon as lurking.

5 Related Work

In [31, 32], the authors developed the first computational approach to lurker mining, focusing on ranking problems. To this purpose, they proposed a topology-driven definition of lurking behavior, based on principles of overconsumption, authoritativeness of the information received, and non-authoritativeness of the information produced. Quantitative and qualitative evaluation results showed how the proposed methods are effective in identifying and ranking lurkers in real-world OSNs.

The same authors also posed a first step toward the definition of delurking strategies in [12], by proposing a targeted influence maximization problem under the linear-threshold diffusion model. In this context, a set of previously identified lurkers is taken as target set of an influence maximization problem, whose objective function is defined upon the concept of *delurking capital*, i.e., the social capital gained by activating lurkers in an online community.

We can also mention research studies that, though not specifically concerning lurking, addressed related problems in OSNs via a game-theoretic approach. For instance, Anand et al. [2] defined a Stackelberg game to maximize the benefit each user gains extending help to other users, hence to determine the advantages of being altruistic. Some interesting remarks relate the altruism of users to their level of capabilities, and indicate that the benefit derived from being altruistic is larger than that reaped by selfish users or free riders. Malliaros and Vazirgiannis [19] also built upon game theory to study the property of users' departure dynamics, i.e., the tendency of individuals to leave the community.

Our proposed approach in this work differs from all the aforementioned studies as it represents both a novel computational approach to lurking and delurking user-behaviors, and a novel application domain in the field of evolutionary games.

6 Conclusion

In this work, we brought for the first time evolutionary game theory into the analysis of lurking behaviors in OSNs. We defined the *Lurker Game* and evaluated it through both a mean-field analysis and by arranging agents on small-world and scale-free networks. Results suggest that *Lurker Game* is suitable to model the dynamics of such a complex phenomenon as lurking, showing a rich behavior depending on the network topology and on the way agents manage their payoff. Remarkably, *Lurker Game* allows us to understand how the adoption of rewarding mechanisms combined with the modeling of hypothetical heterogeneity of users' interests may lead a population towards a cooperative behavior. Further investigations will be mainly devoted to better clarify the interrelation between the two model parameters in *Lurker Game*, also including analysis over other network topologies and larger populations.

References

1. Abramson, G., Kuperman, M.: Social games in social networks. Phys. Rev. E **63** (2001)
2. Anand, S., Chandramouli, R., Subbalakshmi, K.P., Venkataraman, M.: Altruism in social networks: good guys do finish first. Social Netw. Anal. Mining **3**(2), 167–177 (2013)
3. Anderson, A., Huttenlocher, D., Kleinberg, J., Leskovec, J.: Steering user behavior with badges. In: Proceedings of ACM Conference on World Wide Web (WWW) (2013)
4. Anderson, A., Huttenlocher, D., Kleinberg, J., Leskovec, J.: Engaging massive online courses. In: Proceedings of ACM Conference on World Wide Web (WWW) (2014)
5. Ardichvili, A.: Learning and knowledge sharing in virtual communities of practice: motivators, barriers, and enablers. Adv. Dev. Human Resour. **10**, 541–554 (2008)
6. Barabasi, A.L., Albert, R.: Emergence of scaling in random networks. Science **286**, 509–512 (1999)
7. Barabasi, A.L., Albert, R.: Statistical mechanics of complex networks. Rev. Modern Phys. **74**, 47–97 (2002)
8. Barra, A.: The mean field ising model trough interpolating techniques. J. Stat. Phys. **132**(5), 787–809 (2008)
9. Edelmann, N.: Reviewing the definitions of "lurkers" and some implications for online research. Cyberpsychol. Behav. Soc. Netw. **16**(9), 645–649 (2013)
10. Fu, F., Rosenbloom, D.I., Wang, L., Nowak, M.A.: Imitation dynamics of vaccination behavior on social networks. R. Soc.—Proc. B **278** (2011)
11. Galam, S., Walliser, B.: Ising model versus normal form game. Phys. A **389**, 481–489 (2010)
12. Interdonato, R., Pulice, C., Tagarelli, A.: "Got to have faith!": The DEvOTION algorithm for delurking in social networks. In: Proceedings of International Conference on Advances in Social Networks Analysis and Mining (ASONAM), pp. 314–319 (2015)
13. Javarone, M.A.: Is poker a skill game? new insights from statistical physics. EPL **110** (2015)
14. Javarone, M.A.: Statistical physics of the spatial Prisoner's dilemma with memory-aware agents. arXiv:1509.04558 (2015)
15. Javarone, M.A., Atzeni, A.E.: The role of competitiveness in the Prisoner's dilemma. Comput. Soc. Netw. **2** (2015)
16. Kahnwald, N., Khler, T.: Microlearning in virtual communities of practice? An explorative analysis of changing information behaviour. In: Proceedings of Microlearning Conference, pp. 157–172 (2006)
17. Lave, J., Wenger, E.: Situated Learning: Legitimate Peripheral Participation. Cambridge University Press (1991)
18. Lieberman, E., Hauert, C., Nowak, M.A.: Evolutionary dynamics on graphs. Nature **433**, 312–316 (2004)
19. Malliaros, F.D., Vazirgiannis, M.: To stay or not to stay: modeling engagement dynamics in social graphs. In: Proceedings of ACM Conference on Information and Knowledge Management (CIKM), pp. 469–478 (2013)
20. Mobilia, M., Redner, S.: Majority versus minority dynamics: phase transition in an interacting two-state spin system. Phys. Rev. E **68**(4), 046106 (2003)
21. Perc, M., Gomez-Gardenes, J., Szolnoki, A., Floria, L.M., Moreno, Y.: Evolutionary dynamics of group interactions on structured populations: a review. J. R. Soc. Interface **10**(80) (2013)

22. Perc, M., Szolnoki, A.: Social diversity and promotion of cooperation in the spatial Prisoner's dilemma. Phys. Rev. E **77** (2008)
23. Poncela-Casasnovas, J., Gomez-Gardenes, J., Traulsen, A., Moreno, Y.: Evolutionary game dynamics in a growing structured population. New J. Phys. **11**, 083031 (2009)
24. Preece, J.J., Nonnecke, B., Andrews, D.: The top five reasons for lurking: improving community experiences for everyone. Comput. Human Behav. **20**(2), 201–223 (2004)
25. Santos, F.C., Santos, M.D., Pacheco, J.M.: Social diversity promotes the emergence of cooperation in public goods games. Nature **454**, 231–216 (2008)
26. Van Segbroeck, S., Santos, F.C., Lenaerts, T., Pacheco, J.M.: Reacting differently to adverse ties promotes cooperation in social networks. Phys. Rev. Lett. **102**, 058105 (2009)
27. Sun, N., Rau, P.P.-L., Ma, L.: Understanding lurkers in online communities: a literature review. Comput. Human Behav. **38**, 110–117 (2014)
28. Szabo, G., Fath, G.: Evolutionary games on graphs. Phys. Rep. **446** (2007)
29. Szolnoki, A., Perc, M.: Reward and cooperation in the spatial public goods game. EPL **92** (2010)
30. Szolnoki, A., Szabo, G., Perc, M.: Phase diagrams for the spatial public goods game with pool punishment. Phys. Rev. E **83** (2011)
31. Tagarelli, A., Interdonato, R.: "Who's out there?": identifying and ranking Lurkers in social networks. In: Proceedings of International Conference on Advances in Social Networks Analysis and Mining (ASONAM), pp. 215–222 (2013)
32. Tagarelli, A., Interdonato, R.: Lurking in social networks: topology-based analysis and ranking methods. Soc. Netw. Analys. Mining **4**(230), 27 (2014)
33. Tomassini, M.: Introduction to Evolutionary Game Theory (2014)
34. Wang, Z., Szolnoki, A., Perc, M.: Interdependent network reciprocity in evolutionary games. Sci. Rep. **3**, 1183 (2013)
35. Watts, D.J., Strogatz, S.H.: Collective dynamics of small-world networks. Nature **393**, 440–442 (1998)

Part VII
Applications of Networks

The Network of Genetic Admixture in Humans

Hend Alrasheed and Feodor F. Dragan

Abstract Recent advances in the field of genetic data analysis reveal promising findings in the field of human history; especially when combined with proper data analysis tools. Within the field of modern genetics, there is evidence that the human populations have genetically interacted as a result of several events. The genetic admixture contains multiple pieces of DNA that have been passed down subsequently through generations making it combine DNA from different source groups. In this paper, we construct and analyze the network of human genetic admixture. We study the topology of this network, we investigate its δ-hyperbolicity (negative curvature), and, using it, identify the core vertices by proposing the δ-hyperbolicity-neighborhood measure that we assign to each vertex.

1 Introduction

Using networks to describe systems that are composed of elements and the interactions or connections between those elements aids analyzing and understanding them. Therefore, networks in multiple disciplines ranging from computer science to systems biology are being modeled as graphs were vertices represent the different elements and edges represent the different interactions among those elements. Within the field of modern genetics, there is evidence that the human populations have interacted throughout history. This interaction, which may occur as a result of migrations, invasions, and slavery, results in transfer of genetics and accordingly creates admixed populations. The genetic admixture contains multiple pieces of DNA that have been passed down subsequently through generations making it combine DNA from different source groups.

The work in [10] uses DNA from many people around the world (95 populations) to identify the mixed source groups and to decide when did those mixing events

H. Alrasheed (✉) · F.F. Dragan
Department of Computer Science, Kent State University, Kent, OH 44242, USA
e-mail: halrashe@kent.edu

F.F. Dragan
e-mail: dragan@cs.kent.edu

H. Cherifi et al. (eds.), *Complex Networks VII*, Studies in Computational Intelligence 644, DOI 10.1007/978-3-319-30569-1_18

had occurred. Their results are presented on an interactive map in [1]. Their work concludes that many populations are results of genetically mixed groups that mixed throughout the last 4000 years. Furthermore, some of those mixed source populations are geographically very spread. Finally, even though genetic mixing among source groups is often local with respect to time and space, neighboring populations do not necessarily share the same ancestry or history.

Even though it is interesting to analyze the details of the direct genetic admix among populations, it is equally interesting to see how this genetic admix looks like in the organization level by the use of graph-theoretical tools. This global approach of analyzing the genetic admix as a system not only as individual components may increase our understanding of the human history in multiple aspects; for example, the transmission of languages and cultures. In this paper, we construct and analyze the network of human genetic admixture. We investigate the topology of this network by studying the degree distribution, the clustering coefficient, and the different measures of centrality. We also investigate the δ-hyperbolicity of this network and, using it, identify the core vertices. For this we propose the δ-hyperbolicity-neighborhood of each vertex. Then we use this measure to identify the core vertices. Based on our analysis, we find the average distance between a pair of populations across the network are relatively small suggesting a small-world network. We also find that the network comprises a number of sub-networks when edges are pruned based on their weights. Those sub-networks are formed by multiple neighboring populations. Also, we identify key vertices according to a number of centrality measures, and we find that those measures correlate very well. Moreover, we find the core vertices identified based on the δ-hyperbolicity-neighborhood measure correlate to some extent with some of the typical centrality measures such as the betweenness centrality.

2 Data and Network Construction

The data was obtained from the *Genetic Atlas of Human Admixture History* interactive website [1], which is a companion of the work presented in [10]. In this work, the authors study 95 populations (a population or a group is a set of individuals with similar genetic makeup). For each individual population p, they show the set of other source populations that are genetically admixed in the DNA of population p. For example, Fig. 1, which is a screen shot from [1], shows that the Polish population has the following admixing groups: Lithuanian (53.1 %), Norwegian (16 %), Russian (12.9 %), Moroccan (3.7 %), Sardinian (3.7 %), Basque (2.6 %), etc. The percent associated with each source group indicates its contribution to that population such that all admixed populations collectively make 100 %. Overall, we found 2685 distinct edges in this network.

Here we construct and study the network of genetic admixture in human populations. In this network vertices represent the different populations and an edge connects two populations if one participates in the genetic makeup of the other. Each edge has an associated direction and a weight. For a source population u and a

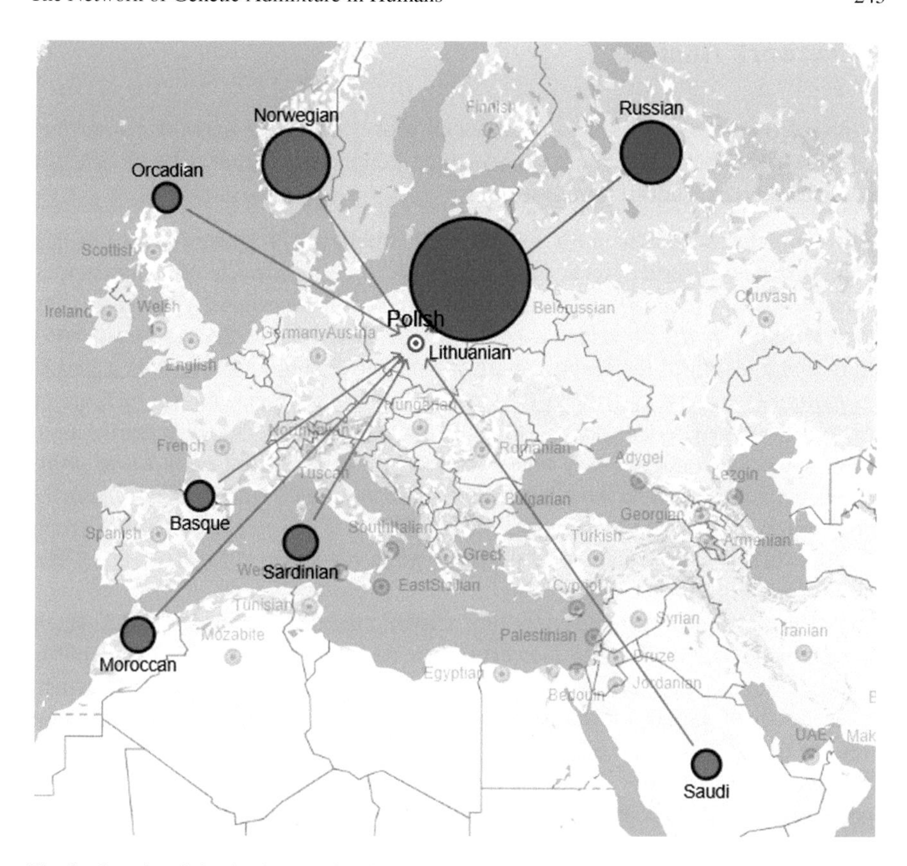

Fig. 1 Genetic mixing in the genetic of the Polish population

population v, the edge e_{uv} is directed towards v. The weight assigned to each edge e_{uv} denoted as w_{uv} is based on the percent of contribution in which it participates in building the DNA of this admixed group. Hence, the larger the weight is, the more significant the contribution. We normalize the weights as follows. The weight of an edge e_{uv} becomes $w_{uv} = (100 - \lambda_{uv})/10$, where λ_{uv} is the percent of the contribution as reported in [1]. This way the smaller the weight is, the larger the contribution, and as a result, the shorter the distance between the two populations. For example, if a source group u represents 50 % of group v's DNA, then the edge leaving u towards v has a weight of 5.

A graph can be expressed by its adjacency matrix a_{uv} where the value a_{uv} is one if vertices u and v are connected and zero otherwise and w_{uv} represents the weight of that edge if one is present. We use this representation throughout this work. For several reasons that will become apparent later on in this text, we will be analyzing the weighted and the unweighted versions of this network. In the weighted network, different edges will have associated weights as described above. In the unweighted network, edges are either present or absent (we ignore w_{uv} in this case). Table 1 gives some overall statistics of the constructed networks.

3 Network Analysis

In this section, we study some fundamental global and local network parameters of the two generated networks: the weighted genetic admixture network and the unweighted genetic admixture network.

Diameter, characteristic path length, and small-world property. According to the distances between vertices in the graph, the *eccentricity* of a vertex u is $ecc(u) = \max_{v \in V}\{d(u, v)\}$. The minimum value of the eccentricity represents the graph's *radius*: $rad(G) = min_{u \in V}\{ecc(u)\}$. The *diameter* of the graph $diam(G)$ refers to the length of the longest shortest path between any two vertices u and v, i.e., $diam(G) = \max_{u,v \in V}\{d(u, v)\}$. Another important distance related measure of graphs is the *characteristic path length* (CPL) which is the average distance between vertex pairs. See Table 1. Many real-world networks exhibit the *small-world property*. A network is said to have this property when it has a small CPL or diameter compared to the size of the network. Let $size(G) = |V| + |E|$ be the size of graph G, a network has the small-world property when $diam(G) \leq \log_2(size(G))$. For our network, $\log_2(size(G)) = 11.44$.

For the unweighted genetic admixture network, the diameter is 4, which is small compared to the network's size. However, since the diameter in graphs is susceptible to outlier vertex pairs [11], we are also interested in the *effective diameter* which represents the maximum distance between a fraction of vertex pairs (in our case 90 %) of the network. The effective diameter for this network is 2, and the CPL is 1.8. Clearly, this network exhibits the small-world property. This indicates that if one population p_1 does not contribute (directly) in the genetic make up of another population p_2, then there is a small chain of population exchanges between the two. For the weighted network, the diameter is ≈36, the effective diameter is ≈19, and the CPL is ≈16. In both networks the diameter is finite which means that all vertices are reachable from one another. In other words, the network of human genetic admixture has one connected component. Also, we find that the network is biconnected.

Weights and network components. One would expect neighboring populations to be genetically admixed; however, it was concluded in [10] that some mixture

Table 1 Basic network parameters

Measure	Directed weighted	Directed unweighted
$diam(G)$	36.38	4
CPL	15.83	1.8
$rad(G)$	19.68	2
$\bar{k}^+(G)$	143.8	14.6
$\bar{k}^-(G)$	273	28.3
$\bar{k}(G)$	416.8	42.9

$diam(G)$: network's diameter; CPL: characteristic path length; $rad(G)$: graph's radius; $\bar{k}^+(G)$, $\bar{k}^-(G)$, $\bar{k}(G)$: average in-degree, average out-degree, average total degree respectively

events include populations that belong to very distant locations. This is evident considering our genetic admixture network that is represented by one connected component. However, this also motivates investigating the sub-networks that this single component comprises. Specifically, we focus on the number of sub-networks, the size of each sub-network, and how the populations in each sub-network are connected. To obtain the set of sub-networks, we use the edge weights as an indication of their importance. Given a threshold number t, where $0 \leq t \leq 100$, first, we fix the threshold weight t and construct a graph $G^t = (V, E^t)$ by pruning those edges with weights less than t. Second, we identify the set of strongly connected components for the directed network G^t; each strongly connected component represents one sub-network.

We start this process with $t = 100\,\%$ (the highest possible weight). At this point, every single vertex in the graph G^{100} represents a strongly connected component on its own. Then we gradually reduce threshold t (obtaining a different set of sub-networks) until we get to a point in which all vertices are in the same component (when $t = 0$). The number of strongly connected components as well as some other properties about each component are listed in Table 2. An interesting observation about the formation of populations into distinct sub-networks is that it is highly affected by the geographic locations of those populations. For an example, see Table 3 in which we provide a list of all sub-networks with size ≥2 along with the geographic location to which the listed populations belong. The geographic regions are as presented in [10]. Note that all populations in a sub-network either belong to the same geographic region or to a region that is close geographically. This indicates that, for some populations, the genetic admixing with neighboring populations is more significant. Another interesting observation is that the small-world property is evident in the sub-networks. For example, G^3, G^2, and G^1 in Table 2.

Table 2 Sub-networks in each G^t that result from pruning edges with weights $<t$

t (%)	$\lvert G^t \rvert$	$\lvert E^t \rvert$	Min # of vertices in a sub-network	Max # of vertices in a sub-network	Diam of largest sub-network
90	95	2	1	1	0
70	95	7	1	1	0
50	94	18	1	2	1
30	88	63	1	3	1
10	36	222	1	33	8
5	18	418	1	61	5
3	12	601	1	78	6
1	2	1099	5	90	4
0	1	2685	95	95	4

t: edge threshold; $\lvert G^t \rvert$, $\lvert E^t \rvert$: number of sub-networks and edges in G^t; Diam of largest sub-network is the longest (unweighted) path that exists between any two vertices

Table 3 A list of populations in some of the sub-networks in G^t where $t = 20$ and the geographic region(s) of the listed populations

No.	Populations	Geographic region
1	GermanyAustria, Finnish, Norwegian, English, Ireland Scottish, Spanish, French	N.W.Europe, E.Europe
2	Cambodian, Dai, Han, HanNchina, Tujia, Miao	S.EastAsia
3	Balochi, Brahui, Sindhi, Pathan	C.SouthAsia
4	BantuSouthAfrica, SanKhomani, SanNamibia	Bantu, San
5	Belorussian, Polish, Lithuanian	E.Europe
6	Ethiopian, EthiopianJew	Ethiopian
7	BantuKenya, Yoruba	Bantu, W.Africa
8	Adygei, Georgian	W.Asia
9	Bedouin, Saudi	S.MiddleEast
10	Daur, Oroqen	N.EastAsia
11	Yi, Naxi	S.EastAsia

Clustering coefficient. The *clustering coefficient* for a vertex v, denoted as $cc(v)$, indicates the likeliness that any two neighbors of v are also neighbors. Given an unweighted graph $G = (V, E)$ and a vertex $v \in V$, let $N(v)$ be the neighborhood of v consisting of all vertices adjacent to v. Also, let $e_{N(v)}$ be the set of edges between every pair of vertices in v's neighborhood. Then $cc(v) = \frac{2|e_{N(v)}|}{|N(v)||N(v)-1|}$. $0 \leq cc(v) \leq 1$. The clustering coefficient $CC(G) \in [0, 1]$ of a graph G is the average of $cc(v)$ taken over all vertices $v \in V$. $CC(G) = 0$ when there is no clustering and $CC(G) = 1$ when the clustering is very high which happens when the network includes a number of sub-networks each of which is highly dense and connected with other sub-networks with very few links.

For our network, the clustering coefficient measures the tendency of two populations that both already genetically admixed with a third population to themselves admix (we ignore the directions here). The average clustering coefficient of the network is about 0.57. We are also interested in exploring the following: if a population p_1 contributes to the genetic admix of another population p_2, what is the probability that population p_2 also contributes to p_1's admix? This question can be answered using the graph's *reciprocity* which is another important property of directed networks [13]. The reciprocity of a given graph, denoted as $R(G)$, is the fraction of edges that point to both directions (vertices) and it is calculated as $R(G) = \frac{|e^*_{uv}|}{|E|}$, where $|e^*_{uv}|$ is the number of bidirectional edges and $0 \leq R(G) \leq 1$. The reciporcity of our genetic admixture network is 0.24 which means that if population p_1 contributes to the DNA of population p_2, then there is a probability of 24 % that p_2 also contributes to the DNA of population p_1. This could be explained by the one direction immigration. Close analysis of those pairs of populations, that admix in only one direction, shows that the admix involves non-neighboring populations.

Degree distribution and the degree centrality. The *degree* of a vertex u (denoted as k_u) in an undirected graph G is the number of edges that have u as one of their endpoints, i.e., $k_u = \sum_v a_{uv}$. If G is directed, then a vertex u has an *in-degree* denoted as k_u^+ that represents the number of edges in E that have u as a source vertex, and an *out-degree* k_u^- that represents the number of edges that have u as a target vertex. The in-degrees and the out-degrees of the vertices in our directed unweighted genetic admixture network fluctuates between 1 and 63, with Papuan and Druze having the highest in-degree and out-degree respectively. In case of weighted networks, the weights of the edges are important to give a more precise characterization of its complexity. Therefore, rather than considering the number of incident edges, we consider their weights. Hence, the degree k_u is defined as $\sum_v a_{uv} w_{uv}$. The in-degree and the out-degree are defined accordingly. The population with the highest weighted in-degree is Papuan and the population with the highest out-degree is Druze. The *average degree* of the graph G, $\bar{k}(G)$, is defined as $\bar{k}(G) = \frac{1}{|V|} \sum_{u \in V} k_u$. See Table 1.

The degree centrality considers the central vertices as the set of vertices with the highest number of connections. The degree centrality is a local measure since it only relies on the number of neighbors [6]. Therefore, we compute the degree distribution $p(k)$ and the cumulative degree distribution $P(k) = \sum_{\ell \geq k} p(\ell)$ which indicates the fraction of vertices with degree k or larger. The cumulative degree distribution often provides some global characteristic of the network. In Fig. 2, we plot the cumulative in-degree and out-degree distributions for our directed unweighted genetic admixture network in a semilogarithmic scale. One could think that vertices with very high in-degrees act like populations that belong to popular geographical locations that may had attracted immigrants or had represented commercial attractions. However, it also may be the case that the high in-degree is just a result of being geographically close to multiple other populations and the genetic admix is just a consequence of the location.

Distances and centrality. The distance $d(u, v)$ between two vertices u and v in a graph G is the number of edges in a shortest (u, v)-path that connects them. When G is a weighted graph, the distance $d(u, v)$ is the sum of the weights of all edges in a shortest (u, v)-path from u to v (direction is important). The centrality measures presented in this section are all based on the set of shortest paths in a graph. A centrality measure rank the vertices according to their importance. Then it identifies

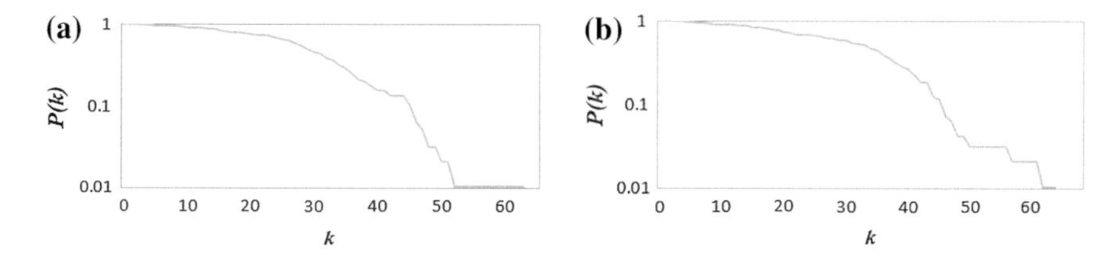

Fig. 2 The cumulative degree distribution $P(k)$ with the in-degree k (**a**) and the out-degree k (**b**). The *horizontal axis* for each chart is the in-degree or the out-degree and the *vertical axis* is the cumulative probability distribution of that degree

the set of vertices that are most significant and accordingly more central. There are multiple centrality measures each of which identifies the key vertices based on a distinct purpose. In this section, we limit our discussion to those measures that are directly based on the notion of distances.

The *betweenness centrality* measure expresses how much effect each vertex has in the communication in the network assuming that all traffic follows shortest paths. Informally, the betweenness centrality of a vertex v refers to the total number of shortest paths between every vertex pair that pass through v. Let $\alpha_{wz}(v)$ be the fraction of shortest paths between w and z that pass through v, i.e., $\alpha_{wz}(v) = \sigma_{wz}(v)/\sigma_{wz}$, where $\sigma_{wz}(v)$ is the number of all shortest paths between w and z that pass through v and σ_{wz} is the number of all shortest paths between w and z. The betweenness centrality $c_B(v)$ of v is $c_B(v) = \sum_{w \in V} \sum_{z \in V} \alpha_{wz}(v)$ [6]. Higher values of this measure indicates higher importance of the vertex. The *closeness centrality* considers the central vertices as the subset of vertices with the minimum total distance to all other vertices. The closeness centrality $c_C(v)$ of a vertex v is defined as $c_C(v) = 1/\sum_{u \in V} d(v, u)$ [6]. The *eccentricity centrality* suggests that the center of the graph includes the vertex (or vertices) that has the shortest distance to all other vertices. For a given vertex v, the eccentricity centrality is $c_E(v) = 1/\max\{d(v, u) : u \in V\}$ [9]. The vertices with the highest eccentricity centrality in fact form the *center* of the network $C(G)$. In other words, $C(G) = \{u \in V : ecc(u) = rad(G)\}$. Tables 4 and 5 list the highest ten populations for the degree, betweenness, eccentricity, and closeness centrality measures for the unweighted and the weighted networks respectively. Note that in the fifth column of Table 4, all the five listed populations have equal eccentricity centrality. For the unweighted network, the Spearman rank correlation coefficient, which tests the association between two sets of ranks, between the betweenness and the closeness centralities is 0.677 with 70 % common populations in the list of top 10 populations. For the weighted network, the Spearman rank correlation between the two measures is about 0.41.

Table 4 Top ten populations with respect to degree, betweenness, eccentricity, and closeness centrality measures for the directed unweighted genetic admixture network

In-degree	Out-degree	Tot-degree	Betweenness	Eccentricity	Closeness
Papuan	Druze	Adygie	Burusho	Papuan	Druze
Maya	Palestinian	Armenian	Papuan	Melanesian	Palestinian
Melanesian	Burusho	Balochi	Druze	Columbian	Burusho
Burusho	Maya	BantuKenya	Melanesian	Lahu	Maya
Uzbekistani	Hazara	BantuSouthAfrica	IndianJew	Hazara	Hazara
IndianJew	Melanesian	Basque	Maya	–	Melanesian
Cambodian	Sardinian	Bedouin	Hazara	–	Papuan
Adygie	Papuan	Belorussian	Palestinian	–	Sardinian
Turkish	Kalash	BiakaPygmy	Adygie	–	Kalash
Pathan	Indian	Brahui	MbutiPygmy	–	Brahui

Table 5 Top ten populations with respect to degree, betweenness, eccentricity, and closeness centrality measures for the directed weighted genetic admixture network

In-degree	Out-degree	Tot-degree	Betweenness	Eccentricity	Closeness
Papuan	Druze	Papuan	Spanish	Burusho	Druze
Melanesian	Palestinian	Burusho	Maya	Armenian	Han
Indian	Burusho	Maya	Han	Pathan	Palestinian
Burusho	Maya	Melanesian	SanKhomani	Maya	Maya
Maya	Hazara	Indian	Moroccan	Hazara	Burusho
Pathan	Indian	Palestinian	Iranian	Melanesian	Balochi
Myanmar	Melanesian	Druze	Cypriot	Papuan	Jordanian
Cambodian	Papuan	Hazara	Pathan	She	Moroccan
IndianJew	Bedouin	IndianJew	Adygei	Han	Brahui
Adygei	Mozabite	MbutiPygmy	EastSicilian	Yakut	Melanesian

4 δ-Hyperbolicity and Network's Core

δ-Hyperbolicity is a measure that captures the notion of negative curvature in abstract metric spaces including graphs. A simple graph $G = (V, E)$ naturally defines a metric space (V, d) on its vertex set V where the distance $d(u, v)$ is defined as the length a shortest (v, u)-path between v and u. In graphs, δ-hyperbolicity measures how close the graph's structure is to a tree structure metrically [8]. Given a graph $G = (V, E)$, x, y, u, and $v \in V$ are four distinct vertices, and the three sums: $d(x, y) + d(u, v)$, $d(x, u) + d(y, v)$, and $d(x, v) + d(y, u)$ sorted in a non-increasing order, the hyperbolicity of the quadruple x, y, u, v denoted as $\delta(x, y, u, v)$ is defined as: $\delta(x, y, u, v) = (d(x, y) + d(u, v)) - (d(x, u) + d(y, v))/2$. The δ-hyperbolicity of the graph is $\delta(G) = \max_{x,y,u,v \in G} \delta(x, y, u, v)$. Generally, the smaller the value of $\delta(G)$ the closer the graph is to a tree metrically and, as a result, the hyperbolicity property is more evident. Even though the δ-hyperbolicity by definition considers the maximum difference between any two largest distance sums for any quadruple, recent research also analyzes the distribution of δ-hyperbolicity of the quadruples [2, 3, 7]. This makes the value of the average δ-hyperbolicity (taken over all quadruples) equally important. The small δ-hyperbolicity property has been found in many real-world networks [2, 3, 7, 12]. However, in many of those networks, this low value is a direct result of the small-world property especially that the inequality $\delta(G) \leq \frac{diam(G)}{2}$ is sharp. For our unweighted network, $\delta(G) = 2$ and the average δ-hyperbolicity of the graph $\delta'(G)$ is 0.24. For our weighted network, $\delta(G) = 15$ and $\delta'(G) = 1.96$.

δ-Hyperbolicity, centrality, and network's core. It was suggested in [4] that the concentration of load on a subset of vertices of the network, for communication assuming shortest path routing, is due to its negative curvature or δ-hyperbolicity. This concentration can be seen as a bend in those shortest paths towards a core of the network defined by its most central vertices. However, the identification of core vertices differ according to the centrality measure used to decide the central vertices. In [4], the core is defined as the subset of vertices with highest betweenness centrality. In [3] the core is defined based on the eccentricity centrality and the betweenness centrality.

Proposition 1 ([3]) *Let G be a δ-hyperbolic graph and x, y be arbitrary vertices of G. If $d(x, y) > 4\delta + 1$, then on any shortest (x, y)-path there is a vertex w with $ecc(w) < \max\{ecc(x), ecc(y)\}$.*

According to the proposition, shortest paths bend towards vertices with smaller eccentricity making the graph's core mostly represented with vertices with the smallest eccentricity ($rad(G)$ or $rad(G) + 1$ in most cases). Then, the betweenness centrality is used to prioritize those vertices according to their participation. In [2], it has been observed that if one constructs a small r-neighborhood where ($r = \delta(G)$) around a vertex v on a shortest path between two vertices x and y, then all shortest paths between x and y include a vertex in this r-neighborhood.[1] Our goal is to identify the core vertices using the δ-hyperbolicity of the network without any presumptions about the centrality of the vertices in the network. Then we analyze the core vertices in terms of their centrality.

For each integer $r \geq 0$, let $N_r(u)$ denotes the *neighborhood* of distance at most r centered at u, i.e., $N_r(u) = \{v \in V : d(u, v) \leq r\}$. We define the δ-hyperbolicity-neighborhood of a vertex u, denoted as $N_\Delta(u)$, as the smallest integer Δ, where $0 \leq \Delta \leq \delta(G)$, such that the majority of vertex pairs (more than 90 %) are covered by that neighborhood. We say a vertex pair (w, z) is covered by the δ-hyperbolicity-neighborhood of a vertex v, if there is at least one vertex $u \in N_\Delta(v)$ such that $d(w, z) = d(w, v) + d(v, z)$. Figure 3 shows that the δ-hyperbolicity-neighborhoods of the majority of vertices when $\Delta = 0$ cover a small percent of vertex pairs (between 3 % and 15 %). An exception is those vertices with higher betweenness; for example, Papuan that covers about 33 % of other vertex pairs. However, when $\Delta = 1$, the δ-hyperbolicity-neighborhood around each vertex covers the majority of vertex pairs. Again some exceptions include French that covers only 27 %. For the details, take a look at Table 6.

[1]Note that the value of r could be higher but never exceeds $6\delta(G) + 2$ [2]. However, for real-networks it was observed in [2] that $r \approx \delta(G)$.

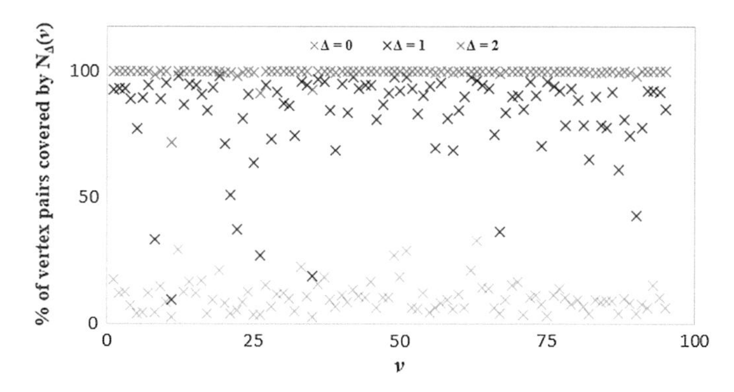

Fig. 3 Percent of vertex pairs covered by the δ-hyperbolicity-neighborhood $N_{\Delta}(v)$ of each vertex v in the directed unweighted genetic admixture network. $1 \le v \le 95$ and $0 \le \Delta \le \delta(G) = 2$

Table 6 Percent of vertex pairs covered by each δ-hyperbolicity-neighborhood in the directed unweighted genetic admixture network

v	Population	$N_0(v)$ (%)	$N_1(v)$ (%)	$N_2(v)$ (%)
a Populations with high coverage				
12	Burusho	29.48	98.53	100
33	Hazara	22.5	96.2	100
63	Papuan	33.08	96.77	100
b Populations with low coverage				
10	Bulgarian	2.8	9.18	72.06
25	French	3.68	27.22	91.58
35	Hungarian	2.79	18.8	93.1

Now we can rank our vertices according to their δ-hyperbolicity-neighborhoods. Each vertex v has two values: (1) Δ, that represents the smallest integer $\Delta \le \delta(G)$ such that the δ-hyperbolicity-neighborhood $N_{\Delta}(v)$ covers more than 90 % of vertex pairs, and (2) the percent of vertex pairs covered by this δ-hyperbolicity-neighborhood. We lexicographically sort all vertices according to those two values. The results are listed in Table 7. The higher the ranking of a vertex, the more it becomes part of the core set.

Discussions. From Fig. 3 and the results listed in Table 7 it is clear that the ranking of vertices obtained according by the coverage of their δ-hyperbolicity-neighborhoods corresponds to some extent with the ranking obtained from the centrality measures; especially the out-degree centrality, the betweenness centrality, the eccentricity centrality, and the closeness centrality. One can see that the majority of the top ten populations in each centrality measure are also present in the top ten list of the vertices with respect to their δ-hyperbolicity-neighborhoods. In contrast, some populations who are not at the top of the ranking with respect to some centrality measures

Table 7 Top ten populations with respect to the δ-hyperbolicity-neighborhoods of vertices in the directed unweighted genetic admixture network

Rank	Population	In-degree rank	Out-degree rank	Betweenness rank	Eccentricity rank	Closeness rank
1	Burusho	4	3	1	2	3
2	Druze	19	1	3	2	1
3	Palestinian	16	2	8	2	2
4	Melanesian	3	5	4	1	6
5	Kalash	17	7	19	2	9
6	Maya	2	4	6	2	4
7	Indian	10	7	15	2	11
8	Papuan	1	7	2	1	7
9	IndianJew	5	9	5	2	22
10	Sardinian	29	6	35	2	8

Here we compare this rank of each vertex with its rank according to the five centrality measures discussed earlier: the in-degree, out-degree, betweenness, eccentricity, and closeness

actually appear as core vertices according to their δ-hyperbolicity-neighborhood. For example, the three populations: Kalash, Indian, and Sardinian all are considered as core vertices according to their δ-hyperbolicity-neighborhood; however, they have lower values for the eccentricity centrality and/or the betweenness centrality measures. This can be justified by the existence of multiple core vertices distributed over multiple cores of the network defined using different centrality measures (or even by the existence of a number of nested cores). Some core vertices are more important with respect to their location and according to the percent of other vertex pairs they cover. This makes those vertices have higher values for the eccentricity centrality and/or the betweenness centrality measures. Still other core vertices, which may have lower eccentricity or betweenness centralities, are important (i.e., essential for communication) for a smaller percent of vertices. This observation motivates investigating the existence of multiple communities that revolve around those different core vertices. Generally, communities in a network are represented by a number of highly dense (with respect to the number of connections) set of vertices; and different communities are linked with fewer connections. Here we use the Louvain method [5] for detecting communities in our unweighted genetic admixture network (we ignore the direction of the edges here). The method identifies three communities (or modules) in our network which admits a modularity of 1.62. The modularity here measures the density of connections inside communities to the density of connections outside communities. See Figs. 4 and 5. Unlike the sub-networks identified earlier in Sect. 3, the modules are represented mostly by non-neighboring populations, and the core vertices are distributed among the different modules.

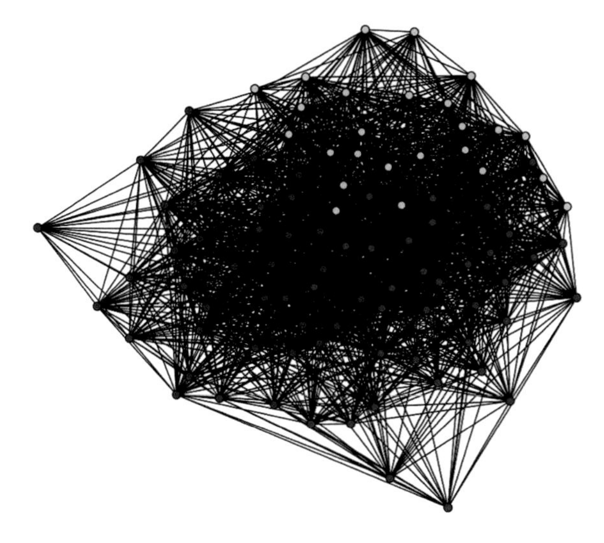

Fig. 4 Populations in three different modules. Module sizes (with respect to the number of vertices) are 36, 34, 25

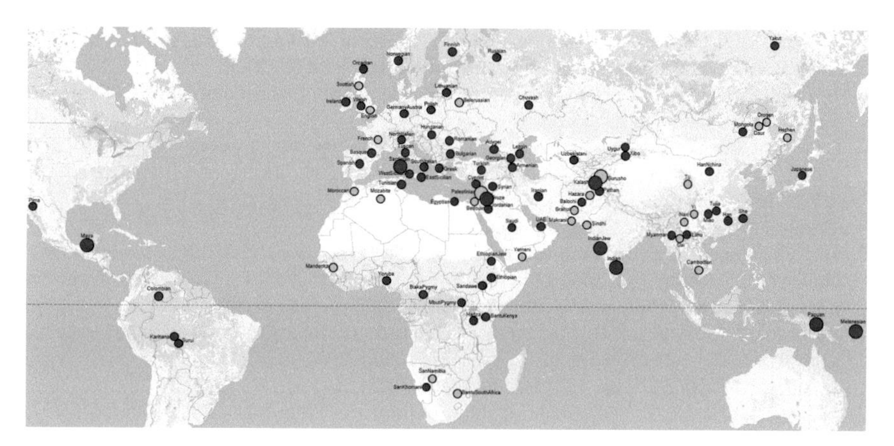

Fig. 5 Each population is assigned to a different module. *Larger circles* indicates the core vertices in each module based on the δ-hyperbolicity-neighborhood of vertices

5 Conclusion

We have studied the genetic admixture network of humans in which vertices represent populations and a link exists between a pair of populations if one participates in the genetic admix of the other. We have considered both the weighted and the unweighted versions of the network. The networks studied were based on data published in [10]. Based on our analysis, we find the average distance between a pair of populations across the network are relatively small suggesting a small-world network. We also find that the network comprises a number of sub-networks when edges are pruned based on their weights. Those sub-networks are formed by multiple neighboring populations. Also, we identify key vertices according to a number of centrality measures, and we find that those measures correlate very well. Finally, we propose a method of identifying core vertices based on the δ-hyperbolicity of the network. This network is dynamic; i.e., the connections among populations are based on a specific time frame. It is interesting to capture different admixing statistics based on various time frames and compare how the dynamicity of this network changes with time.

References

1. http://www.admixturemap.paintmychromosomes.com
2. Albert, R., DasGupta, B., Mobasheri, N.: Topological implications of negative curvature for biological and social networks. Phys. Rev. E **89**(3), 032811 (2014)
3. Alrasheed, H., Dragan, F.F.: Core-periphery models for graphs based on their δ-hyperbolicity: an example using biological networks. In: Complex Networks VI, pp. 65–77. Springer (2015)
4. Baryshnikov, Y.: On the curvature of the internet. In: Workshop on Stochastic Geometry and Teletraffic. Eindhoven, The Netherlands (2002)
5. Blondel, V.D., Guillaume, J.-L., Lambiotte, R., Lefebvre, E.: Fast unfolding of communities in large networks. J. Stat. Mech: Theory Exp. **2008**(10), P10008 (2008)
6. Brandes, U., Erlebach, T.: Network analysis: methodological foundations, vol. 3418. Springer Science & Business Media (2005)
7. De Montgolfier, F., Soto, M., Viennot, L.: Treewidth and hyperbolicity of the internet. In: 2011 10th IEEE International Symposium on Network Computing and Applications (NCA), pp. 25–32. IEEE (2011)
8. Gromov, M.: Hyperbolic Groups. Springer (1987)
9. Hage, P., Harary, F.: Eccentricity and centrality in networks. Soc. Netw. **17**(1), 57–63 (1995)
10. Hellenthal, G., Busby, G.B., Band, G., Wilson, J.F., Capelli, C., Falush, D., Myers, S.: A genetic atlas of human admixture history. Science **343**(6172), 747–751 (2014)
11. Leskovec, J., Chakrabarti, D., Kleinberg, J., Faloutsos, C.: Realistic, mathematically tractable graph generation and evolution, using kronecker multiplication. In: Knowledge Discovery in Databases: PKDD 2005, pp. 133–145. Springer (2005)
12. Narayan, O., Saniee, I.: Large-scale curvature of networks. Phys. Rev. E **84**(6), 066108 (2011)
13. Newman, M.E., Forrest, S., Balthrop, J.: Email networks and the spread of computer viruses. Phys. Rev. E **66**(3), 035101 (2002)

Network Science and Narratives: Basic Model and Application to Victor Hugo's Les Misérables

Semi Min and Juyong Park

Abstract Propelled by the recent advances in digitization of books and computational methods for automated text analysis, we are witnessing a promising opportunity for a serious scientific study of narratives. The importance of such an endeavor stems from the fact that a good story, albeit often fictional and artificial, is composed of highly believable characters who interact and experience a sequence of events together in a realistic world setting, and thus a better understanding of narratives may yield new insights for comprehending various real social phenomena as well as literary fiction. Here we present the basic scientific framework for modeling narrative as complex networks, which allows us to study how the narrative structure is reflected in the network of characters and how they allow us to understand the dynamics of narrative progression. This paper contains the fundamental network model of narratives and its properties that serves as the starting point for a more comprehensive future work.

1 Introduction

Advances in quantitative methodologies for the modeling and analyses of large-scale heterogeneous data in recent years have made possible the understanding of various complex social, technological, and biological systems from novel perspectives. And the horizon is expanding from such traditional complex systems to other fields including culture and humanities to find answers to new and long-standing problems, helped by the advent of large-scale data sets. For example, massive digitization of books (e.g., Project Gutenberg and Google Books) have allowed scholars to perform high-throughput analyses of language and literature, collective memory, the adoption of technology, censorship, and historical epidemiology [1]. The global

S. Min · J. Park (✉)
Graduate School of Culture Technology, Korea Advanced Institute of Science & Technology, Daejeon 34141, Republic of Korea
e-mail: juyongp@kaist.ac.kr

S. Min
e-mail: minsm86@kaist.ac.kr

H. Cherifi et al. (eds.), *Complex Networks VII*, Studies in Computational Intelligence 644, DOI 10.1007/978-3-319-30569-1_19

nature of web and the social media are also enabling many transnational studies of cultural phenomena: Park et al. studied complex network of western classical composers constructed from a comprehensive recordings data from ArkivMusic [2] and revealed its basic properties and growth dynamics [3]. Schich et al. analyzed the network of notable individuals and showed the emergent processes in cultural history [4].

As evidenced by these previous works, a data analysis framework that has attracted significant attention in recent years is network science. Network science attempts to uncover the underlying principles of a complex system from the patterns and the nature of the connections or interactions among its components [5]. By focusing on the interconnected nature of things, network science has made great strides in enabling a deep understanding of not only the systems easily recognizable as a network such as the Worldwide Web [6] and the Internet backbone [7], but also those that have been intensively studied in other traditional domains such as biology, management, and sociolgy [8–10].

In this paper we utilize the network framework to study a system that has not yet been widely studied by the framework, but one that we believe can benefit hugely: Narratives. Narratives, or stories as they are called more colloquially, have long been studied by narratologists who tried to reveal the patterns in their structure, as they realized the ubiquity and importance of narratives as the form by which we perceive and communicate our experience and surroundings. The connection between narratives and networks can be easily seen, for example, from the New Oxford American Dictionary's definition of a narrative as "a spoken or written account of connected events." Therefore the connection between the elements of the events—people (characters), things (devices), environment, etc.—are very important in the construction of a well-written, engaging narrative. It is this importance of the connections between the elements that render narratives an appropriate topic for network science. While still in its infancy, there have been a number of notable works in the past several years that do highlight the importance of networks for understanding narratives. The community structure of the co-appearance network of characters from Victor Hugo's Les Misérables [11], the network of characters from William Shakespeare's Hamlet constructed from word exchange [12], and the social networks of characters from from 19th-century British novels and serials from conversations [13] are such examples.

There have historically been many attempts to characterize the plot structure of narrative plot by extracting common and oft-observed patterns, including Aristotle's three-act plot structure theory. According to the theory, Act One is the setup where the central theme and question are raised, followed by two major turning points that form Acts Two and Three before the narrative concludes with a climax and a final resolution. Variant forms of it such as a four-act plot structure theory exist as well [14, 15]. While these types of theories still find use, there are naturally many ways in which they can be extended or modified to be able to help us understand the huge body of literature in existence. It should be not only possible but also necessary: given the large number of narratives, it's easy to understand that the themes may be different from narrative to narrative, and each narrative may contain more than three events and many more characters, resulting in the level of complexity that requires dividing the narrative into more than merely three or four smaller parts. This is what we wish to achieve in this paper: We attempt to establish a network science-based framework for narrative analysis that more accurately identifies the complexity of a narrative from the narrative itself. Note that this paper presents the fundamental modeling philosophy of our approach and some very basic results, and a more extensive work will be presented elsewhere in the future. We illustrate the framework using the English translation of Victor Hugo's Les Misérables [16], freely available on Project Gutenberg [17]. Based on the Paris Uprising of 1832 CE, it is considered a classic that vividly conveys the social conditions of a tumultuous era and insights into the human psyche via richly developed characters around the main character Jean Valjean, a fugitive who transforms into a force for good in a chaotic France while being haunted by the shadow of his criminal past [18]. The resulting complex web of characters and plots render Les Misérablesan appropriate material for the application of the methods we develop here.

2 Network of Characters: Topology and Growth

Our work begins from the the realization that it is very often the characters that, through their actions and interactions, advance a narrative. This makes it appropriate to represent the structure of narrative using character appearance and interactions. Since there can be varying numbers of characters appearing together at various points in the narrative, we represent the narrative as a set of interacting **character timelines** shown in Fig. 1. A character timeline is a record of the character's appearance in each narrative unit (e.g., a chapter in a novel or a scene in a play) of the narrative. What we consider as an appearance of a character and an interaction is up to the narrative format and the modeler; while it is straightforward in a dialog-driven script for a movie or a play since a character's appearances and dialogs are very clearly marked,

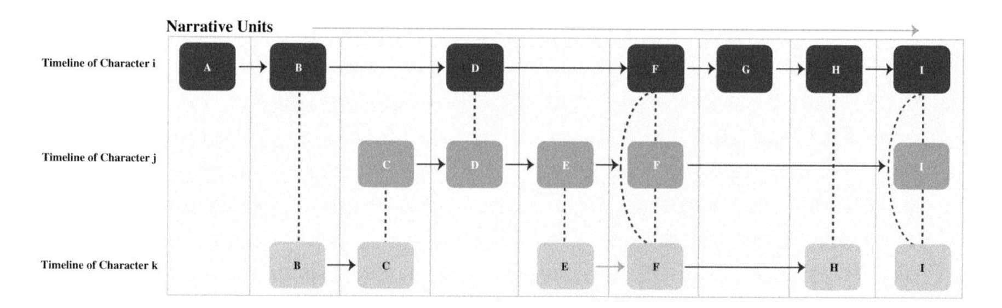

Fig. 1 Character-centric diagrammatic representation of our narrative flow model. According to the model, a narrative is a set of character timelines, the record of the characters' appearances in each narrative unit such as a chapter. We build the network of characters by connecting the characters that appear in a common narrative unit

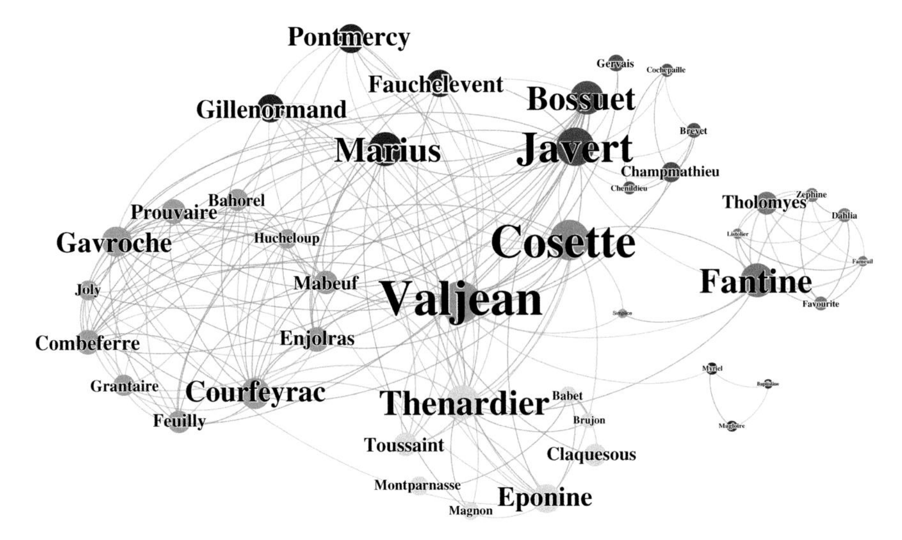

Fig. 2 The character network of Les Misérables. The node radius is proportional to its degree (the number of its neighbors). The network shows many characteristics of a social network such as small-world property and high clustering

there can be ambiguities in a novel. Here, we count the appearance of a character in a chapter as a single appearance, and consider tow characters to be interacting when they appear together in a common chapter of Les Misérables. This implies that an interaction could mean not only an explicit conversation or contact, but also experiencing a common event or situation.

In a well-written story, it is often the adventures, tribulations, and successes or failures of the characters that constitute its core content. The state of the character therefore is intimately tied to the state of narrative progression. Furthermore, since interactions between characters is an essential part of a character's experience and development, understanding the dynamics of the character relationships is key to understanding the structure of the narrative itself. We thus start by constructing the network of characters based on Fig. 1. The network of Les Misérables contains 63 characters after very minor ones are excluded. Drawing an edge between two characters if they have appeared in a chapter together (Fig. 1 or Ref. [11]) results in $m = 504$ edges. The network is shown in Fig. 2. In it, 25.8 % of the character pairs are connected, the mean geodesic length is 1.85, the network diameter is 4 (between the pair of Babet and Geborand, and 17 other pairs of relatively minor characters), and the clustering coefficient is 0.77.[1]

From the perspective of the narrative flow, what interests us is how the network grows over time and what we learn about the narrative from the patterns. This is

[1]Although our network appears denser than typical social networks [19, 20], this is likely due to the fact that most characters of the novel are involved in some common plot while the rest of the story world is pushed into the background.

because the network is essentially coupled to the narrative flow: Starting from an empty network in the beginning of the narrative, the network grows as new characters are introduced and interact with others. In this sense, we can say that the temporal growth of the network is intimately connected to the concept of the so-called narrative stages. A common classification of narrative stages is *Exposition*, *Rising Action*, *Climax*, *Falling Action*, and *Resolution* [21]. The Exposition stage introduces the characters and the space they inhabit. Once the motives and allegiances of the characters are presented, in the Rising Action the characters begin to struggle against each other until all conflicts are resolved throughout the last three stages. Based on this stage distinction, therefore, we can assume that n and m would not simply increase linearly in time but nonlinearly in accordance with the nature of the stages. In Fig. 3 we show the growth of n and m along the narrative time measured in chapters. As expected, the growth is not linear, especially for the number of nodes n. After the first batch of characters are introduced at the beginning of the narrative, there are specific points in the narrative where many new characters are introduced simultaneously (noted *S1*, *S2*, and *S3* in Fig. 3) that suggest they are the Exposition stages. An inspection of the actual story confirms this:

- Stage *S1*: Fantine's friends are introduced as her happy days are depicted.
- Stage *S2*: Valjean's former fellow prison inmates appear to testify during the trial of the fake Valjean.
- Stage *S3*: "The Friends of ABC" (young progressives) are introduced, shown debating on various social issues of the day.

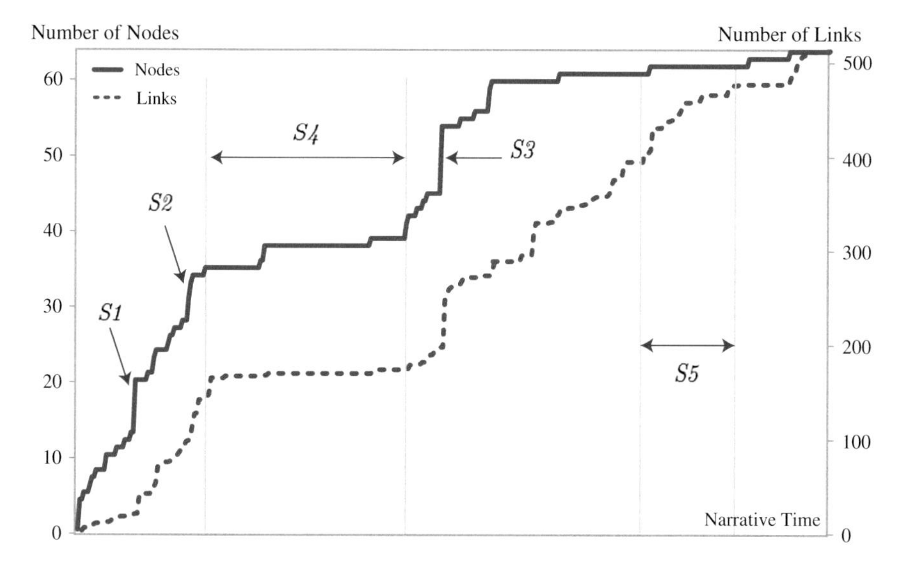

Fig. 3 The growths of the number of characters n and the number of edges m in Les Misérables. The growths are nonlinear, indicating that different stages in narratives serve different roles in the network growth via character introduction and formation of new connections

There is also a stretch of chapters (*S4*) where the network shows little growth. This part largely coincides with Volume 2 ("Cosette") of the novel composed of those chapters that contain no narrative progression (i.e. the author digresses to discuss the battle of Waterloo, religion, the vagrant children of Paris, etc.) or that show no network growth, being mainly about Valjean and Cosette's flight from the pursuit of Thenadier while avoiding people in general. Finally, near the end of the narrative at *S5*, it is the number of edges m that lead the growth of the network while n shows little increase. This—new edges being created between existing nodes without the addition to new ones—implies a convergence of the characters into a common environment: this part in fact describes the scene at the barricade where nearly all major characters (who have been introduced before) converge.

The network growth pattern shown Fig. 3 is aggregate. To understand the narrative structures based on the character network in more detail, we need to study the centrality of each character. We introduce three measures: Appearance a, unweighted degree k, and weighted degree w. Appearance a is the simplest, equal to the number of narrative units (chapters in our analysis of Les Misérables) in which a character appears. We show the final histogram of a in Fig. 4a. It has a skewed distribution with many characters appearing in a handful of chapters and a few characters appearing in many chapters, for instance Marius appears in 122 chapters, Valjean in 121, and Cosette in 97, whereas the mean and the median are 19.3 and 9 respectively, nearly an order of magnitude smaller than the most frequent characters. In Fig. 4d we show the temporal growth of each character's cumulative appearance. Although Marius and Valjean are similar in the total appearances (122 and 121, respectively), how these values are reached are very different. Valjean first appears in the beginning of the novel, then with regularity until there is a noticeable absence between chapters 160 and 233 (indicated by a plateau). During Valjean's absence, Marius, making his first appearance in chapter 170, takes the center stage in the novel and appears in almost every chapter until he overtakes Valjean in appearance. This is a direct reflection of the structure of Les Misérables: the first part is mainly about Valjean (with Marius absent), the second part is mainly about Marius (with Valjean absent), and the final part features both as major characters.

The degree k, the number of connections in the network, on the other hand, differs in interesting ways from a. The three highest-degree nodes are Valjean ($k = 43$), Cosette ($k = 41$), and Javert ($k = 39$), whereas Marius is down to $k = 34$. The degree therefore captures the nature of the social sphere around a character that appearance alone cannot tell: Valjean is a well-travelled character linking many different spheres of the story world, whereas Marius associates with a narrow pool of characters (namely the young fellow rebels) and his love interest Cosette. The weighted degree, the total number of times a character meets others, shown in Fig. 4e, f is in a sense a combination of the two. Here Valjean is again the leading character, followed by Marius, Cosette, Thenadier, and Javert.

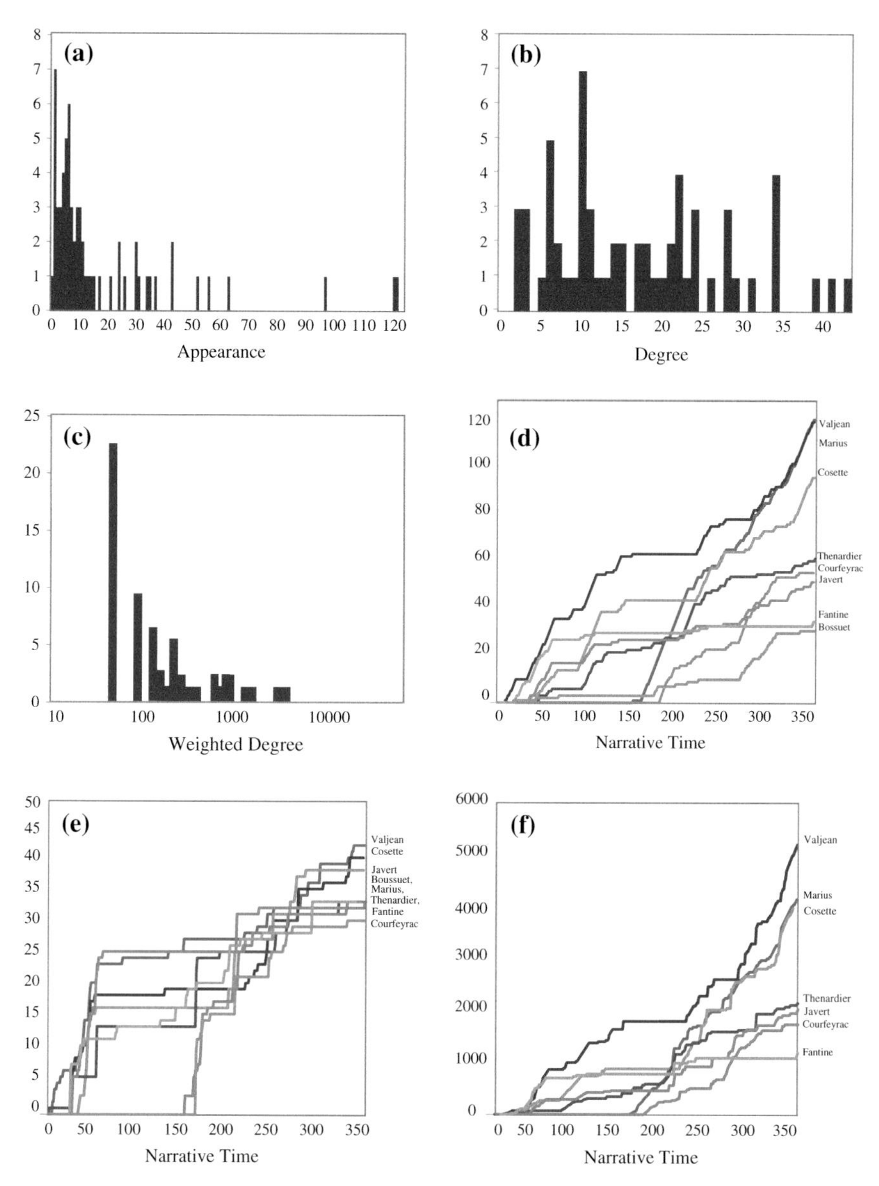

Fig. 4 Histograms of (**a**) appearance, (**b**) unweighted degree, and (**c**) weighted degree of the characters of Les Misérables. The histograms are relatively skewed, with some characters having high values and many having small values. The three most frequently appearing characters are Marius (122), Valjean (121), and Cosette (97), while the three highest-degree characters are Valjean (43), Cosette (41), and Javert (39), and the three highest-weighted degree characters are Valjean (5203), Marius (4148), and Cosette (3977). The discrepancies indicate the differences in the characteristics of their social networks. (**d**)–(**f**) show the growth of these quantities for each character, showing the different stages of the narrative in which the characters are very actively depicted. **a** Appearance distribution. **b** Unweighted degree distribution. **c** Weighted degree distribution. **d** Appearance. **e** Degree. **f** Weighted degree

3 Conclusion

In this paper we proposed a network framework for studying a narrative. By modeling a narrative flow as a growing dynamic network of characters, we demonstrated that some very basic measurements of the network reveal interesting pattern of the network structure. For instance, the growth of the network size showed a range of interesting behaviors such as a sudden increase in the number of nodes and characters that corresponded to the different characteristics of narrative stages. The role of an individual character in the narrative flow was measured via multiple centrality measures, which revealed different aspects of the character's role and position in the network. While this work is the first part of a longer forthcoming research that makes explicit use of the narrative text to more fully capture the complexity in storytelling, we believe that this work contains some useful and fundamental ideas for bringing the methods of complex networks towards understanding narratives.

Acknowledgments The authors would like to thank Kyungyeon Moon, Wonjae Lee, and Bong Gwan Jun for helpful comments. This work was supported by the National Research Foundation of Korea (NRF-20100004910 and NRF-2013S1A3A2055285), BK21 Plus Postgraduate Organization for Content Science, and the Digital Contents Research and Development program of MSIP (R0184-15-1037, Development of Data Mining Core Technologies for Real-time Intelligent Information Recommendation in Smart Spaces).

References

1. Michel, J.B., Shen, Y.K., Aiden, A.P., Veres, A., Gray, M.K., Pickett, J.P., Hoiberg, D., Clancy, D., Norvig, P., Orwant, J., et al.: Science **331**(6014), 176 (2011)
2. ArkivMusic. http://www.arkivmusic.com/. Accessed 2015-10-08
3. Park, D., Bae, A., Schich, M., Park, J.: EPJ Data Sci. **4**(1), 1 (2015)
4. Schich, M., Song, C., Ahn, Y.Y., Mirsky, A., Martino, M., Barabási, A.L., Helbing, D.: Science **345**(6196), 558 (2014)
5. Han, J., Kamber, M., Pei, J.: Data Mining: Concepts and Techniques. Elsevier (2011)
6. Albert, R., Jeong, H., Barabási, A.L.: Nature **401**(6749), 130 (1999)
7. Choi, J.H., Barnett, G.A., Chon, B.S.: Glob. Netw. **6**(1), 81 (2006)
8. Grimm, V., Revilla, E., Berger, U., Jeltsch, F., Mooij, W.M., Railsback, S.F., Thulke, H.H., Weiner, J., Wiegand, T., DeAngelis, D.L.: Science **310**(5750), 987 (2005)
9. Jeong, H., Tombor, B., Albert, R., Oltvai, Z.N., Barabási, A.L.: Nature **407**(6804), 651 (2000)
10. Borgatti, S.P., Foster, P.C.: J. Manag. **29**(6), 991 (2003)
11. Newman, M.: Networks: An Introduction. Oxford University Press (2010)
12. Moretti, F.: New Left Review (2011)
13. Elson, D.K., Dames, N., McKeown, K.R.: In: Proceedings of the 48th Annual Meeting of the Association for Computational Linguistics, pp. 138–147. Association for Computational Linguistics (2010)
14. Field, S.: Screenplay: The Foundations of Screenwriting. Delta (2007)
15. Vogler, C.: The Writer's Journey. Michael Wiese Productions (2007)
16. Hugo, V.: Les Misérables, vol. 5. Lassalle (1862)
17. Gtenberg, T.P.: https://www.gutenberg.org. Accessed 2015-10-08
18. Welsh, A.: Nineteenth-Century Fiction, pp. 8–23 (1978)

19. Wasserman, S., Faust, K.: Social Network Analysis: Methods and Applications, vol. 8. Cambridge University Press (1994)
20. Marsden, P.V.: Ann. Rev. Sociol. 435–463 (1990)
21. Freytag, G.: Freytag's Technique of the Drama: An Exposition of Dramatic Composition and Art. Scholarly Press (1896)

Mental Lexicon Growth Modelling Reveals the Multiplexity of the English Language

Massimo Stella and Markus Brede

Abstract In this work we extend previous analyses of linguistic networks by adopting a multi-layer network framework for modelling the human mental lexicon, i.e. an abstract mental repository where words and concepts are stored together with their linguistic patterns. Across a three-layer linguistic multiplex, we model English words as nodes and connect them according to (i) phonological similarities, (ii) synonym relationships and (iii) free word associations. Our main aim is to exploit this multi-layered structure to explore the influence of phonological and semantic relationships on lexicon assembly over time. We propose a model of lexicon growth which is driven by the phonological layer: words are suggested according to different orderings of insertion (e.g. shorter word length, highest frequency, semantic multiplex features) and accepted or rejected subject to constraints. We then measure times of network assembly and compare these to empirical data about the age of acquisition of words. In agreement with empirical studies in psycholinguistics, our results provide quantitative evidence for the hypothesis that word acquisition is driven by features at multiple levels of organisation within language.

1 Introduction

Human language is a complex system: it relies on a hierarchical, multi-level combination of simple components (i.e. graphemes, phonemes, words, periods) where "each unit is defined by, and only by, its relations with the other ones" [1, 2]. This definition [1] might explain some of the success of complex network modelling of language for investigating the cognitive processes behind the so-called human mental lexicon (HML) [2]. Psycholinguists conjecture [1, 3, 4] that words and concepts are stored within the human mind in such mental repository, which allows word retrieval according to multiple relationships (i.e. semantic, phonological, etc.). One can imagine the HML as an extensive database, where words are stored together with

M. Stella (✉) · M. Brede
Institute for Complex Systems Simulation, University of Southampton, Southampton, UK
e-mail: massimo.stella@inbox.com

H. Cherifi et al. (eds.), *Complex Networks VII*, Studies in Computational Intelligence 644, DOI 10.1007/978-3-319-30569-1_20

their linguistic patterns (e.g. synonym relations, etc.) on which a distance metric can be imposed, allowing for comparisons across entries.

In the last fifteen years, different layers of the HML have been investigated using tools from network theory. Motter et al. [5] constructed a semantic network of synonyms, where words appearing as synonyms in a dictionary were connected. The resulting network exhibited small-worldness (i.e. higher clustering coefficient and similar mean shortest path length compared to random graphs). It also displayed a heavy-tailed degree distribution with scaling exponent $\gamma \simeq 3.5$. The authors attributed both the presence of network hubs and the small-world feature to *polysemy*, i.e. a given word having more meanings depending on context and thus gathering more links. Sigman and Cecchi [6] showed that polysemic links create shortcuts within semantic networks, thus reducing path lengths between semantically distant concepts. This is relevant to cognitive processing because the semantic topology correlates with performance in word retrieval in memory tasks [1, 3, 7, 8]. It is conjectured that words within the HML are recollected together with a set of additional properties (e.g. being animated, etc.) [7]. Empirical evidence supports the hypothesis that adjacent words in a semantic network inherit features from their neighborhood, so that words closer on the network topology can be processed in a correlated way, thus reducing memory effort [1, 3, 7]. Semantic networks were further analysed by Steyvers and Tenenbaum in [9]. By proposing a network growth model based on preferential attachment, the authors investigated the role of word learning variables (e.g. frequency and age of acquisition) on shaping the structure of semantic networks. They showed that higher frequency words tend to have more semantic connections and tend to be acquired at earlier stages of development , thus highlighting an interplay between network topology and language learning.

Complex networks were also proposed as a suitable tool for analysing the phonological layer of the HML. Vitevitch suggested phonological networks (PNs) [4] as complex networks in which words are connected if phonologically similar, i.e. if they differ by the addition, substitution or deletion of one phoneme. Experimental evidence showed that the resulting network degree and local clustering coefficient both correlated positively with speech errors and word identification times, indicating that also the topological properties of a word in the phonological network plays an important role in its cognitive processing [10]. In [11, 12], we checked that artificial corpora, made of uncorrelated random words, could not reproduce specific features of the English phonological network. By means of percolation experiments we showed that the real PN actually inherits some features (e.g. a degree distribution with a heavy tail) from its embedding space, but it also displays some patterns that are extremely hard to match with random word models (e.g. the PN's empirical core-periphery structure). By proposing a family of null models that respect the spatial embedding, we identified two constraints possibly acting on phoneme organisation: (i) a maximum size of phonological neighborhoods (above which word confusability [4, 10] becomes predominant) and (ii) a tendency to avoid local clustering (which correlates with word confusability [10]).

To the best of our knowledge, until now there has been no theoretical framework modelling both the semantic and the phonological aspects of the mental lexicon in terms of a multiplex network. Multiplexes represent a novel and quite prolific research field [13, 14]: in a multiplex the same set of nodes can be connected differently in different layers of networks. Historically, the idea of context-dependent links originated from the social sciences [13]. However, it is only in the last five years that these multi-layered networks were successfully applied in a wide collection of different scenarios, such as robustness of infrastructure, science of science and game-theoretic dilemmas, among many others (for further references see [14]).

Exploring the multiplexity of the English language to study lexicon formation is the main idea of this study. We specifically focus on the interplay between semantics and phonological factors in the assembly of the repertoire over time. In detail, we build a three-layer multiplex network, where each layer represents a given linguistic network and where the same set of nodes is replicated across all layers. We focus our analysis on a minimalistic network growth model where the lexicon is assembled over time and real words get inserted, one at a time, according to a given ordering, either based on exogenous features (e.g. word frequency) or multiplex features of the HML. Our main aim is to quantify the influence of each ordering in the assembly times of the empirical multiplex, in order to assess the impact of word features on lexicon growth. For this purpose, we test our experiments with empirical data of the age of acquisition of English words, obtained from [15]. Our results highlight the presence of an interplay between the phonological and the semantic layers in structuring the mental lexicon.

This paper is structured as following: in Sect. 2 we report on the dataset we adopted for the multiplex construction and we compare it to datasets of commonly spoken English; in Sect. 3 we introduce the model of lexicon growth; the results are discussed in Sect. 4, conclusions and future work directions are reported at the end.

2 Multiplex Construction

We build a linguistic multiplex of three unweighted, undirected graphs/layers, comprising an intersection of $N = 4731$ words and based on the following interactions:

1. Free Word Associations (based on the Edinburgh Associative Thesaurus [16]);
2. Synonyms (based on WordNet 3.0 [17]);
3. Phonological Relationships (based on WordNet 3.0 and manually checked, automatic phonological transcriptions into the IPA alphabet [18]).

The thus constructed linguistic multiplex includes one phonological layer and two semantic layers. With synonym relationships and word associations we chose to include two semantic layers, mainly because of large structural differences in the topology of these networks. Free associations capture also those linguistic patterns that cannot be expressed in terms of other semantic relationships (e.g. opposites, synonyms, etc.). These relationships are still of primary importance for cognitive processes [3, 16]. In fact, experimental evidence indicates that such links act as pointers for word retrieval [1, 3]. Their greater generality is what differentiates associations from synonymy relationships, which have been extensively investigated in the linguistic literature [3, 8, 9].

Representativeness of the Data

A network representation of language should be indicative of real patterns in the mental lexicon, therefore the linguistic multiplex should be based on commonly used words. Unfortunately WordNet 3.0 does not include frequency counts, therefore we tested our data through the word frequencies from the Opensubtitles dataset [19], i.e. a lexicon based on more than $1.4 \cdot 10^8$ word counts from TV series subtitles.

The word length distributions reported in Fig. 1 indicate that our smaller-size word sample contains more shorter words when compared to WordNet. Furthermore, Fig. 1 shows that words contain less phonemes than orthographic characters, on average. For instance, a word in our sample contained 4.78 ± 0.03 phonemes and 5.39 ± 0.03 orthographic characters. Given this difference, we are using both phonetic and orthographic word lengths in our growth experiments in Sect. 4, as proxies for word acquisition through hearing and reading, respectively.

In Fig. 1 the word frequency distribution of our 4731 sampled words is compared against the whole Opensubtitles repository and against the 4731 words from Opensubtitles with the highest frequencies. Interestingly, the whole dataset exhibits a heavy tail behaviour. The cumulative probabilities $P(F \geq z)$ of finding a word with frequency F greater than or equal to z tell us that higher frequency words in our dataset are more likely than in the whole Opensubtitles but also less likely than in the frequency ranked subsample. Furthermore, excluding extremely frequent words, our sample reproduces the same power-law like behaviour of the whole Opensubtitles

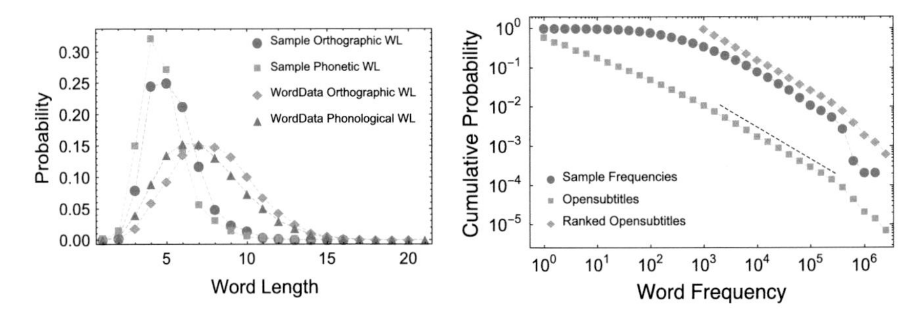

Fig. 1 *Left* Orthographic and phonetic word length distributions for our sample (*blue dots* and *golden squares*, respectively) and for roughly 29000 words phonetically transcribed within WordNet 3.0 (*green diamonds* and *red triangles*). Opensubtitles is not used in the word length distributions because it does not have phonological transcriptions. *Right* Empirical probability distributions of word frequency within our data sample (*blue dots*), the Opensubtitles repository (*golden squares*) and a ranked subsample of the Opensubtitles list of the same size as our sample (*green diamonds*). The *dashed black line* gives a power-law with exponent $\gamma = 1.83 \pm 0.03$

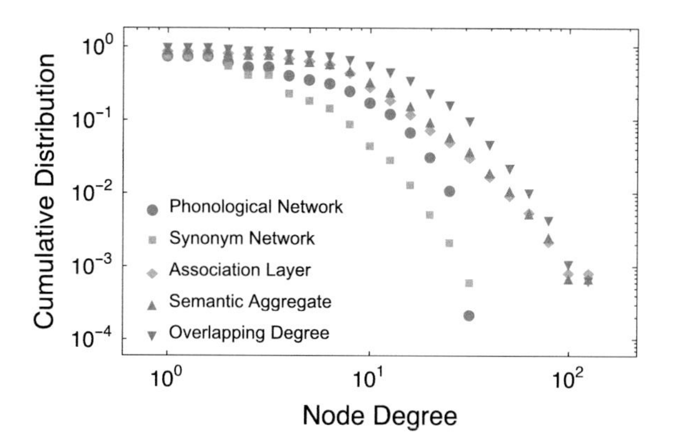

Fig. 2 Cumulative degree distribution $P(K \geq k)$ for the phonological network (*blue dots*), the synonym network (*golden squares*), the free association network (*green diamonds*), the semantic aggregate network associations+synonyms (*red triangles*) and the overlapping network (*purple triangles*)

dataset, for mid-and high frequencies. Because of this over-representation of higher frequency words, we can reasonably assume that our data is a good representation of commonly used English words.

Multiplex Network Structure

We begin the analysis of the multiplex by investigating the cumulative degree distributions $P(K \geq k)$ [20] of individual layers and of the multiplex, reported in Fig. 2. The degree distributions span different orders of magnitudes and display different behaviors. There is a considerable fraction of hubs within the association network, which displays a heavy tail degree distribution. The phonological network displays a cut-off around degree $k \approx 30$ while the synonym network shows a degree distribution that can be approximated by an exponential. We also investigate the multiplex *overlapping degree* o_i [13], i.e. the sum of degrees $k_i^{[\alpha]}$ of node i on each layer α:

$$o_i = \sum_{\alpha=1}^{3} k_i^{[\alpha]}. \quad (1)$$

Interestingly, the overlapping degree seems to have a more pronounced exponential decay compared to the degree of the *aggregated semantic layers* (i.e. a network where any link is present if it is present in at least one of the original layers). This reveals negative degree correlations on lower degrees, in our linguistic multiplex. A scatter plot highlights the presence of hub nodes in the semantic aggregate that have low degrees ($k \leq 15$) in the phonological layer. For this reason, locally combining layer topologies is of interest for our assembly experiments, cf. Sect. 4.

Table 1 reports some network metrics [20] for the individual layers and the aggregated semantic multiplex, compared to configuration models with the same degree distributions. All the three layers display the small-world feature, in agreementwith

Table 1 Metrics for the multiplex with $N = 4731$ nodes: edge count L, average degree $\langle k \rangle$, mean clustering coefficient CC, assortativity coefficient a, giant component node count $GCSize$, network diameter D and mean shortest path length $\langle d \rangle$

Network	L	$\langle k \rangle$	CC	a	$GCSize$	D	$\langle d \rangle$
Phonological	15447	6.5	0.24	0.61	3668	22	6.7
Phonological CM	15447	6.5	0.004(1)	0.0048(4)	4580(10)	10	4.3(5)
Synonym	7010	3.0	0.23	0.26	2989	15	5.9
Synonym CM	7010	3.0	0.002(1)	−0.02(3)	3396(9)	13	5.1(2)
Association	20375	8.6	0.1	−0.11	3664	7	3.6
Association CM	20375	8.6	0.09(2)	−0.005(1)	3658(8)	7	3.5(3)
Semantic Aggregate	26056	11.	0.18	−0.06	4298	9	3.6
Sem. Agg. Combined CMs	27374	11.6	0.01(2)	0.015(1)	4320(10)	8	3.5(2)
Multiplex Aggregate	40983	17.3	0.15	0.018	4689	9	3.4
Mult. Agg. Combined CMs	42787	18.1	0.012(5)	0.024(4)	4713(6)	8	3.2(1)

Error bars on the last digit are reported in parentheses and are based on 20 repetitions. For instance, 3.0(4) means 3.0 ± 0.4. CM aggregates are obtained by combining CM layers and therefore differ in degree from their empirical counterparts

previous results [5, 11]. The current literature suggests that small-worldness might be related to language robustness to individual word retrieval failure (e.g. in aphasia [3]) while also enhancing network navigability [10]. It is noteworthy that the phonological layer displays a network diameter almost three times larger than the mean path distance. Since its configuration model (CM) counterpart does not reproduce such pattern, this is an indication of a strong core-periphery structure within the network [11]. Further, all the individual layers are disconnected and have a giant component (GC). The GC size is hardly matched by CMs for the phonological and synonym layers while there is good agreement for the association layer. Interestingly, the two semantic layers display different organisational features: the synonym layer is more disconnected but more clustered than the association one; and while synonyms display an assortative mixing by degree associations are disassortative [20], instead. Therefore, in the association layer there are hub words surrounded by many poorly connected nodes while in the synonym layer large neighborhoods tend to be directly connected with each other. Indeed, because of these different topological features we will keep these layers distinct within our linguistic multiplex. Assortative mixing is also strongly present at the phonological level, but note that in this case high assortativity is a feature inherited from the embedding space of phonological networks [11, 12]. Configuration model aggregates are formed by aggregating the individual configuration model layers. Only 0.5 % of the edges in the association layer overlap with the synonym layer. The empirical networks display a higher edge overlap when compared to the configuration models (4.5 % of the edges overlap across the real layers versus the 0.1 % of the CMs).

3 Simulated Network Assembly

In [11], we suggested a network growth procedure as a null model for phonological networks (PNs), in which an artificial PN was built from randomly assembled strings of phonemes, satisfying some empirical constraints (e.g. phoneme frequencies). In this work, we extend that model by adopting a multiplex perspective.

Let us model the mental lexicon as a network which grows over time. Our model is *localist* [9], i.e. in it each concept is partially associated with an individual node/word in the network. Concepts are acquired to the lexicon by inserting single nodes/words. However, a given concept is represented in its full meaning by a word/node together with its links, since they retain further information about the concept itself (e.g. a neighborood can translate into a semantic context [9] or it can provide information about word confusability [4]). In the following we will use "words" to identify single nodes and "concepts" to identify jointly a node and its local connectivity. We follow an approach similar to Steyvers and Tenenbaum [9].

At each time step, a node/word is tentatively inserted into the lexicon. Then, we check for phonological similarities between the new word and the others already in the network, i.e. we check for links on the phonological level. If the new node/word receives at least one connection (i.e. it becomes *active* in the multiplex jargon [14]) on the phonological network, then it is accepted to the lexicon. Otherwise, if the node/word does not receive any connection, we reject it with probability f, putting it back to the list of not yet included words. Words are suggested from this list according to a given multiplex or exogenous criterion and until all words have been accepted. We measure the average assembly time T, i.e. the time it takes until a full network comprising all 4731 words has been built. The rejection probability f is the only free parameter of the model, but acceptance/rejection of words also depends strongly on the ordering in which they are suggested. There are many possibilities of different orderings that could be considered. We tested several of them and then selected a sample of those experiments that provided a wide pool of different results:

1. random ordering as a baseline reference case (Rand. Order);
2. phonologically shorter words first (Short Pho., e.g. "a", "ad", "ash", ...);
3. orthographically shorter words first (Short Wor., e.g. "a", "ad", "be", ...);
4. more frequent words first (Freq., e.g. "a", "in", "have", ...);
5. higher degree words in the association layer first, where hubs are the most recollected words in semantic memory (Asso., e.g. "man", "water", "sex", ...);
6. higher degree words in the synonym layer first, notice the difference with the association layer in the ranking (Syno., e.g. "take", "hold", "get", ...);
7. higher degree words in the semantic multiplex aggregate first, association hubs prevail over the synonyms (As.+Sy., e.g. "man", "water", "sex", ...);
8. empirical age of acquisition [15] (AoA, e.g. "momma", "potty", "water", ...);
9. random phonological/random semantic neighbors, i.e. select a word at random on the phonological level, select one of its neighbors on the semantic aggregate at random, avoiding repetitions (R. Ph./Ag.);

10. random phonological/frequent semantic neighbors, i.e. select a phonological word at random, select one of its neighbors on the semantic aggregate at random but proportionally to its frequency (R. Ph./F. Ag.);
11. frequent phonological/frequent semantic neighbors, i.e. select a phonological word at random but proportionally to its frequency, similarly select one of its neighbors on the semantic aggregate (F. Ph./Ag.).

In our model the growth dynamics is driven by the phonological layer. Although this could be made more realistic, our choice is motivated by two empirical observations. Firstly, there is widely accepted empirical evidence showing that phonological memory (i.e. the growing set of phonological transcriptions that are checked for connections, in our model) plays a critical role in concept acquisition [21–25]. Furthermore, there are recent empirical studies in children that strongly emphasise that lexical acquisition is heavily influenced by the phonology of the words, at least at early stages of the lexicon's assembly [25]. Psycholinguists conjecture that this lead of phonology in the lexicon growth might occur because children could find it easier to produce and understand words containing phonemes already presented in their phonological inventory [1, 22]. This empirical bias is what our model tries to capture by checking for phonological similarities before word acceptance/rejection. However, it is also true that semantics and other external features do influence the lexicon's growth and our model does account for this interplay through the orderings of word insertion. In fact, there is also evidence that, after an initial state in which phonological learning is predominant, lexical learning lets children learn novel words whose sounds are not present in their inventories [24, 25]. Our model captures also this aspect, since even novel words that do not have phonological similarities can be probabilistically accepted. The second motivation behind adopting the phonological network as a check for linguistic relationships is that detecting phonological similarities is straightforward: it can be done on a quantitative basis (i.e. check for phoneme strings having edit distance one). Conversely, detecting semantic relationships (i.e. are two words synonyms?) can be extremely difficult without any external source of information (e.g. a dictionary or an experiment).

Beyond the type of links we check, another key element of our model is the "activation" requirement, i.e. the fact that a word has to receive at least one connection in order not to undergo the probabilistic rejection/acceptance stage. Being connected to any other node is the simplest requirement one can think of in terms of local connectivity, which is pivotal in the activation spread [7]. We have made this modelling choice mainly in the interests of meaningful parsimony. While we do not mean to preclude a possible role for other growth dynamics, we have to start from a simple, yet meaningful, dynamics that minimises the number of free parameters. It has to be underlined that our chosen model represents, at best, a highly simplified abstraction of the cognitive processes driving real lexicon growth. We chose to follow simplicity, mainly because of how little is known about the evolution of the real, large-scale human mental lexicon [1, 9]. Other viable approaches that might fall in the same simplicity category as our model should also be explored in the future.

Interestingly, the same word is used in both the multiplex and the psycholinguistics jargons: an "active" node in a multiplex is one having at least one link [14], the "activation" in psycholinguistics is a theoretical stimulus signal that spreads through connections across the semantic and/or phonological layers of the mental lexicon when words are to be identified and retrieved [3, 7, 10]. Indeed, our model accepts preferentially "active" (in the multiplex jargon) and potentially "activable" words (in an activation spread model scenario). Our focus on local connectivity was inspired by previous models of lexicon growth [9, 26], which conjecture that memory search processes might be sensitive to the local connectivity of concepts.

4 Results and Discussion

In Fig. 3 we report the normalized assembly times T/T_{random} for different orderings for several values of the rejection parameter f. These values are rescaled to the random case. Interestingly, such rescaling shows that inserting words with our selected orderings always decreases the assembly times compared to the random scenario. This effect becomes more evident when the rejection probability f is large. For instance, when the probability of rejecting each inactive word is $f = 0.8$, inserting words ordered according to their phonetic length (shorter first) fully assembles the network in roughly 78 % of the time necessary in the random case and this distinction is statistically significant. Notice that a-priori, we chose orderings loosely inspired by a least memory effort principle [1] so that this general trend is expected. Nonetheless, there is an interesting variety of behaviors that need further analysis.

Intuitively, inserting shorter phonetic-length words first is the optimal case in terms of minimum assembly time. Orthographic word length gives slightly higher assembly times. Inserting words according to their frequency gives results that are very close to semantic measures such as the degree rankings in the semantic layers/aggregate and to multiplex features. All these orderings show a trend close to the one where words are inserted within the growing lexicon according to their age of acquisition.

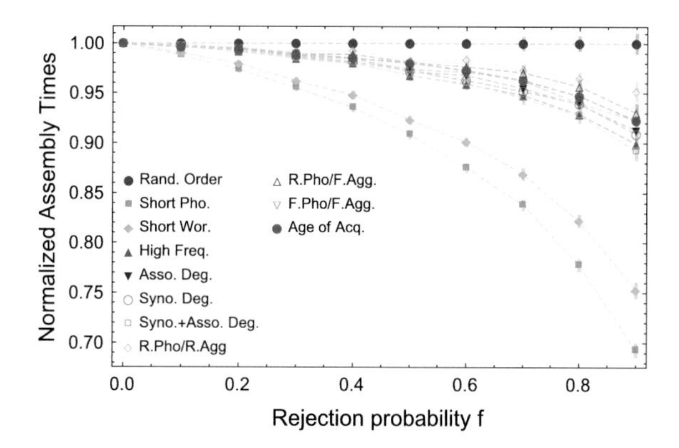

Fig. 3 Normalized assembly times for different orderings at different rejection probabilities f. These normalized times indicate the average time necessary for the network to get assembled through a given ordering rescaled to the random reference case. Error bars indicate standard errors and are evaluated over 20 different runs

Since assembly times are the quantitative proxies of our model for the likelihood of the mechanisms underlying lexicon growth, we adopt the times based on the age of acquisition as another reference point for testing the influence of the other orderings.

We start from the distributions of the assembly time for each ordering, at several values of the rejection probability f. We then quantify the overlap of the interquartile range of the age of acquisition case with the other scenarios. We consider the overlap of interquartile ranges rather than the overlap of the whole distributions because interquartile ranges represent a more robust measure of scale against fluctuations on extreme values in small, skewed empirical distributions as ours [27]. Also, interquartile ranges are easy to compute and visualise by commonly used box plots [27]. An example is reported in Fig. 4, where a box plot for the interquartile ranges of all our orderings are reported for $f = 0.8$. For instance, in that case the frequency ordering does not give results compatible with the empirical case (even though it is very close to the ordering with the degree in the semantic aggregate). Further, considering only the semantic degrees gives a slightly stronger overlap, but is not yet compatible with the age of acquisition case. Ordering words by their phonological and the semantic network degrees gives the closest results to the empirical age of acquisition scenario. We interpret this result as a quantitative proof of the importance of the multiplex structure of human language in shaping organisational features of the human mental lexicon. Locally navigating across the linguistic multiplex with a word frequency bias gives the best, highest overlapping results, within the framework of our theoretical model.

In Fig. 5 we checked the performance of the multiplex-based ordering versus f. Let us underline that during a given assembly f is kept fixed. However, when the probability of rejecting unconnected words is low, the orderings based only on either frequency or the semantic degrees perform relatively well. We can think of this stage as the real lexical learning phase [3, 23, 25, 26], which happens later in language development and where novel words are inserted within the lexicon according

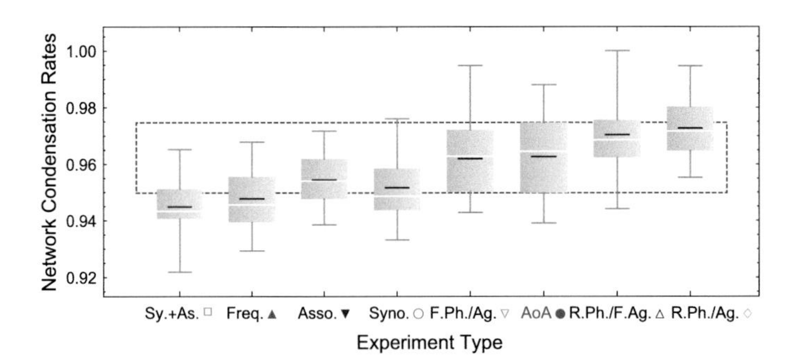

Fig. 4 Normalized assembly times of different orderings for a rejection probability equal to 0.8. The age of acquisition is highlighted in *red*. Whiskers represent distribution extremes while interquartile ranges are represented by *orange boxes*. White dashes indicate medians while *black dashes* represent means, instead. Interquartile overlaps represent the fraction of *orange boxes* falling within the ranges of the age of acquisition scenario

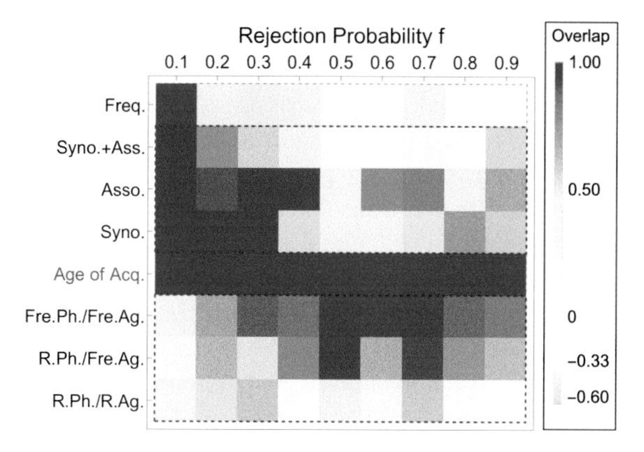

Fig. 5 Heat map of the normalized overlaps of the interquartile ranges of the assembly times relative to the age of acquisition case. The colours indicate: *red* = a perfect overlap (see the age of acquisition row); *white* = the absence of overlap; *blue* = the respective interquartile ranges are quite far. Orderings based on multiplex features are highlighted, the semantic ones on *top* and those based on multiplex neighborhoods on the *bottom*

to their semantic information and almost independent of phonological similarities. Larger values of the rejection probability f correspond to scenarios where the frequency and semantic degree orderings give results significantly different from the age of acquisition case. We can interpret this stage as a phonological learning phase [21–23], where words are inserted to the lexicon strongly based on their phonological similarities and where the phonological and the semantic layers are strongly interdependent. Therefore, our model highlights an interesting shift from one strongly semantic to a strongly multiplex stage, depending on the f parameter. This is a first quantitative finding about the importance of a multiplex modelling of the human mental lexicon. In fact, partial knowledge as frequency or phonological information only is unable to reproduce the same patterns across the whole parameter space.

5 Conclusions and Future Work

Here we proposed a simplified model of lexicon growth which is based on a representation of the English HML on several levels via the framework of multiplex networks. Motivated by empirical evidence and technical advantages in checking for phonological links we focus on the phonological level for the growth dynamics.

Numerically estimated assembly times identify a higher likelihood of a lexicon growth encapsulating information from the multiplex structure of free associations, synonyms and phonologically similar words, compared to assembly based on information from single layers or only word frequencies or lengths. In fact, assembly times can be thought of as proxies for the likelihood of the mechanisms underlying lexicon growth. When words are acquired without strong phonological biases (as in later stages of children's linguistic development) then orderings based on frequency and on semantic local centralities (i.e. node degree) are in good agreement with the empirical case. On the other hand, when words are acquired with stronger biases, as it happens in earlier stages of children's linguistic development, orderings based on the multiplexity of the English language provide results closer to the real scenario.

There are many interesting questions that this preliminary work opens. The first is a more extensive investigation of the multiplex features of the English language, e.g. a more detailed structural investigation of multiplex reducibility, layer overlap, cartography, clustering, efficiency and robustness to word retrieval failure [14]. Another interesting research direction would be trying to generalize our model by basing acceptance on the formation of more than one connections, or rather on links created also on other multiplex layers different from the phonological one, possibly by using the empirical semantic connections as a reference. This generalisation would be more realistic but also more cumbersome in adding more parameters to a model, which, already in this simple version, is capable of displaying an interplay between lexical and phonological learning.

From a complex systems perspective, it would be interesting to explore further the "multiplexity" of the English language, namely the interplay between phonological and semantic features, also by comparing the model against real data from children. Last but not least, a multiplex analysis for languages different from English could represent an interesting theoretical framework for testing both distinctive and universal features of human language.

Acknowledgments MS acknowledges the DTC in Complex Systems Simulation, University of Southampton for financial support. The authors acknowledge Dr. Srinandan Dasmahapatra, Nicole Beckage and the reviewers for providing insightful comments.

References

1. Aitchison, J.: Words in the Mind: An Introduction to the Mental Lexicon. Wiley (2012)
2. Baronchelli, A., Ferrer-i Cancho, R., Pastor-Satorras, R., Chater, N., Christiansen, M.H.: Trends Cogn. Sci. **17**(7), 348 (2013)
3. Beckage, N.M., Colunga, E.: Towards a Theoretical Framework of Analyzing Complex Linguistic Networks. In: Mehler, A., Blanchard, P., Job, B., Banisch, S. (eds.) Springer (2015)
4. Vitevitch, M.S.: J. Speech Lang. Hear. Res. **51**(2), 408 (2008)
5. Motter, A.E., de Moura, A.P., Lai, Y.C., Dasgupta, P.: Phys. Rev. E **65**(6), 065102 (2002)
6. Sigman, M., Cecchi, G.A.: PNAS **99**(3), 1742 (2002)
7. Collins, A.M., Loftus, E.F.: Psychol. Rev. **82**(6), 407 (1975)
8. de Deyne, S., Storms, G.: Behav. Res. Methods **40**(1), 213 (2008)
9. Steyvers, M., Tenenbaum, J.B.: Cogn. Sci. **29**(1), 41 (2005)
10. Vitevitch, M.S., Chan, K.Y., Goldstein, R.: Cogn. Psychol. **68**, 1 (2014)
11. Stella, M., Brede, M.: J. Stat. Mech. **2015**, P05006 (2015)
12. Stella, M., Brede, M.: Accepted in Lecture Notes in Computer Science (2015)
13. Battiston, F., Nicosia, V., Latora, V.: Phys. Rev. E **89**(3), 032804 (2014)
14. Boccaletti, S., Bianconi, G., Criado, R., Del Genio, C.I., Gómez-Gardeñes, J., Romance, M., Sendina-Nadal, I., Wang, Z., Zanin, M.: Phys. Rep. **544**(1), 1 (2014)

15. Kuperman, V., Stadthagen-Gonzalez, H., Brysbaert, M.: Behav. Res. Methods **44**(4), 978 (2012)
16. Kiss, G., Armstrong, C.A., Milroy, R.: Medical Research Council (1972)
17. Miller, G., Fellbaum, C., Tengi, R., Wakefield, P., Langone, H., Haskell, B.: WordNet. MIT Press, Cambridge (1998)
18. Brondsted, T. (2008). http://tom.brondsted.dk/text2phoneme/
19. Barbaresi, A.: Language-classified open subtitles (laclos). Ph.D. thesis, BBAW (2014)
20. Newman, M.: Networks: an Introduction. Oxford University Press (2010)
21. Ferguson, C.A., Farwell, C.B: Language. pp. 419–439 (1975)
22. Schwartz, R.G., Leonard, L.B.: J. Child Lang. **9**(02), 319 (1982)
23. Stoel-Gammon, C., Cooper, J.A.: J. Child Lang. **11**(02), 247 (1984)
24. Hoff, E., Core, C., Bridges, K.: J. Child Lang. **35**(04), 903 (2008)
25. Wiethan, F.M., Nóro, L.A., Mota, H.B.: CoDAS, vol. 4, pp. 260–264. SciELO, Brasil (2014)
26. Beckage, N.M., Aguilar, A., Colunga, E.: Proceedings of CogSci2015 (2015)
27. Moore, D.S., McCabe, G.P.: Introduction to the Practice of Statistics. WH Freeman/Times Books/Henry Holt & Co (1989)

Empirical Analysis of Crypto Currencies

Manoj Kumar Popuri and Mehmet Hadi Gunes

Abstract Analysis of the currency networks is not easy as the transactions are not centralized but rather take place over a large number of banks and commercial entities. Digital crypto currencies, however, require a public ledger to work and provide an opportunity for analysis of currency transactions. A crypto currency is a medium of exchange using cryptography to secure the transactions and to control the creation of new units. In this paper, we analyze two of the popular crypto currencies, i.e., Bitcoin and Litecoin. We construct network of transactions from public transaction ledger. We investigate the structure of currency transaction network by measuring the network characteristics.

1 Introduction

Currency is a medium of exchange, which arose out of need to address the inefficiency of barter. Digital currency is a form of currency that is electronically created and stored [14]. Crypto currencies are often decentralized digital cash systems and there is no single overseeing authority [13]. The first public crypto currency was Bitcoin, proposed in 2008 by Satoshi Nakamoto, a pseudonym [10]. Even though the system went online in January 2009, Bitcoin had very few users and didn't have real world value for a year. Since its inception, over 48 million transactions took place. The market value of Bitcoins in circulation peaked at about 14 billion USD on May 12, 2013, and as of Dec 1, 2015 is about 5.63 billion USD.

The Bitcoin system operates as an online peer-to-peer network, and anyone can join the system by installing the client application. Instead of having a bank account maintained by a central authority, each user has a unique address that consists of a pair of public and private keys. Existing coins are associated to the public key of the owner, and outgoing payments have to be signed by the owner using the

M.K. Popuri · M.H. Gunes (✉)
University of Nevada, Reno, USA
e-mail: mgunes@unr.edu

M.K. Popuri
e-mail: mpopuri@unr.edu

H. Cherifi et al. (eds.), *Complex Networks VII*, Studies in Computational Intelligence 644, DOI 10.1007/978-3-319-30569-1_21

corresponding private key. After validation of transaction with the owner's public key, the successful transactions are formed into blocks.

The transactions of all the crypto currencies are available to anyone by installing the client and connecting to peer to peer network. Such detailed information is rarely available in financial systems, making the crypto currency networks a valuable source of empirical data involving monetary transactions. Due to the anonymity of the crypto currencies and potentially unlimited number of pseudo identities a user could generate, however, it is hard to determine which observed phenomena are specific to the system and which results can be generalized.

An earlier study by Daniel et al. analyzes the Bitcoin transaction network to investigate the movement of money and observe the dynamics of the network [7]. In their analysis of Bitcoin data on May 7th 2013, they observe 17 million transactions among 13 million addresses where only a million of them had nonzero balance. According to their analysis there is a strong correlation between the balance and the indegree of individual nodes. They found that the Bitcoin network is gradually increasing since 2010 with some fluctuations, e.g., the boom in the exchange rate in 2011. According to their analysis both the in-degree and out-degree are highly heterogeneous with power law distributions. They also found that Bitcoin network is disassortative except for only a brief period in the initial deployment where the number of nodes were few.

The study of networks has emerged in diverse disciplines as a means of analyzing complex relational data [12]. Network analysis has been applied to physical phenomena [15], biological systems [6], transportation systems [1], social networks [11], software systems [3], linguistics [2] and academy [5].

In this paper, we compare two most popular crypto currencies as a network, by analyzing their transaction ledger. We map the transaction network of Bitcoin and Litecoin digital currencies from their public ledger and analyze the complex network of each digital currency. In our network, the nodes are the addresses of Bitcoin users and the edges are the transaction between two users.

2 Bitcoin Network

We downloaded the Bitcoin ledger and decoded the data collected from the wallet. Bitcoin network is a growing network where the number of unique addresses created increases exponentially. The major increase in the number of unique addresses occurred after the first boom in 2011 and the second one when the Bitcoin market value crossed 1000 USD. The network we are analysing is comprising of N = 49,390,594 nodes, total incoming transactions E_{in} = 151,933,127, and total outgoing transactions E_{out} = 151,857,042. We also divide the transaction data by year to study the evolution of the network over the years.

Degree

The degree distribution captures the underlying structure of a network by summarizing the degree characteristics of the nodes. Figure 1 present the in degree and out

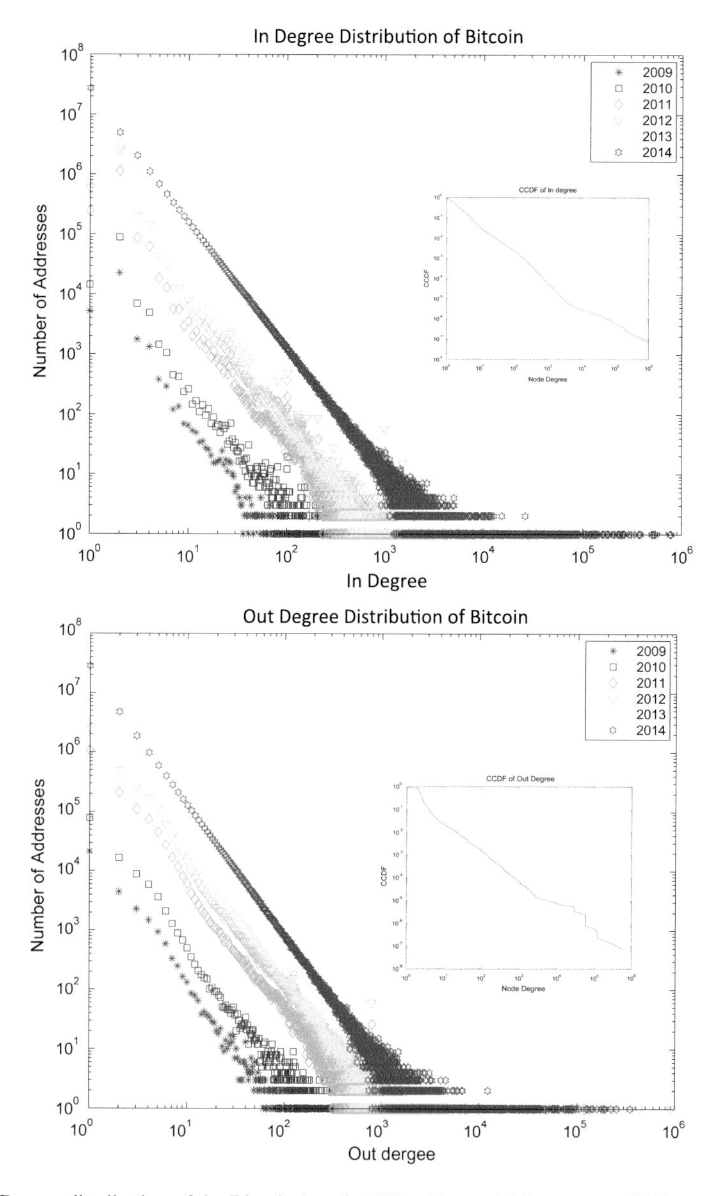

Fig. 1 Degree distribution of the Bitcoin (yearly PDF with overlaid aggregate CDF)

degree distributions of the Bitcoin transactions, respectively. While *probability distribution functions* show yearly distributions, overlaid *cumulative distribution function* shows the distribution for all transactions. We find that the degree distributions of yearly transactions as well as all transactions follow power law distribution, which makes Bitcoin network a scale free network, for both in degree and out degree. The power laws of the overall degree distributions are $\alpha_{in} \sim -2.21$ and $\alpha_{out} \sim -2.10$.

Table 1 Degree characteristics of yearly Bitcoin transactions

		In degree				Out degree			
$Year$	$Nodes$	$Edges$	Max	Avg	α_{in}	$Edges$	Max	Avg	α_{out}
2009	32,699	98,611	1,257	3.01	1.94	95,499	1,528	2.92	1.78
2010	122,167	374,712	1,826	3.07	2.00	478,271	5,8829	3.19	1.83
2011	1,610,899	5,198,488	118,016	3.23	2.13	5,461,888	59,297	3.39	1.88
2012	3,780,767	14,570,562	913,847	3.85	2.14	14,130,630	570,898	3.73	1.90
2013	5,082,351	16,338,332	16,969	3.21	2.19	16,442,626	69,919	3.23	1.92
2014	38,761,711	115,352,422	636,092	2.98	2.21	116,241,889	1,765,959	3.01	2.10

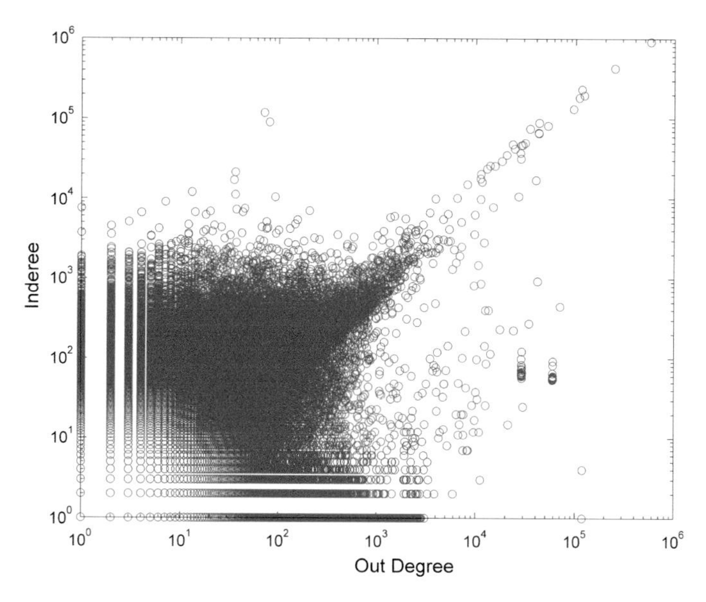

Fig. 2 Degree correlations of Bitcoin

Table 2 Network characteristics of yearly Bitcoin transactions

Year	2009	2010	2011	2012	2013	2014
Assortativity	−0.30	−0.14	−0.03	−0.025	−0.017	−0.019
Clustering	0.00	0.22	0.21	0.10	0.055	0.04

Table 1 presents the characteristics of yearly Bitcoin transactions. We observe that from 2009 to 2011 the degree distribution slope increases considerably and thereafter has been increasing but slightly.

Assortativity

We computed the nearest neighbour degree function $K_n^{in}(K_{out})$, which measures the in degree K_{in} of the nodes with respect to out degree K_{out}. Figure 2 presents the *degree correlations* for the Bitcoin network. In the graph, we observe that there is a disassortative behaviour between the In and out Degrees of the nodes. That is, the nodes with high out degree tend to connect to the node with low in degree.

As a summary measure *assortativity coefficient* is calculated as the Pearson correlation coefficient of degree between pairs of linked nodes. Positive values of r indicate a preference to link between nodes of similar degree, while negative values indicate preference to link between nodes of different degree. Table 2 presents the yearly assortativity coefficients of the Bitcoin transactions. We observe that the in-out degree correlation coefficient is negative, except for only a brief period in the initial phase. After mid-2010, the degree correlation coefficient stays between r $\approx$ -0.012 and r ≈ -0.015 suggesting that the network is disassortative. In general, for large scale-free networks, assortativity vanishes as the network size increases [9] and a similar behavior is observed in the Bitcoin network.

Clustering

We also measured the *average clustering coefficient*, which measures local density of edges. Table 2 presents clustering coefficients of the yearly Bitcoin transactions. We observed that in 2009 clustering is 0, indicating that there were no triangles among users. Then, between 2010 and 2011, clustering is high, fluctuating around 0.22. This can be due to few early adopters transferring money between their multiple accounts to test the network. As the number of users increase in the subsequent years, the clustering coefficient reduces from 0.10 in 2012 to around 0.04 in 2014, which is still much higher than a random network of similar size.

Richest Bitcoin Addresses

We traced the top 100 richest addresses in the Bitcoin and analysed for unique patterns. The total Bitcoins in circulation are 14,917,575 BTC with a market value of 377.93 USD as of Dec 1, 2015. The top 100 richest nodes in Bitcoin hold 19.88 % of wealth as shown in Fig. 3. We noticed couple of interesting behaviours among the richest Bitcoin users. For instance, the richest node transfers his/her bitcoins to four new addresses and then on the same day transfers all coins back into a single new address, which becomes the new richest address.

Figure 4 shows the in and out degree of the top 100 users. We observe that the incoming transactions to the richest people are through mining nodes, which indicates that most of the richest nodes are miners. We also observe that approximately 73 % of the richest people have 0 out degree, which means that they just accumulate money without spending it.

Anonymity

Even though Bitcoin data is anonymous, an active attacker can observe the IP address of a transaction request and match it to an actual user [4]. Hence, some users might be interested in hiding their IP address when communicating with the network. Anonymizer technologies allow one to hide a user's IP address and are widely used. Tor is currently the most popular anonymizer network with millions of users [8].

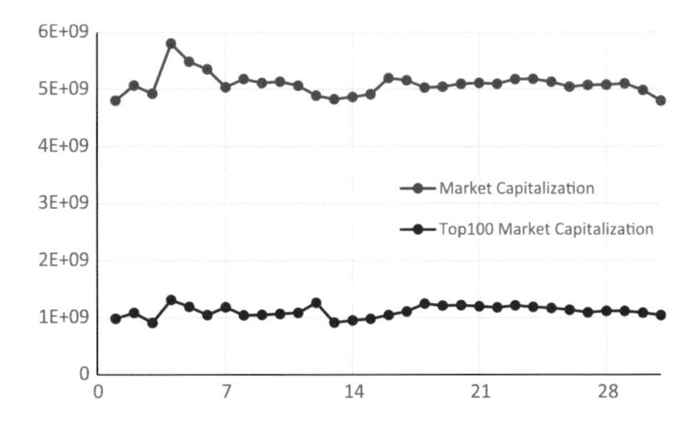

Fig. 3 Richest users during Nov 2015

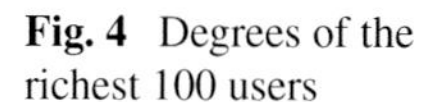
Fig. 4 Degrees of the richest 100 users

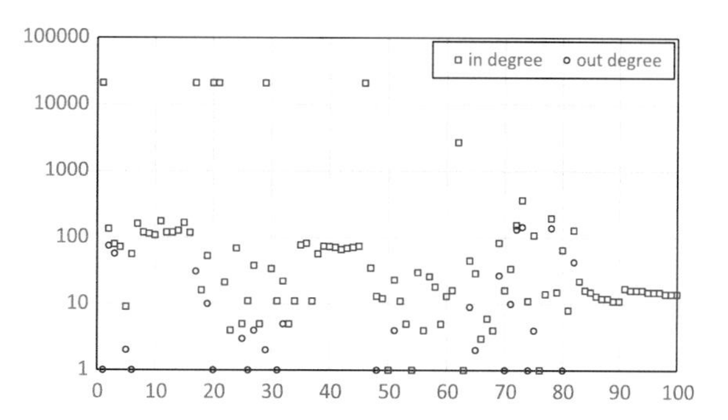

Fig. 5 Anonymity among MyWallet users

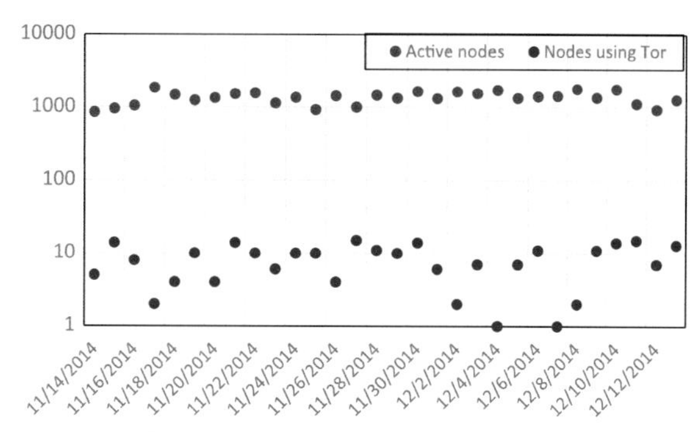

To analyze the percentage of anonymous Bitcoin users, we compared the IP addresses connected to the Bitcoin with the IP addresses of Tor exit nodes every hour. We analyzed the IP addresses for 30 days to find the ratio of users connecting to the Bitcoin anonymously as shown in Fig. 5. We observed that among 800 to 2000 connects to MyWallet at a given time only up to 20 nodes are using Tor anonymizer.

3 Litecoin Network

Litecoin is refereed to as the silver form of Bitcoin where the protocol is designed so that custom hardware cannot be used for mining. Even through Litecoin market value is 1 % of Bitcoin, the Litecoin network has a total $N = 6{,}990{,}919$ unique addresses, total $E_{in} = 56{,}205{,}576$ incoming transactions, and total $E_{out} = 52{,}456{,}092$ outgoing transactions Fig. 6

Degree

We calculated the in degree and out degree distributions of the network in Fig. 7. Unlike Bitcoin network, the Litecoin network growth is continuous. The degree distributions of aggregate transactions show a power law pattern with an exponent of $\alpha_{in} \sim -2.14$ for in degree and $\alpha_{out} \sim -2.01$ for out degree.

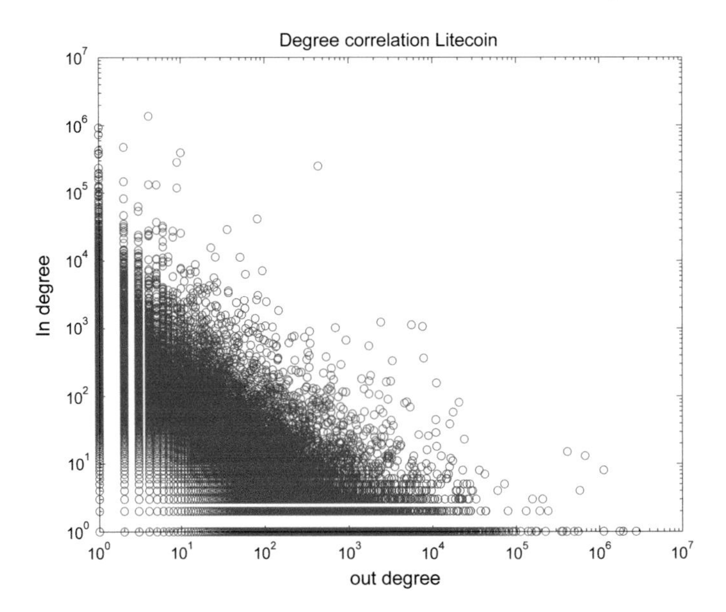

Fig. 6 Degree correlations of Litecoin

Table 3 presents yearly Litecoin network characteristics. The in degree and out degree power law exponents are more stable than the Bitcoin network.

Assortativity
We compute the *degree correlation*, i.e., the in degree K_{in} of the nodes with out degree K_{out}, for the network in Fig. 6. We find that the in-out degree correlation is dissortaative as the nodes with high degree have low in degree. The distribution is different from Fig. 2 for Bitcoin where the very high degree nodes connected to other very high degree nodes.

Table 4 presents *assortativity coefficient* of yearly Litecoin transactions. We find that the in-out degrees of yearly transactions is dissasortative in 2011 and 2012 but over the time become non-assortative in 2014.

Clustering
We also measured the average clustering coefficient in Table 4. We observed that, in the initial phase clustering is high. After the initial phase the clustering coefficient reduces from 0.33 in 2012 to around 0.032 in 2014.

Richest Litecoin Addresses
The total Litecoin in circulation are 43,455,110 LTC with a market value of 0.00959 USD as of Dec 1, 2015. The 48.89 % of the total market capitalization of the Litecoin is hold by the richest 100 people. We observed that the behaviour of the top 100 addresses in the Litecoin network are similar to the Bitcoin's richest users. We find that among the 100 richest nodes 82 % of the nodes have 0 out degree as shown

Table 3 Degree characteristics of yearly Litecoin transactions

$Year$	$Nodes$	In Degree				Out Degree			
		$Edges$	Max	Avg	α_{in}	$Edges$	Max	Avg	α_{out}
2011	22,400	754,734	170,892	33.69	1.81	63,163	4037	2.81	2.21
2012	545,576	10,391,318	1,124,344	19.04	1.90	2,484,673	395,841	4.55	2.13
2013	2,546,672	25,208,855	2,765,143	9.89	2.02	17,876,786	734,660	7.01	2.00
2014	6,735,643	19,850,699	360,129	2.94	2.21	32,031,470	1,373,967	4.75	2.12

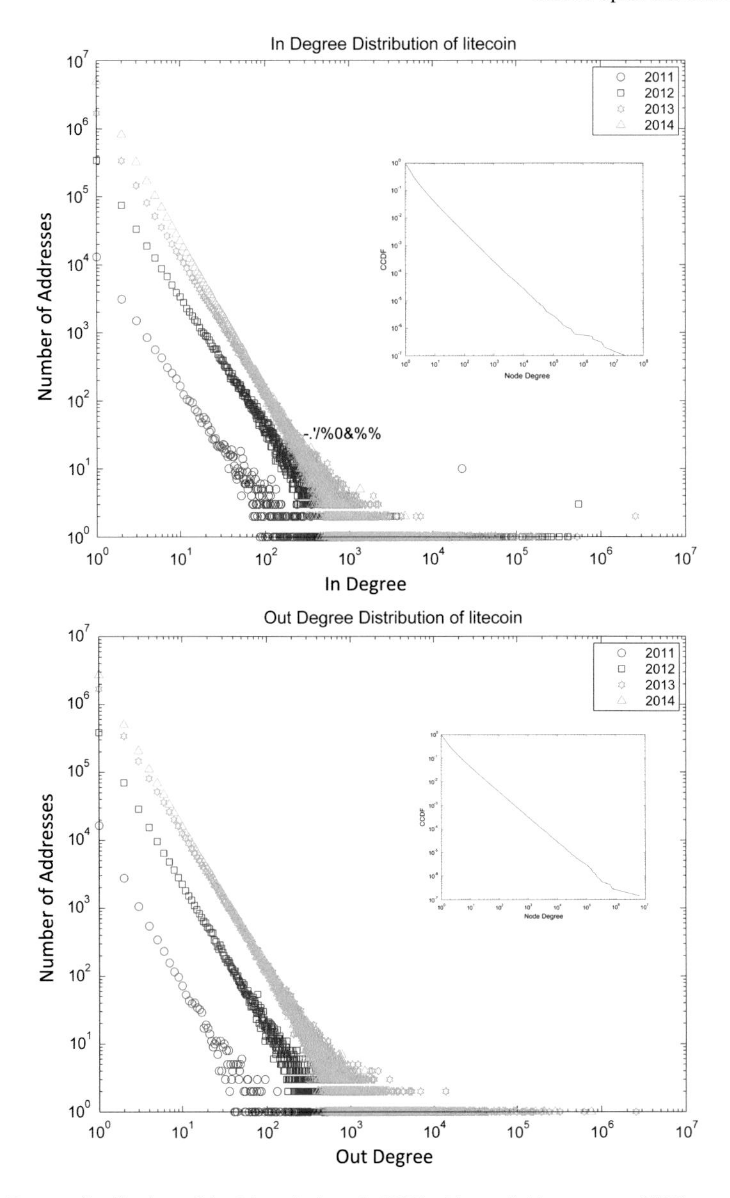

Fig. 7 Degree distribution of the Litecoin (yearly PDF with overlaid aggregate CDF)

in Fig. 8. We observe an interesting pattern among the richest Litecoin users where more than two thirds of the 100 richest nodes simply transfer their Litecoins into a new account while paying a small transaction fee. This can be an indication that those accounts belong to a single user.

Table 4 Network characteristics of yearly Litecoin transaction networks

Year	2011	2012	2013	2014
Assortativity	−0.036	−0.027	−0.015	−0.000
Clustering	0.33	0.18	0.062	0.038

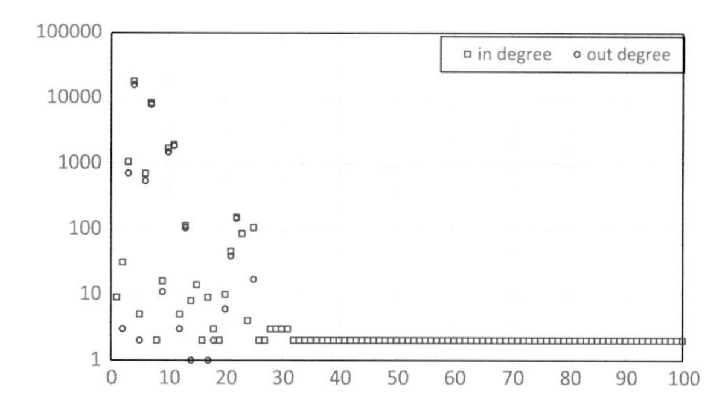

Fig. 8 Degrees of the richest 100 users

4 Conclusions

We have performed a detailed analysis of the two popular digital currencies, i.e., Bitcoin and Litecoin. After becoming popular after 2011, Bitcoin is characterized by a dissasortative degree correlation and power law in- and out-degree distributions. Litecoin network has disassortative degree correlation and power law in- and out-degree distributions since inception in 2011. The characteristics of richest nodes in Bitcoin and Litecoin are similar. We also found that majority of the richest nodes are interested in just accumulating money.

Acknowledgments This material is based upon work supported by the National Science Foundation under grant number EPS- IIA-1301726.

References

1. Cheung, D.P., Gunes, M.H.: A complex network analysis of the United States air transportation. In: IEEE/ACM ASONAM, pp. 699–701. Washington, DC, USA (2012)
2. Crosley, G., Gunes, M.H.: Using complex network representation to identify important structural components of Chinese characters. In: Complex Networks, pp. 319–328 (2014)
3. Dittrich, A., Gunes, M.H., Dascalu, S.: Network analysis of software repositories: identifying subject matter experts. In: Complex Networks, pp. 187–198 (2013)
4. Erdin, E., Zachor, C., Gunes, M.H.: How to find hidden users: a survey of attacks on anonymity networks. IEEE Commun. Surv. Tut. **17**(4), 2296–2316 (2015)
5. Kardes, H., Sevincer, A., Gunes, M.H., Yuksel, M.: Six degrees of separation among US researchers. In: IEEE/ACM SONAM, pp. 654–659 (2012)
6. Komurov, K., Gunes, M.H., White, M.A.: Fine-scale dissection of functional protein network organization by statistical network analysis. PLoS ONE **4**(6), e6017 (2009)

7. Kondor, D., Psfai, M., Csabai, I., Vattay, G.: Do the rich get richer? an empirical analysis of the bitcoin transaction network. PLoS ONE **9**(2), e86197 (2014)
8. Li, B., Erdin, E., Gunes, M.H., Bebis, G., Shipley, T.: An overview of anonymity technology usage. Comput. Commun. **36**(12), 1269–1283 (2013)
9. Menche, J., Valleriani, A., Lipowsky, R.: Asymptotic properties of degree-correlated scale-free networks. Phys. Rev. E **81**(4), 046103 (2010)
10. Nakamoto, S.: Bitcoin: A peer-to-peer electronic cash system. White Paper **1**(2), 1–9 (2008)
11. Naruchitparames, J., Gunes, M.H., Louis, S.J.: Friend recommendations in social networks using genetic algorithms and network topology. In: IEEE CEC, pp. 2207–2214 (2011)
12. Newman, M.: Networks: An Introduction. Oxford University Press Inc, New York, NY, USA (2010)
13. Ober, M., Katzenbeisser, S., Hamacher, K.: Structure and anonymity of the bitcoin transaction graph. Future Internet **5**(2), 237–250 (2013)
14. Shoaib, M., Ilyas, M., Khiyal, M.S.H.: Official digital currency. In: ICDIM, pp. 346–352 (2013)
15. Tian, G., Gunes, M.H.: Complex network analysis of ozone transport. In: Complex Networks, pp. 87–96 (2014)

Circadian Patterns on Wikipedia Edits

Y. Gandica, R. Lambiotte, T. Carletti, F. Sampaio dos Aidos and J. Carvalho

Abstract Cyclic behaviour and circadian patterns emerging from the editing activity of Wikipedia are hereby considered. Such patterns affect many human activities, mobility routes, energy storage and synchronization, among others. Because the editing of Wikipedia is the result of a voluntary process made by many independent human beings, the question about the signature of such circadian patterns on such data is not straightforward. We however show in this work that Wikipedia editing presents well defined periodic patterns with respect to daily, weekly and monthly activity. In addition, we also show the periodic nature of the number of inter-event in time. The results of our work shed some light on the activity scheduling present in our society, contributing to the circadian patterns understanding.

1 Introduction

The success of research in digital social patterns hinges on the access to high quality data. Even though the availability of recorded data and its accessibility are rapidly increasing, many data sets are not freely available for research. Wikipedia (WP) is an important exception, as not only it is considered a robust and trustworthy source of

Y. Gandica (✉) · R. Lambiotte · T. Carletti
Department of Mathematics and Namur Center for Complex Systems—naXys,
University of Namur, rempart de la Vierge 8, 5000 Namur, Belgium
e-mail: ygandica@gmail.com

R. Lambiotte
e-mail: renaud.lambiotte@unamur.be

T. Carletti
e-mail: timoteo.carletti@unamur.be

F.S. dos Aidos · J. Carvalho
CFisUC, Department of Physics, University of Coimbra, 3004-516 Coimbra, Portugal
e-mail: aidos@teor.fis.uc.pt

J. Carvalho
e-mail: joaoclcarvalho@gmail.com

H. Cherifi et al. (eds.), *Complex Networks VII*, Studies in Computational Intelligence 644, DOI 10.1007/978-3-319-30569-1_22

information [1] but it also is accessible by anyone with a connection to the internet. This platform also offers to anyone all its past editing record.

The high quality of the WP encyclopedia is the result of a collective effort by millions of volunteers in an apparently disorganized process of editing, acceptance, and rejection, which works as an effective and robust peer review procedure [2]. Some editors spend a lot of time, either editing a single page on a specific subject, or editing several pages over a diversity of subjects. It has been shown [3] that some statistical aspects of WP editing can be reproduced by using three simple mechanisms: preferential attachment, that represents the editors' ownership feeling, displayed by a strong tendency of users to improve and defend their previous contributions [2]; a fitness parameter, that describes the greater or smaller predisposition for users to edit, which may be caused by the authority of editors who are experts on the page topic, or merely by their personality traits [4–6], and an aging factor [7, 8], describing the time-dependent behavior, with an initial high motivation to edit, followed by a tendency to decrease the editing activity due to theme completeness, personal saturation, blockage [9], and/or any other possible personal cause. From the last mechanisms, if we analyze the data coming from editors with long editing activity, the condition of fitness is satisfied and the system is in the regime before time saturation. In this sense, we can study the general behavior when authors are mostly influenced by the ownership feeling over the WP page.

In this communication we propose to explore if, under this condition, the data coming from WP editing exhibits circadian patterns. By circadian patterns we mean any kind of regular activity, in terms of days, weeks or even months and seasons. Human beings are constrained by biological circadian cycles. However the society duties generate other uniformities in our daily activities. Currently, one important aspect of circadian sequence studies comes from research on human bursty behaviour. Some authors have suggested that circadian patterns are one of the main causes for the fat-tailed nature of inter-event distributions [10, 11].

The issue of circadian patterns is, by itself, of great interest. Beyond obvious concerns for medicine and biology, this subject has implications regarding mobility routes, energy storage and synchronization, among others. Some previous studies from data gathering of mobile calls, emails, web-pages, etc., have shown that they exhibit circadian patterns. The existence of such patterns in WP editing, however, cannot be taken for granted, due to the voluntary character of the editing process. Still, we show in this work that WP editing does present well defined circadian patterns. As a result, the WP database can be used to address several interesting current issues regarding the cyclic nature of human activity.

This article is structured as following: in Sect. 2 we explain the source of the data sample and in Sect. 3 the data cyclic nature is studied. In Sect. 4 we show the circadian patterns on data, while in Sect. 5 the circadian patterns of the inter-event probabilities are shown. Finally in Sect. 6 we extract our conclusions.

2 Data

Our data sample is a database of WP edits, of pages written in English in the period of about 10 years ending in January 2010; this dump containing 4.64×10^6 pages [13]. For each entry we have the WP page name, the edit time stamp and the identification of the editor who did the changes. We analyze the information separately for each page and for each editor, and we reduce the impact of outliers by eliminating the pages or editors with less than 2000 edits. This number is a good compromise between having enough pages/editors and the pages being frequently updated or the editor being reasonably active in this time interval. We removed from the data the edits made by WP-bots, which are programs that go through the WP, carrying out automatically repetitive and mundane tasks to maintain the WP pages (as software programs, their edit pattern is different from the humans'). Moreover, we only considered the editors who logged in with a username before editing, in order to univocally identify the editor; in this way, we discarded the entries associated to IP numbers.

3 Cyclic Nature

A natural way to study periodicity on data is by means of a Fourier power spectrum. For this purpose, we selected the 100 most active editors and computed the Fourier spectrum of each editor's time activity. Results are reported in the upper panels of Fig. 1, where we show the power spectrum for three representative editors. In general, editors have the main power peak at $\sim 1.157 \times 10^{-5}$ Hz corresponding to a period of 24 h and a second peak at $\sim 2.315 \times 10^{-5}$ Hz, matching a 12 h period, a harmonic from the main frequency. In the bottom panels of the same figure we show the result of the same procedure applied to pages. We found that, in general, pages lack predominant power peaks. They appear only in pages related to records (for example births/deaths counting) or companies, which in general update their pages daily. In the bottom panels of Fig. 1 we show one sample of a regular page (center), one of a company (right) and one of death counting (left).

4 Circadian Event Patterns

In order to better illustrate these circadian patterns, we show in Fig. 2 the average number of edits on a day for the four most active editors as a function of the time of day at which the edit is recorded. To build this plot we divided the whole time span of our sample, about ten years, into days and for each day we computed the average number of edits done in windows whose duration is 30 min. We can notice that these editors are characterized by a high editing activity. Closely inspecting the data we found that these editors handle several pages and play the role of administrators. It is

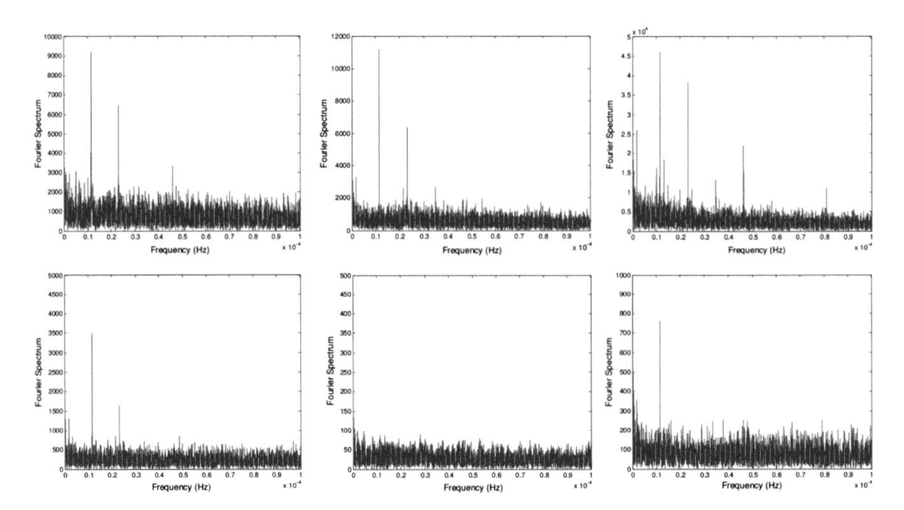

Fig. 1 Fourier power spectrum computed using a fast Fourier transform for three editors (*upper panels*) and for three pages (*bottom panels*) showing the regular behavior found in each group. In general, the highest power peak was found for editors, corresponding to a period of about 24 h ($f \sim 0.1 \times 10^{-4}$ Hz), and a second, smaller, peak corresponding to a period of about 12 h ($f \sim 0.2 \times 10^{-4}$ Hz), a harmonic of the first frequency. Peaks were not found for pages that are not related to special events or companies, which in general update their pages daily. We show two examples (*left* and *right panels*) of these special cases in this figure

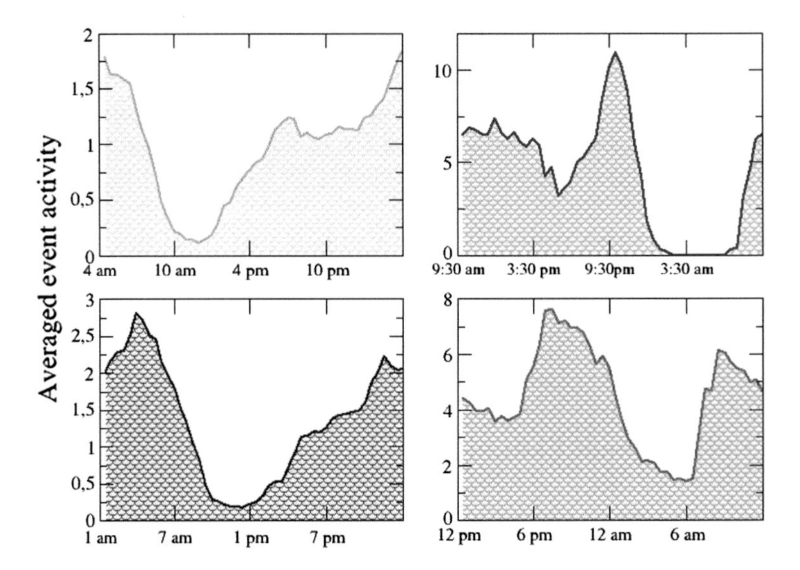

Fig. 2 Distribution of editing pattern along the day, averaged over the data time span (about ten years), for four editors. The time is given by the WP servers and the editors can be in different time zones, which can also be different from the WP servers time. The sleeping/resting hours are visible and also the one or two peaks of more active editing. (Figure taken from [14] with the permission of the authors)

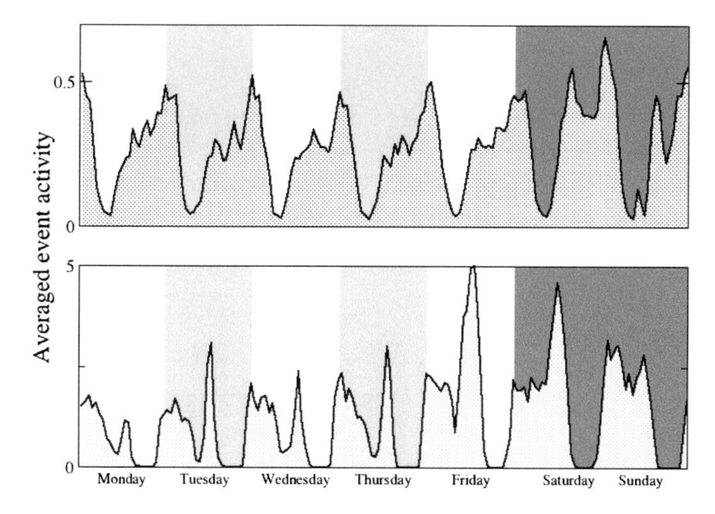

Fig. 3 Editing activity as a function of the week day, averaged over the data time span (about ten years), for two editors. Circadian patterns are clearly visible. The highest activity peak can switch between morning and evening, depending on the week day. In general, the week-ends have different editing activity. (Figure taken from [14] with the permission of the authors)

also clear that each editor has a well defined activity schedule with peaks at particular times.

In Fig. 3 we report the weekly editing activity. The circadian patterns are clearly visible. Once again each editor has a characteristic activity pattern. The highest activity peak can switch between mornings and evenings, depending on the day. In the process of WP editing, the change of activity patterns on week-ends is clear.

Finally, as some countries are strongly influenced by seasonal changes, we are interested in understanding whether WP editors are also influenced by the life style adopted in each season. In Fig. 4 we show the activity over the year, plotting the total number of edits for the whole data time span. We show the total number of edits in each month (stars) and in each day of the year (filled color) for the most active WP editors. We can see that the editors in the two left panels have two peaks, in June and in December, while the editors in the right panels only display a peak in December.

5 Daily Inter-event Patterns

The circadian patterns in the editing probability are somehow expected. Any regular activity done by a human being has a probability that depends on the time of day, the day of the week and time of the year. The editing of WP is also affected by these probabilities, as was shown in Figs. 2 and 3. When addressing circadian patterns, researchers concentrate mainly in the event probability. However, another relevant question is how are the inter-event times affected by the human circadian nature.

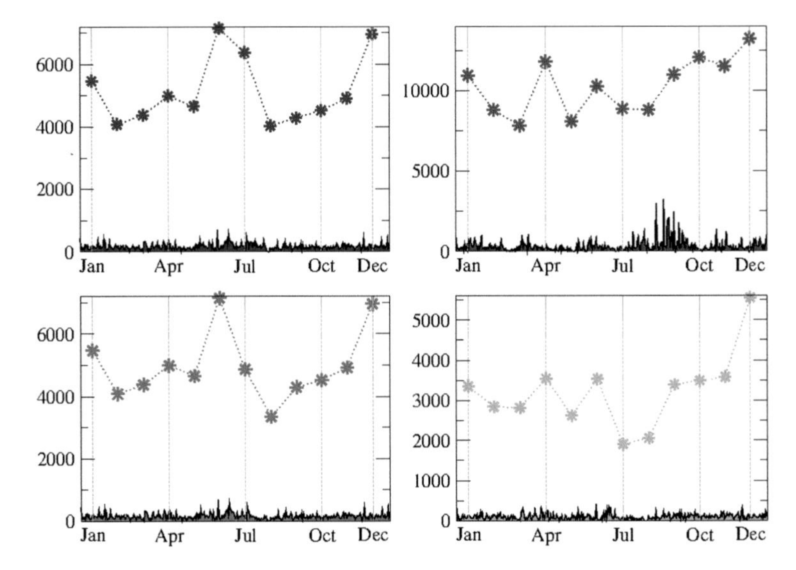

Fig. 4 Number of edits along the year, per month (shown by *stars*) and per day (with *filled color*), for four of the most active editors, over the data time span (about ten years)

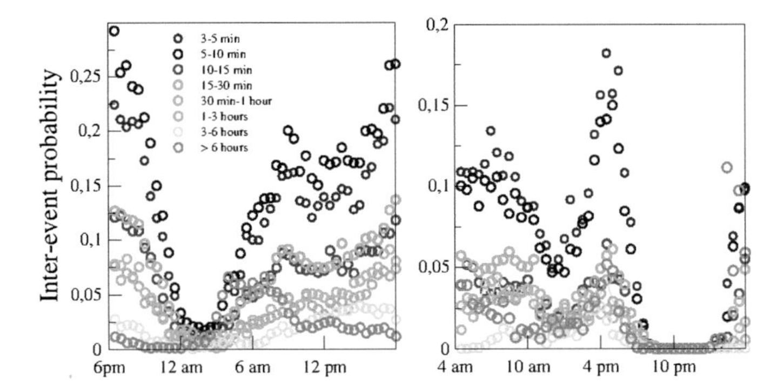

Fig. 5 Distributions of inter-event for edits starting at different hours of the day, averaged over the data time interval (about 10 years). Each color corresponds to a specific interval of inter-events. There is also a structure regarding the inter-event activity depending on the hour in a day. The editors are the same as in Fig. 3

To answer this question we show in Fig. 5 the inter-event time probability along the day. Each color in the plot represents the probability to have inter-event times inside a particular range of values. It is apparent that each range of the averaged fraction of inter-events follows a similar pattern, according to the specific editor's circadian cycle. This means that the activities in different time scales are affected in a similar way by the intrinsic circadian pattern of these editors.

6 Conclusions

Cyclic nature and circadian patterns have been studied over the editing activity of WP. The fast Fourier transform was used to test cyclic behavior in data for several editors on one WP page and over the activity of one editor over several WP pages. We found that, in general, the editors have the main power peak at $\sim 1.157 \times 10^{-5}$ Hz corresponding to a period of 24 h and a second peak at $\sim 2.315 \times 10^{-5}$ Hz (12 h period), a harmonic of the main frequency. On the other hand, page editing does not show predominant power peaks. These peaks appear in pages only if they are related to records (for example births/deaths counting) or companies, which in general update their pages daily. Clear circadian patterns were visible when plotting the averaged editing activity along the day and along the week. Along the year the intensity of activity seems conditioned by holidays. In addition, circadian patterns over inter-event activity were also reported, showing that the activities in different time scales are affected in a similar way by the intrinsic circadian cycle of the editors. It was compelling to check how the human activity patterns obtained from this online data-base are conditioned by social constraints. The only biological constraint that we found, however, was a crucial one: resting time.

Acknowledgments The work of Y.G, R.L and T.C. presents research results of the Belgian Network DYSCO (Dynamical Systems, Control, and Optimization), funded by the Interuniversity Attraction Poles Programme, initiated by the Belgian State, Science Policy Office. The scientific responsibility rests with its author(s).

References

1. Giles, J.: Internet encyclopaedias go head to head. Nature **438**, 900 (2005)
2. Halfaker, A., Kittur, A., Kraut, R., Riedl, J.: A jury of your peers: quality, experience and ownership in Wikipedia. In: WikiSym09, article no. 15
3. Gandica, Y., Carvalho, J., Dos Aidos F.S.: Wikipedia editing dynamics. Phys. Rev. E **91**, 012824 (2015)
4. Kimmons, R.M.: Understanding collaboration in Wikipedia. First Monday **16**, 12 (2011)
5. Bianconi, G., Barabási, A.-L.: Competition and multiscaling in evolving networks. EPL **54**, 436 (2001)
6. Dorogovtsev, S.N., Mendes, J.F.F.: Scaling properties of scale-free evolving networks: continuous approach. Phys. Rev. E **63**, 056125 (2001)
7. Vicent, N.L., Gómez, Z., Kappen, H.J., Kaltenbrunner, A.: World Wide Web **16**, 645 (2013)
8. Dorogovtsev, S.N., Mendes, J.F.F.: Evolution of networks with aging of sites. Phys. Rev. E **62**, 1842 (2000)
9. Javanmardi, S., Lopes, C., Baldi, P.: Modeling User Reputation in Wikis. Stat. Anal. Data Min.: ASA Data Sci. J. **3**, 126 (2010)
10. Malmgren, R.D., et al.: Poissonian explanation for heavy tails in e-mail communication. PNAS **105**(47), 18153 (2008)
11. Malmgren, R.D., et al.: On universality in human correspondence activity. Science **325**, 1696 (2009)

12. Hang-Hyun, J., et al.: Circadian pattern and burstiness in mobile phone communication. New J. Phys. **14**, 013055 (2012)
13. http://wwm.phy.bme.hu/
14. Gandica, Y., Carvalho, J., Aidos, F.S., Lambiotte, R., Carletti, T.: On the origin of burstiness in human behavior: the wikipedia edits case (2016)

Lyric-Based Music Recommendation

Derek Gossi and Mehmet H. Gunes

Abstract Traditional music recommendation systems rely on collaborative filtering to recommend songs or artists. This is computationally efficient and performs well method but is not effective when there is limited or no user input. For these cases, it may be useful to consider content-based recommendation. This paper considers a content-based recommendation system based on lyrical data. We compare a complex network of lyrical recommendations to an equivalent collaborative filtering network. We used user generated tag data from Last.fm to produce 23 subgraphs of each network based on tag categories representing musical genre, mood, and gender of vocalist. We analyzed these subgraphs to determine how recommendations within each network tend to stay within tag categories. Finally, we compared the lyrical recommendations to the collaborative filtering recommendations to determine how well lyrical recommendations perform. We see that the lyrical network is significantly more clustered within tag categories than the collaborative filtering network, particularly within small musical niches, and recommendations based on lyrics alone perform 12.6 times better than random recommendations.

1 Introduction

Due to the proliferation of music streaming and subscription services in recent years, there has been increasing interest in determining how various songs and artists are connected to one another, and ultimately to a given listener. The goal in analyzing a musical network in this context is often to recommend songs or artists to a person who has a list of songs or artists that they are already known to enjoy listening to. This is a difficult challenge in many ways, especially when one considers the unique subtleties present in musical expression. Often, preferences for a given listener span a wide range of genres and styles, making labeling the data solely in this manner

D. Gossi
Mathematics & Statistics, University of Nevada, Reno, Nevada, USA

M.H. Gunes (✉)
Computer Science and Engineering, University of Nevada, Reno, Nevada, USA
e-mail: mgunes@unr.edu

H. Cherifi et al. (eds.), *Complex Networks VII*, Studies in Computational Intelligence 644, DOI 10.1007/978-3-319-30569-1_23

insufficient in the recommendation task. Up to this point, the methodology that has been employed for the recommendation task in an industrial setting has been largely focused on linking listeners together by preferences using a collaborative filtering method, i.e. if user 1 likes songs A, B, and C, and user 2 likes songs A, B, and D, the recommender might recommend song D to user 1, and song C to user 2. However, this methodology is an indirect approach, as it does not use actual musical or lyrical content. In particular, collaborative filtering recommendation systems do not scale well to new or existing entries suffering from a lack of user ratings.

A growing area of the literature in musical analysis is focused on using audio and lyrical features to classify artists and songs [4–7]. This methodology goes beyond user preference lists and attempts to find relationships amongst the songs and artists. This, ideally, would lead to stronger recommendation engines, as well as a more thorough understanding of the music. The idea is to find factors beyond genre that influence a given listener's probability of enjoying a song. These could be related to tempo, mood, production style, use in a dance setting, or lyrical sentiment. While analysis of the full audio wave data has the most potential to improve recommendation systems, the success of existing algorithms is limited when given the task of classifying this complex unstructured data. Another option is to utilize song lyrics to connect various songs and artists, where lyrics are available. This presents its own unique challenges, as accurate lyrical analysis involves being able to decipher subtleties such as irony, hyperbole, and ambiguity. Even disregarding the complexities of lyrical analysis, the lyrics in conjunction with the music provide another layer of complexity. For example, a given set of lyrics over a slow and maudlin musical background may be open to a completely different interpretation than the same set of lyrics over a fast-paced and energetic musical background. However, even considering these challenges, lyrical analysis has proved in the literature to be a worthwhile endeavor.

Research in the area of lyrical analysis has grown in recent years with the increasing availability of large datasets to train algorithms. Work in this area has tended to focus on classification of lyrical content into categories such as mood using labeled training data [3]. More recent work has separated lyrical analysis from standard text analysis, by showing the importance of rhyme, repetition, and meter. However, the research beyond standard text analysis tools has been fairly minimal. The release of the Million Song Dataset (MDS), by far the largest dataset available to use in research of this type, has begun a new wave of analysis [2]. What has been missing from the literature is a complex network approach to lyrical analysis where the topology of the network is defined by artists linked together by lyrical similarity. This paper uses lyrical network and compares clustering methods on these networks. While certain genres such as pop and rock have been notoriously difficult to separate, even for human analyzers, with the aid of the MSD there may be new possibilities for genre clustering and defining the edges.

2 Methodology

Lyrics Dataset: Lyrical data is provided by the musiXmatch Dataset (MXD) [1], which provides lyrics for 237,662 tracks and 22,821 unique artists (implying 10.4 average number of tracks per artist in the dataset), which are each directly linked to the MSD. This is, of course, a small subset of the full MSD. The remaining tracks were omitted due to either copyright restrictions, a given track not containing lyrics, or duplication. The lyrics for each track are provided in a BOW format, and are stemmed using a modified version the Porter2 stemming algorithm. Thus, words that are similar in a statistical sense, e.g. kneel, kneeled, and kneeling, are mapped to the same stem (in this case, "kneel"), and are treated as identical terms in the analysis. The total dictionary of terms used as features in this analysis is limited to the top 5,000 words present in the dataset, which accounts for approximately 92 % of the complete set of unique words. This limited dictionary is chosen because many of the term features outside this list are noisy and unusable, or used too infrequently in the dataset to be of much statistical value.

Term Frequency Matrix: A vector space model (VSM) is implemented to represent the lyrical data. While the BOW format loses valuable information about the location of words relative to each other within a given song, it is a convenient statistical tool as it allows us to define a song as a vector along a finite dictionary of terms, where each component in the vector represents the frequency of a given word. Once each song is vectorized in this manner, we have a sparse term frequency matrix of size $n \times 5000$, where $n = 237{,}662$ songs in the dataset. Further, we create a similar reduced term frequency matrix of artists by adding frequency vectors across a given artist's song catalog within the dataset, resulting in a sparse $n_a \times 5000$ matrix, where $n_a = 22{,}821$ unique artists in the dataset. This "summed" artist matrix represents the full dictionary of words used by a given artist in the dataset.

Term Frequency-Inverse Document Frequency (TF-IDF) Weighting: The term frequency matrix of artists, while modeling the lyrical data of each artist as elements in a common vector space, suffers from the fact that "unimportant" and "important" words are weighted similarly. This is remedied by first eliminating statistically unimportant stop words such as *the, is, at, which,* and *on*. Secondly, term frequency-inverse document frequency weighting (TF-IDF) is utilized to minimize the importance of common words occurring frequently in the dataset. Formally, we multiply the raw term frequency by the inverse document frequency $IDF(w) = \log\left(\frac{|A|}{|a \in A : w \in a| + 1}\right)$ where A is the set of all artists in the dataset, and $a \in A$. Calculating TF-IDF weights for each element in the term frequency matrix results in an adjusted sparse matrix in the same vector space as the term frequency matrix.

Pairwise Similarity Matrix: Pairwise similarity between artists is calculated using cosine similarity. If x and y are artist TF-IDF vectors, then their cosine similarity is $C\left(\underline{x}, \underline{y}\right) = \frac{\underline{x}\underline{y}^T}{||\underline{x}||\,\left\|\underline{y}\right\|}$. Computing the cosine similarity for each artist vector results in an $n_a \; x \; n_a$ pairwise similarity matrix where 0 implies the two vectors are orthogonal and completely dissimilar, and 1 implies two artists are identical, in terms of frequency and types of words used in the lyrics.

Threshold Selection: From the similarity matrix, we are able to define which pairs of artists are "similar" to each other in the lyrical sense. Rather than applying a similarity threshold directly to similarity matrix to obtain the set of edges, we utilize a *k nearest neighbor* approach to emulate what would be seen in a traditional collaborative filtering recommendation network. The k nearest neighbors of a given artist are the k artists which would be recommended to a user given a known preference to the artist in question. This fixes the outdegree for every node in the dataset at k, while the indegree is of unknown distribution. This resembles real-world limitations of recommendation networks, as explored in [3]. The level k is chosen to be 10 for this analysis, a level deemed sufficient enough to collect a useful range of connections while not going beyond the practical limitations of a real-world recommendation network. However, it should be noted that there is no true empirical justification for this choice of k.

Collaborative Filtering Network: We compare the topology of the network defined by lyrical similarity to that of a traditional collaborative filtering approach. The Echo Nest Taste Profile Subset (ENTPS) [1], provided as part of the Million Song Dataset Challenge, includes data on the number of times a given user has listened to a song in the MSD. The data includes 1,019,318 unique users with 48,373,586 user/song/count triples. With this dataset, a traditional collaborative filtering network can be defined by utilizing memory-based filtering using songs rather than users, i.e. item-based collaborative filtering. We begin by vectorizing data for a given item—in this case a song—where the *i* component in the vector represents the number of plays by user i. We compute pairwise cosine similarity for each song in the dataset where the song vectors belong to the vector space of users. Once the pairwise similarity matrix is generated, we compute the edges of the network using the *k nearest neighbors* approach, with k = 10, in a similar manner to the lyrics network. The lyrics network and the collaborative filtering network are then reduced to 18,290 unique artists shared by both datasets, implying 80.2 % of the original 22,821 unique artists in the lyrics dataset also have user play count data in the ENTPS. With each node having outdegree of 10, each of the two artist networks has 18,290 nodes and 182,900 edges.

Tag Data: To enhance analysis of the recommendation networks, we link user-generated tag data to artists in the network by utilizing the Last.fm dataset [1]. The Last.fm data is linked to the MSD and includes 522,366 unique tags and 505,216 tracks with at least one tag. We significantly reduce the tag set to the most general and descriptive tag categories. The 500 top tags in the dataset are grouped into relevant tag categories representing a unique musical genre, mood, or gender. Table 1 presents the unique tags in the Last.fm dataset that are used as tag categories. Overall, 23 tag categories are considered. 15 of the 23 tag categories represent musical genre. Four mood categories are considered. For simplicity, several of the genre groups include groups of similar, but technically different, genres. For example, soul, R&B, and funk are grouped together, while in actuality they are distinct genres.

Table 1 Last.fm tag categories

Category	Type	Unique tags included
Rock	Genre	Rock, Classic R., Hard R., Progressive R., Pop R., Soft R., Rock n Roll
Pop	Genre	Pop, Pop Rock
Alternative	Genre	Alternative, Alternative Rock
Indie	Genre+	Indie, Indie Rock, Indie Pop
Electronic	Genre	Electronic, Electronica, Electro, House, Trance, Techno, Progressive Trance
Dance	Genre+	Dance, party, club
Jazz	Genre	Jazz, Jazzy
Folk	Genre	Singer-Songwriter, Folk, Acoustic, Folk Rock, Singer Songwriter,
Metal	Genre	Metal, Heavy M., Death M., Progressive M., Black M., Power M., Gothic M., Melodic M., Doom M., Thrash M., Metalcore, Nu M.
Soul	Genre	Soul, RnB, Funk, R&B, RB, R and B
Hip Hop	Genre	Hip-Hop, Hip Hop, Rap, Hiphop
Punk	Genre	Punk, Punk Rock
Blues	Genre	Blues, Blues Rock
Country	Genre	Country, Classic Country
Reggae	Genre	Reggae
Latin	Genre	Latin, Spanish, Latino
Christian	Genre	Christian, Worship
Relaxing	Mood	Chillout, Mellow, Chill, Relax, Relaxing, Calm, Chill Out
Romantic	Mood	Love Songs, Love Song, Sensual, Sex, Sexy
Positive	Mood	Fun, Happy, Upbeat, Energetic, Uplifting, Feel Good, Energy, Positive
Negative	Mood	Sad, Melancholy, Melancholic, Dark, Moody, Bittersweet
Male	Gender	Male Vocalists, Male Vocalist, Male Vocals, Male
Female	Gender	Female Vocalists, Female Vocalist, Female, Female Vocals, Female Vocal

Subgraph Analysis: To determine how each network is structured within certain musical communities, subgraphs are generated and analyzed for each of the 23 tag categories. Subgraphs are generated by limiting the artist set within a given subgraph to only artists featuring a tag from the tag category associated with that subgraph. By analyzing the number of edges remaining in each subgraph compared to the number of edges "leaving" the subgraph, we can see if recommendations within each network tend to stay within certain tag categories, or certain niches within the full artist set. Formally, we compare the actual number of edges in each subgraph to the number of edges that would remain in the subgraph given that the 10 outgoing edges from each node in the subgraph are distributed to random nodes in the network. The ratio to the actual number of edges in the subgraph to the expected randomly distributed number of edges in the subgraph will indicate how much recommendations tend to remain in a given tag category. The expected number of edges remaining in the subgraph given random edge distribution is calculated as follows. Given every node in the graph has outdegree of k, say node s is inside the subgraph S of full node set N. Then the expected number of edges from s going to other nodes in S is $\sum_{i=1}^{k} i \frac{\binom{|S|-1}{i}\binom{|N \setminus S|}{k-i}}{\binom{|N|-1}{k}}$. If edges are chosen randomly, each $s \in S$ selects edges independently. Expected number of edges remaining in the subgraph S is $|S| \sum_{i=1}^{k} i \frac{\binom{|S|-1}{i}\binom{|N \setminus S|}{k-i}}{\binom{|N|-1}{k}}$. For large N, this is approximately $|S| K \frac{|S|}{|N|}$.

Comparison of the Recommendation Task: We compare the ranked recommendation lists for both networks, assuming that the collaborative filtering network represents the "true" rankings. For each artist, we consider the ranked list of the top 1,000 most similar artists using both lyrical similarity and user similarity metrics, and calculate the difference between the two ranked lists used a Rank Biased Overlap (RBO) metric, calculated as follows. Let $CF_i^{(j)}$ represent the set of the first i elements in the ranked list of the collaborative filtering network for artist j. Let $L_i^{(j)}$ be defined in an identical manner for the lyrical network. The RBO of the lyrical and collaborative filtering rankings for artist j is $RBO^{(j)} = \frac{1}{1000} \sum_{i=1}^{1000} \frac{CF_i^{(j)} \cap L_i^{(j)}}{i}$. The mean RBO for the full artist set is the mean of the $RBO^{(j)}$ across all artists j. We also compare the RBO of the collaborative filtering rankings with a set of random rankings for each artist, to see if using lyrical similarity rankings is how much more accurate than simply recommending a set of randomly ordered artists in the dataset.

3 Results

Comparing the general topology of the two networks in Table 2, several differences stand out. The average clustering coefficient of the lyrics network is significantly greater than that of the collaborative filtering network and the average shortest path is also greater. This indicates that the lyrics network, when compared to the collaborative filtering network, tends to be clustered more around certain communities of the network, which would also increase the average shortest path as there are less bridges between communities.

While the average outdegree of both networks is fixed at 10, the indegree distributions are displayed in Figs. 1 and 2. Neither network displays a power law in its indegree distribution, with clear curvature in the log-log plots. The lyrics network

Table 2 Network topology comparison

Network	Diameter	Average shortest path	Clustering coefficient
Lyrics Network	10	4.52	0.217
CF Network	6	4.22	0.119

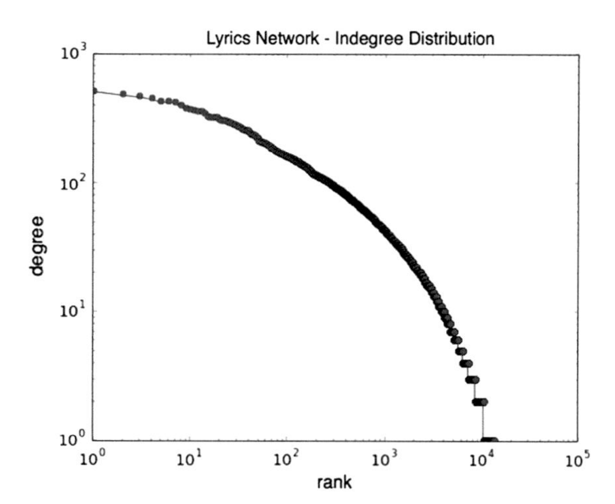

Fig. 1 Lyrics network

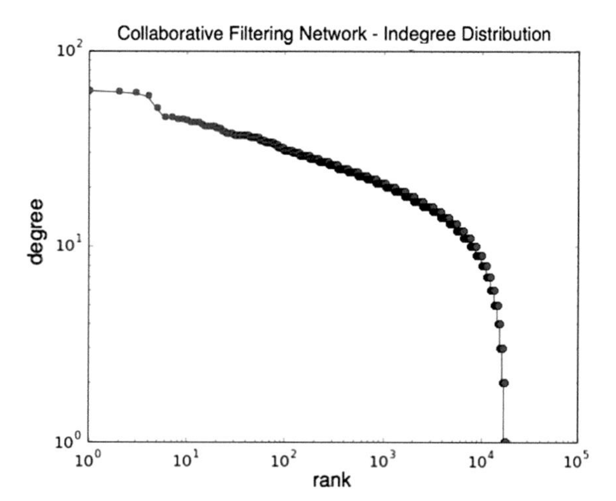

Fig. 2 Collaborative filtering network

is significantly biased than the collaborative filtering network, with the top 10 % of nodes receiving 65.1 % of the possible edges. In comparison, the top 10 % of nodes in the collaborative filtering network only receive 22.6 % of the possible edges.

Results of the subgraph analysis shown in Tables 3 and 4 indicate that the lyrics network is significantly more clustered within certain tag categories than the collaborative filtering network, indicating that users tend to listen to music across a broad spectrum of categories. However, users do not randomly listen to music across tag categories, as every tag category subgraph does have more edges than the expected number of edges given random edge preference. In addition, in both networks as the

Table 3 Lyrics network subgraph analysis

Subgraph	Nodes	Percentage of nodes (%)	Edges	Random edges	Actual/random
Rock	7,622	42	61,658	31,763	1.9
Pop	6,277	34	45,870	21,542	2.1
Relaxing	5,425	30	37,959	16,091	2.4
Alternative	5,126	28	35,333	14,366	2.5
Positive	5,052	28	34,657	13,954	2.5
Male	4,529	25	27,533	11,215	2.5
Romantic	4,416	24	30,058	10,662	2.8
Dance	4,404	24	23,320	10,604	2.2
Indie	4,192	23	23,987	9,608	2.5
Negative	4,135	23	28,643	9,348	3.1
Female	3,617	20	15,538	7,153	2.2
Folk	3,588	20	22,603	7,039	3.2
Electronic	3,536	19	12,981	6,836	1.9
Soul	3,123	17	16,494	5,333	3.1
Metal	2,535	14	14,377	3,514	4.1
Punk	2,282	12	10,715	2,847	3.8
Hip Hop	2,088	11	10,703	2,384	4.5
Jazz	1,955	11	7,040	2,090	3.4
Blues	1,752	10	8,128	1,678	4.8
Country	1,249	7	6,094	853	7.1
Latin	1,117	6	6,675	682	9.8
Reggae	818	4	1,391	366	3.8
Christian	711	4	1,960	276	7.1

size of the node set in a subgraph decreases, the ratio of actual edges to expected random edges increases. This makes intuitive sense, as smaller tag categories indicates more specific niches—these specific niches would tend to have more unique lyrical sets, and listeners of a specific niche of music likely would not venture outside this niche as much as a listener of popular genres or categories.

In the lyrics network, Folk, Metal, Country, Hip Hop, Blues, Latin, and Christian music are particularly strong communities, as would be expected. In the collaborative filtering network, Metal, Country, Latin, Reggae, and Christian music display strong community preference, with Indie music also displaying above average community structure given the size of the artist set. In terms of mood category, negative music is more clustered than positive music lyrically, with ratios of 3.1 and 2.5, respectively, but neither is significantly clustered in terms of user preference. Music with male vocals is slightly more clustered lyrically than music with female vocals.

Table 4 Collaborative filtering network subgraph analysis

Subgraph	Nodes	Percentage of nodes (%)	Edges	Random edges	Actual/random
Rock	7,622	42	39,382	31,763	1.2
Pop	6,277	34	27,180	21,542	1.3
Relaxing	5,425	30	20,880	16,091	1.3
Alternative	5,126	28	20,561	14,366	1.4
Positive	5,052	28	18,124	13,954	1.3
Male	4,529	25	14,849	11,215	1.3
Romantic	4,416	24	14,428	10,662	1.4
Dance	4,404	24	14,198	10,604	1.3
Indie	4,192	23	15,045	9,608	1.6
Negative	4,135	23	12,928	9,348	1.4
Female	3,617	20	9,764	7,153	1.4
Folk	3,588	20	10,357	7,039	1.5
Electronic	3,536	19	10,161	6,836	1.5
Soul	3,123	17	8,159	5,333	1.5
Metal	2,535	14	7,450	3,514	2.1
Punk	2,282	12	5,711	2,847	2.0
Hip Hop	2,088	11	4,598	2,384	1.9
Jazz	1,955	11	3,280	2,090	1.6
Blues	1,752	10	2,799	1,678	1.7
Country	1,249	7	2,484	853	2.9
Latin	1,117	6	1,921	682	2.8
Reggae	818	4	901	366	2.5
Christian	711	4	1,083	276	3.9

Table 5 Recommendation performance comparison

Ranking compared to CF	RBO	Multiple
Lyrical ranking	0.0649	12.6
Random ranking	0.0052	1

Table 5 displays a comparison of collaborative filtering recommendations with lyrical similarity recommendations utilizing the mean RBO metric. This is compared to the mean RBO of collaborative filtering recommendations and random recommendations to provide a baseline. While the lyrical recommendations have a weak mean RBO of 0.0649, it is 12.6 times superior to random recommendations.

4 Conclusions

When actual user data is unavailable, which especially holds true for many new and emerging songs or artists, it may be advantageous to consider content-based recommendation methods in determining the initial recommendations to and from this artist or song. This paper shows that even a purely lyric-based method provides significant information about the tags the artist or song might have associated with it. Lyrical analysis may be especially successful for niche genres such as Country, Metal, Blues, Hip Hop, and Christian—where the lyric network in this analysis was successful at differentiating these genres from others. At the very least, lyrical analysis could verify whether or not a given recommendation in the lack of user data is "very bad," as it could determine how far a user is venturing from their typical listening history.

By adding more elements to the vector space than just TF-IDF vectors for lyrics, such as direct factors for lyrical sentiment, repetition, and variety of word choice one could likely refine the results of this analysis. Adding factors for audio signal content would also produce stronger results.

Acknowledgments This material is based upon work in part supported by the National Science Foundation under grant number EPS- IIA-1301726.

References

1. Bertin-Mahieu, T., Ellis, D.P.W., Whitman, B., Lamere, P.: The million song dataset. In: Proceedings of the ISMIR (2011)
2. The Million Song Dataset Challenge. Kaggle Inc. (2012). http://www.Kaggle.com
3. Cano, P., Celma, O., Koppenberger, M.: The topology of music recommendation networks. Universitat Pompeu Fabra, February 2008
4. Hu, X., Downie, J.S., Ehmann, A.F.: Lyric text mining in music mood classification. University of Illinois at Urbana-Champaign (2009)
5. Xia, Y., Wong, K., Wang, L., Xu, M.: Sentiment vector space model for lyric-based song sentiment classification. Assoc. Comput. Linguist. 133–136 (2008)
6. Maxwell, T.: Exploring the music genome: lyric clustering with heterogeneous features. University of Edinburgh (2007)
7. Macrae, R., Dixon, S.: Ranking lyrics for online search. In: 13th International Society for Music Information Retrieval Conference (ISMIR 2012), pp. 361–366 (2012)
8. Aiolli, F.: A preliminary study on a recommender system for the million song dataset challenge. University of Padova, Italy (2012)

Identifying Key Opinion Leaders in Evolving Co-authorship Networks—A Descriptive Study of a Proxy Variable for Betweenness Centrality

Johannes Putzke and Hideaki Takeda

Abstract Many researchers identify influentials in a network by their betweenness centrality. Whereas betweenness centrality can be calculated in small, static, connected networks, its calculation in complex, large, evolving networks frequently causes some problems. Hence, we propose a proxy variable for a node's betweenness centrality that can be calculated in large, evolving networks. We illustrate our approach using the example of Key Opinion Leader (KOL) identification in an evolving co-authorship network of researchers who have published articles about PCSK9 (a protein that regulates cholesterol levels).

1 Introduction

The analysis of complex networks has become one of the main research topics in contemporary computer science. The analysis of evolving networks has been a particular focus of attention (see [1] for a literature review). In this context, one main research question has been to identify the most important nodes (hubs, influentials) in a network. One of the most prominent measures of a node's importance in a network is a node's *betweenness centrality* [12].

Whereas a node's betweenness centrality can be calculated in small, static, connected networks, the calculation of a node's betweenness centrality in complex, large, evolving networks frequently causes some problems. For example, betweenness centrality can be calculated in connected graphs only. However, the early evolutionary stages of a network are often characterised by a few edges and nodes. Consequently, the corresponding graph consists of many disconnected components (compare [17]), and the measure of betweenness centrality is either undefined (at the whole network level) or can be calculated separately for each of the unconnected components only. Furthermore, calculating betweenness centrality is computationally too costly

J. Putzke (✉) · H. Takeda
National Institute of Informatics, 2-1-2 Hitotsubashi, Chiyoda, Tokyo 101-8430, Japan
e-mail: putzke@nii.ac.jp

H. Takeda
e-mail: takeda@nii.ac.jp

H. Cherifi et al. (eds.), *Complex Networks VII*, Studies in Computational Intelligence 644, DOI 10.1007/978-3-319-30569-1_24

to allow for dynamic analyses in large networks (compare [10, 18] for the execution times of betweenness centrality calculations on commodity machines).[1]

Nevertheless, many application scenarios require some knowledge about the nodes' betweenness centrality in the (early) evolutionary stages of a network. Therefore, in this paper we propose a proxy variable for a node's betweenness centrality that can also be calculated in the early evolutionary stages of a network and that allows for dynamic analyses.

We illustrate our approach using the example of key opinion leader (KOL) identification in PCSK9 research (see Sect. 3, Data). KOLs are physicians and researchers who influence the treatments prescribed by their peers. Pharmaceutical enterprises spend considerable time and effort identifying KOLs and maintaining a good relationship with them. However, to best build relationships, pharmaceutical enterprises have to identify KOLs in the early stages of the emergence of a new research field and track their importance over time. We suppose that KOLs can be identified through their embeddedness in a co-authorship network. In the network, authors serve as nodes, and a tie is assumed between two authors who have co-authored a publication (compare, for example, [20, 21]).

To summarise, our paper has two research objectives. Our main objective is to answer the research question of whether there is a proxy variable for the KOLs' betweenness centrality that can also be calculated in the (early) evolutionary stages of a co-authorship network. However, to answer this question, we first have to identify who the researchers/KOLs are that have the highest betweenness centrality in the PCSK9 co-authorship network.

The remainder of this paper is structured as follows. The next section reviews the related literature. Section 3, Data, introduces the dataset used for our analyses. Section 4 presents our analyses. The last section, Discussion, addresses the theoretical and managerial implications of our work, notes the limitations of this study, and points to further research.

2 Related Work

In this literature review, we particularly focus on two streams of thought. The first is literature about evolutionary network analysis in computer science. An extensive literature review about this kind of work can be found in a recent paper by Aggarwal and Subbian [1]. Hence, a review about this kind of work is beyond the scope of this paper.

The second stream of thought is research that analyses scientific co-authorship by the means of network analysis (e.g. [20, 21, 23]). In this context, it is particularly interesting to highlight papers that examine the evolution of a co-authorship network. For example, Barabási et al. [5] analyse in a seminal paper the small world properties

[1] The execution times are too high even when applying improved algorithms to calculate betweenness centrality (e.g. [3, 6, 30]).

of an evolving co-authorship network (i.e. they examine whether the network has a larger clustering coefficient [28] than expected for a random network and a small average separation/shortest-path-length).

Backstrom et al. [2] examine how communities/groups develop and evolve in networks using data from DBLP.[2] They are particularly interested in determining who will join which community in the future and how people and topics move between communities.

Also, Franceschet [11] uses data from DBLP for his analyses, in which he distinguishes between the *author-paper affiliation network* and the (author) *collaboration network*. Whereas the author-paper affiliation network is a bipartite graph with two types of nodes (authors and papers (and an edge from an author to a paper if the author has written the paper)), the

> collaboration network is an undirected graph obtained from the projection of the author paper affiliation network on the author set of nodes. Nodes of the collaboration network represent authors and there is an edge between two authors if they have collaborated in at least one paper

(p. 1995). Like Franceschet [11], we focus on the authors' collaboration network in this paper.

> Although the collaboration network is a coarser representation with respect to the affiliation network ... [it] is highly informative since many collaboration patterns can be captured by analyzing this form of representation

([11, p. 1995]). For example, Franceschet [11] analyses the temporal evolution of the connectivity of the collaboration network, the distribution of the number of scholar collaborators, network clustering, the average separation distance among scholars, as well as assortativity by the number of collaborators. He finds that the network is a widely connected small world. Furthermore, he finds the distribution of collaboration among scholars to be highly skewed and concentrated (i.e. there are a few highly productive collaborators responsible for a relatively high share of collaborations). However, he finds the network to be resilient to the removal of these highly productive collaborators.

Liu and Xia [17] examine the structure and evolution of the co-authorship network in the interdisciplinary field of "evolution of cooperation". They illustrate how small clusters evolve into a giant component that can be considered as a small-world network.

Whereas most of the studies presented above focus on network topology and macro-level network properties (such as diameter, distance, components, clustering coefficient etc.) [29], Yan and Ding [29] take a different approach by studying micro-level network properties (i.e. centrality measures). Specifically, they examine how authors' centrality measures change over time. Yan and Ding's [29] study is probably the one the most related to our research.

---

[2] DBLP is a database of computer science publications; http://dblp.org, accessed on July 22nd 2015.

Finally, Yang et al. [31] infer a node's future centrality in a network by a *Node Prominence Profile* (NPP). They base the NPP on the principles of preferential attachment [4] and triadic closure [14]. However, in their study, Yang et al. [31] focus on degree centrality. In this work, we intend to infer a node's betweenness centrality from a single, easily available proxy variable.

3 Data

To analyse KOLs in an evolving complex network, we analysed the co-authorship network of researchers who published an article about PCSK9.[3] In the network, authors serve as nodes of the network, and a tie is assumed between two authors who have co-authored at least one publication. To obtain this network, we searched the PubMed/MEDLINE database[4] for all articles that contain the phrase "PCSK9" in any search field. PubMed/MEDLINE is a comprehensive scholarly bibliographic database maintained by the United States National Library of Medicine that also has been used in comparable research (e.g. [23]).

The decision to analyse the PSCK9 network was taken as we wanted to analyse a network with several thousand nodes only to ensure a high data quality and a clear boundary specification of the network (compare [26] for additional advantages of studying a collaboration network of this size). In total, 952 articles were retrieved from the Pubmed/MEDLINE database. These articles were reported to have been written by 4213 authors. (Two of the articles did not provide any authorship information). Since PubMed did not provide a unique researcher ID for many authors, we manually checked the data for inconsistencies. Indeed, many names that were recognised as belonging to different individuals by our system in fact belonged to the same individual. In most of these cases, an author used her or his initial for some of the publications but not others (e.g. we assumed that the names "Abdiche, Yasmina" and "Abdiche, Yasmina N" belong to the same individual; compare [29] for this procedure). After data cleansing, 3905 authors remained in the database.

The average author of these 3905 authors has written 1.742 articles about PCSK9. This number seems rather low. For example, Yan and Ding [29] find in their co-authorship analyses that the average author in the field of "library and information science" has written 2.4 articles. Newman [22] finds numbers between 2.55 and 11.6 articles for different research domains. The low number in the case of PCSK9 research can be explained by the fact that PCSK9 is a rather new, narrow research field and authors also publish papers about different research subjects.

In contrast, the average PCSK9 article has been written by 7.14 authors. This number seems rather high. In comparison, Yan and Ding [29] report that the average

[3]PCSK9 is a protein which regulates LDL cholesterol levels. By blocking PCSK9, cholesterol levels can be brought substantially down. Hence, drugs can be developed that reduce the risk of cardiovascular diseases by blocking PCSK9.

[4]http://www.ncbi.nlm.nih.gov/pubmed, accessed on June 4th 2015.

Table 1 Number of articles per year

Year	Number of articles	Year	Number of articles
1993	1	2009	47
2003	2	2010	61
2004	14	2011	69
2005	18	2012	120
2006	32	2013	140
2007	41	2014	197
2008	58	2015	152

article in the field of library and information science has 1.8 authors. Newman [22] finds numbers between 1.99 (for the field of theoretical high-energy physics) and 8.96 authors per paper (for the field of high energy physics in general). The high number of 7.14 in the case of biomedical research can be explained by the fact that experimental bio-medical research requires a large group of collaborators (similar to those of experimental high-energy physics). For example, large scale clinical trials are conducted by more than 100 people. In this context, the PubMed/MEDLINE database contains a single PCSK9 paper with 186 authors. (However, this number is still low compared with experimental high-energy physics for which Newman [22] reports a single paper with 1,681 authors.)

The articles about PCSK9 were written between 1993 and 2015. However, only a single article about PCSK9 was published in 1993 and none was published between 1994 and 2003. Table 1 lists the number of articles published per year.

4 Analyses

Using this data sample, we calculated a series of network statistics (including betweenness centrality) for all authors in the network using the R libraries "sna" [8] and "igraph" [9], as well as the software Gephi 0.8.2 beta.[5] During the analyses, we aggregated the network from year to year, which means we assumed that a tie between two authors who co-authored a paper in the past endured till the year of analysis (and was not resolved).[6] Consequently, we studied the cumulative network structure in one-year intervals (i.e. the network from 1993 to 2003, the network from 1993 to 2004, the network from 1993 to 2005, and so on) (compare [17]). This approach is also taken by most comparable studies (e.g. [26]).

However, also in 2015 (the last year of our analyses), the network consisted of several components, and it is not meaningful to compare the authors' betweenness

[5]http://gephi.org, accessed on July 14th 2015.

[6]In the analyses, we left out the years 1994–2002, since no papers about PCSK9 were published then.

Table 2 Top-Authors by betweenness centrality

Author	Betweenness centrality	Author	Betweenness centrality
Seidah Nabil G	0.088853334	Wasserman Scott M	0.020952266
Rader Daniel J	0.040243062	Hovingh G Kees	0.020169608
Robinson Jennifer	0.037968118	Boileau Catherine	0.020161673
Humphries Steve E	0.031357795	Zelcer Noam	0.019938673
Boerwinkle Eric	0.029708067	Jukema J Wouter	0.019929378
Kathiresan Sekar	0.027206458	Park Sahng Wook	0.019409024
Lambert Gilles	0.026681915	Horton Jay D	0.017328605
Thompson John R	0.026023458	Cariou Bertrand	0.015712718
Konrad Robert J	0.023464245	Stein Evan A	0.013969696
Davis Harry R Jr	0.022314041	Ballantyne Christie M	0.013736185
Kastelein John J P	0.021815021	Averna Maurizio	0.013704782
Chen Wei	0.021738187	Rabes Jean-Pierre	0.013482347
Hubbard Brian	0.021491985		

centrality scores between components of different size.[7] Therefore, we decided to focus our further analyses on all authors that belong to the main component of the network in 2015.[8] The main component of the network comprised 2,836 authors (i.e. 72.62 % of all authors).[9] Between these authors were 43,183 edges (i.e. 91.21 % of all edges). Focusing on the main component of a network for further analyses is a valid approach taken by many comparable studies (e.g. [19, 29]).

Table 2 shows the 25-top-authors of this main component by their betweenness centrality. Interviews with managers responsible for PCSK9 from the pharmaceutical industry confirmed that most of these authors are among the most influential people in PCSK9 research.[10]

The main research question of this paper was whether these authors can be identified in the early evolutionary stages of a network by taking a variable as a proxy for the authors' betweenness centrality in the future.

In this paper, we propose the number of an author's unclosed triads as a proxy for her or his betweenness centrality. Figure 1 illustrates an unclosed triad. In the unclosed triad, author A has published a paper jointly with node B and another paper jointly with node C. However, node B and node C have not published any

[7] Although Freeman [13] proposed a standardised measure of betweenness centrality that can theoretically be used for comparing centrality scores between components of different size, we think that it is, for example, not meaningful to compare the maximal betweenness centrality of a node in a component with three actors to that of a node in the main component of a co-authorship network.

[8] The main component of a network is also sometimes referred to as the "giant component".

[9] There were no other meaningful big components in the network. For example, the second (third) biggest component in the network comprised 1.23 % (0.69 %) of all authors.

[10] These influential people include basic researchers as well as researchers conducting clinical trials. Hence, some context knowledge is helpful for reading the tables.

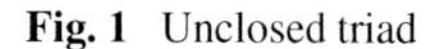

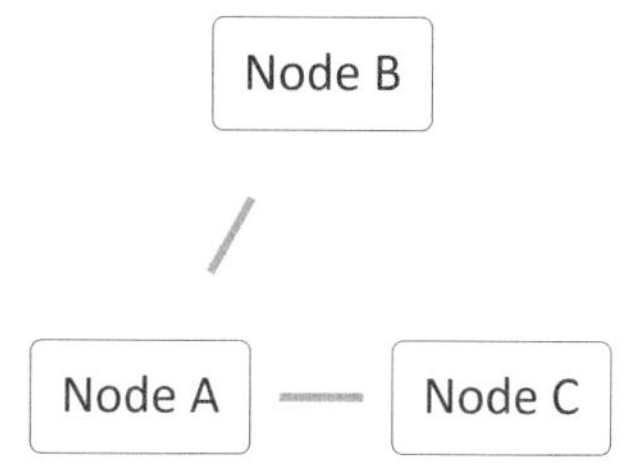

Fig. 1 Unclosed triad

Table 3 Spearman correlation between the number of an author's unclosed triads and this author's betweenness centrality in 2015

Year	Spearman's rho	Year	Spearman's rho
1993–2002	-	2009	0.5767765
2003	0.3581052	2010	0.5976887
2004	0.4428995	2011	0.6691048
2005	0.4435424	2012	0.7276214
2006	0.5192145	2013	0.7779198
2007	0.5859543	2014	0.9017711
2008	0.5787019	2015	0.9876267

paper together. Hence, node A lies between nodes B and C and should have a high betweenness centrality in the final network.[11]

Table 3 illustrates Spearman's [25] rank correlation coefficients between the author's betweenness centrality in 2015 and the number of an author's unclosed triads in the respective years. It is defined as

$$\rho = \frac{\sum(rk(BC_i) - \overline{rk_{BC}})(rk(UT_i) - \overline{rk_{UT}})}{\sqrt{\sum(rk(BC_i) - \overline{rk_{BC}})^2}\sqrt{(rk(UT_i) - \overline{rk_{UT}})^2}} \tag{1}$$

where

$rk(BC_i)$ is the rank of author i's betweenness centrality,
$rk(UT_i)$ is the rank of author i's number of unclosed triads,
$\overline{rk_{BC}}$ is the mean rank of all betweenness centrality scores, and
$\overline{rk_{UT}}$ is the mean rank of all number of unclosed triads.

We decided to use Spearman's rho instead of Pearson's correlation coefficient since we expected the association between betweenness centrality and the number of an author's unclosed triads to be monotonic, but did not want outliers or a nonlinear

[11] We suppose that node A will have a high betweenness centrality in the final network, although node B and node C are more likely to co-author a paper in the future than two random nodes if both have co-authored a paper with node A. In the literature, this fact has been termed the "forbidden triad" [14].

Table 4 Spearman correlation between an author's degree centrality and this author's betweenness centrality in 2015

Year	Spearman's rho	Year	Spearman's rho
1993–2002	–	2009	0.1547840
2003	0.3348258	2010	0.2336706
2004	0.5012240	2011	0.3004783
2005	0.4919271	2012	0.3598940
2006	0.3827627	2013	0.3772301
2007	0.4472469	2014	0.4388881
2008	0.3589474	2015	0.4671028

relationship to bias the results. However, we are aware that we lose some information by treating interval-scaled variables as ordinal.

Since there was only a single publication by two authors between 1993 and 2002, a calculation of Spearman's rho for these years is not meaningful. Also the correlation coefficient for 2003 should be viewed with caution as there were only two publications about PCSK9 in 2003. However, for the remaining years, Spearman's rho indicates a strong to very strong correlation between the author's betweenness centrality in 2015 and the number of the author's unclosed triads in the respective years.

As betweenness centrality and degree centrality are thought to be highly correlated (e.g. [24]), one might argue that an author's degree centrality is a more convenient proxy variable for an author's betweenness centrality than the number of this author's unclosed triads. However, we calculated the correlation between an author's degree centrality and this author's betweenness centrality in 2015 (see Table 4).

By comparing Tables 3 and 4, one can see that the correlation coefficients between an author's number of unclosed triads and this author's betweenness centrality are much higher than those between an author's degree centrality and this author's betweenness centrality.[12] Hence, we can conclude that the number of an author's unclosed triads is a better proxy variable for this author's betweenness centrality than her or his degree centrality.

Although we found that the number of an author's unclosed triads is a good proxy for her or his betweenness centrality in the PCSK9 co-authorship network, this might possibly be a peculiarity of our dataset.

To ensure that the number of an actor's unclosed triads is generally a good proxy for her or his betweenness centrality, we generated a) 100 random Erds-Renyi networks, b) 100 scale-free (Barabási-Albert) networks with a power exponent of $\gamma = 2$, and c) 100 scale-free networks with a power exponent of $\gamma = 3$ [4] using the R package

[12]This is true for all years except 2004 and 2005. However, there were only very few publications in these years (14 and 18 respectively, compare Table 1), and the high correlation coefficients between an author's degree centrality and this author's betweenness centrality for those two years can be explained by chance. Furthermore, the differences in the correlation coefficients between the number of an author's unclosed triads and betweennness centrality and author's degree centrality and betweenness centrality for the years 2004 and 2005 are not very large (0.4428995 vs. 0.5012240 and 0.4435424 vs. 0.4919271).

Table 5 Spearman correlation between the actor's betweenness centrality and the actor's number of unclosed triads

	Erds-Renyi networks	Scale-free networks ($\gamma = 2$)	Scale-free networks ($\gamma = 3$)
Spearman's rho	0.99749	0.99999	0.99999

igraph [9]. All of these networks consisted of 2836 nodes. Afterwards, we calculated for each network the correlation between the actors' betweenness centrality and the actors' number of unclosed triads. Table 5 illustrates the mean correlation coefficients over all 100 networks for the three different types of networks. As the mean correlation coefficients are near 1 in all three types of networks, we can conclude that the number of an actor's unclosed triads is an appropriate proxy variable for her or his betweenness centrality in these artificial networks.

However, we also wanted to ensure that the number of an actor's unclosed triads is a good proxy variable for her or his betweenness centrality in not only these artificial networks but also real world networks. Therefore, we examined the correlation between a node's betweenness centrality and the suggested proxy variable using four well-known real world networks that stem from different domains and have different sizes: (1) Jeong and colleagues' [15] "protein interaction network" ($\rho = 0.971$), (2) Watts and Strogatz' [28] 'power grid" ($\rho = 0.858$), (3) Padgett's "Florentine families network" [7] ($\rho = 0.854$), and (4) Knuth's [16] "Les Miserables dataset" ($\rho = 0.980$). Also, in these four real world networks, the strong correlations indicate that the suggested proxy variable is a good indicator for a node's betweenness centrality. Nevertheless, we cannot infer whether the accuracy of the proxy variable changes with the size of the network based on this data.

For the final step of our analyses, we examined what percentages of the top nodes in the co-authorship network and the four real world networks are correctly identified as such using the proposed heuristic. Figure 2 depicts the number of nodes selected as top nodes (in percent) on the x-axis. The y-axis depicts how many nodes are correctly identified as top nodes using the proposed approach.

For example, if one aims to identify the top 15 % of nodes (by their betweenness centrality) in the co-authorship network, the proposed approach correctly identifies 83 % of these top nodes. Another example is the line for the "Florentine families network" in Fig. 2.[13] If one intends to identify the top 5 % of the Florentine families (i.e. the top family), the proposed proxy variable correctly identifies this one family (i.e. the proxy variable identifies 100 % of the families one intended to identify). Also if one intends to identify the top 12.5 % of the families (i.e. two families), the proposed proxy variable correctly identifies these two families (i.e. 100 %). However, if one intends to identify the top three families by their betweenness centrality (i.e. 18.75 % of all families), the proposed proxy variable only correctly identifies two (i.e. 66.6 % of the families one intended to identify). Finally, when the aim is to

[13]The "Florentine families network" is a very small network (with 16 nodes only).

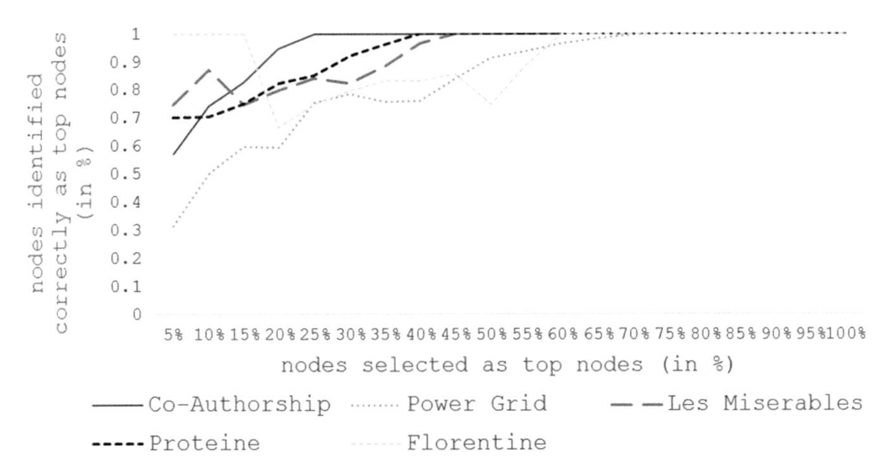

Fig. 2 Nodes selected as *top* nodes and nodes correctly identified as *top* nodes

identify the top four families by their betweenness centrality, the proposed proxy variable correctly identifies three (i.e. 75 %), and so on.

In general, across all networks, the proposed approach identifies a sufficiently large number of top nodes for a variety of application scenarios (such as, for example, KOL identification for marketing campaigns in the co-authorship network).

5 Discussion

In this paper, we aimed at two research objectives. First, we identified KOLs in PCSK9 research by their embeddedness in a co-authorship network. Specifically, we identified them by using their betweenness centrality in the co-authorship network. Second, we proposed a proxy variable for the betweenness centrality of these nodes (i.e. the number of an author's unclosed triads).[14]

We think that both points in themselves are important contributions to practice and literature. Pharmaceutical enterprises spend considerable time and effort identifying KOLs. In this paper, we illustrated an easy and cheap alternative to identify KOLs on the basis of co-authorship data. The proposed method can also be easily conducted with search terms other than "PCSK9".

Furthermore, the proposed proxy variable may serve as an indicator for the nodes' betweenness centrality in a variety of settings where betweenness centrality cannot feasibly be calculated. Since the collaboration network of scientist is a prototype

[14]Encouraged by a literature review and interviews with marketing managers from the pharmaceutical industry, we assumed that authors with a high betweenness centrality have a high influence as well. Although we think that this is a reasonable assumption for co-authorship networks, we want to be clear that structural importance and dynamic influence of nodes do not necessarily have to be the same.

example of a complex evolving network [5], our findings also seem applicable to a variety of other networks as well.

Of course, as with any empirical study, this study is subject to some limitations. We do not consider most of these limitations to void the results, so long as readers remain aware of them as they draw their conclusions. In fact, the limitations suggest some future research. There are four specific limitations to discuss.

First, since PubMed/MEDLINE did not provide unique researcher IDs for all researchers, there might be some problems in distinguishing some of them. Either some researchers might have the same name or some authors might change their name (e.g. after marriage) and be recognised as two different nodes in a network. However, we manually checked the data for inconsistencies and think that the remaining error is of the order of a few percent (compare [22]). Therefore, we do not think that this methodological limitation will significantly affect our results (compare also [5]). Nevertheless, future research should conduct related analyses with a dataset in which each author has a unique researchers' ID.

Second, by selecting all articles about PCSK9 research indexed by the PubMed database, we obtained a clear boundary specification of the network. However, this boundary specification is rather artificial (compare [27]). For example, some of the authors might have collaborated on other articles, and we neglected these co-authorship links in our analyses. Therefore, future research should analyse co-authorship networks with a different boundary specification.

Third, we neglected the fact that KOLs retire and stop publishing papers [5]. Hence, we might possibly have identified some KOLs in our analyses that are not active anymore. Future research could collect additional data on this aspect and explicitly consider the retirement of KOLs in the analyses.

Fourth, during our analyses, we focused on the author collaboration network (compare [11]) and used the author-paper affiliation network (i.e. a bipartite graph) for calculating some descriptive statistics only (such as the number of authors per paper, or the number of papers per author). Future research could analyse the author-paper affiliation network in more detail.

Our hope is that our research will assist others in conducting these types of studies and form the basis for substantial future research into identifying KOLs in co-authorship networks, as well as the use of the number of unclosed triads as a proxy variable for betweenness centrality.

Acknowledgments This work was supported by a fellowship within the FITweltweit programme of the German Academic Exchange Service (DAAD).

References

1. Aggarwal, C., Subbian, K.: Evolutionary network analysis: a survey. ACM Comput. Surv. **47**(1), 10 (2014). doi:10.1145/2601412
2. Backstrom, L., Huttenlocher, D., Kleinberg, J., Lan, X.: Group formation in large social networks: membership, growth, and evolution. In: Paper presented at the 12th ACM SIGKDD International Conference on Knowledge Discovery and Data Mining, Philadelphia, PA, USA

3. Bader, D.A., Kintali, S., Madduri, K., Mihail, M.: Approximating betweenness centrality. In: Algorithms and Models for the Web-Graph, pp. 124–137. Springer, (2007)
4. Barabási, A.L., Albert, R.: Emergence of scaling in random networks. Science **286**(5439), 509–512 (1999). doi:10.1126/science.286.5439.509
5. Barabási, A.L., Jeong, H., Neda, Z., Ravasz, E., Schubert, A., Vicsek, T.: Evolution of the social network of scientific collaborations. Physica A **311**(3–4), 590–614 (2002)
6. Brandes, U.: A faster algorithm for betweenness centrality. J. Math. Sociol. **25**(2), 163–177 (2001)
7. Breiger, R.L., Pattison, P.E.: Cumulated social roles—the duality of persons and their algebras. Soc. Netw. **8**(3), 215–256 (1986). doi:10.1016/0378-8733(86)90006-7
8. Butts, C.T.: Social network analysis with sna. J. Stat. Softw. **24**(6), 1–51 (2008)
9. Czárdi, G., Nepusz, T.: The igraph software package for complex network research. InterJournal, Complex Syst. **1695** (2006)
10. Ediger, D., Jiang, K., Riedy, J., Bader, D., Corley, C., Farber, R., Reynolds, W.N.: Massive social network analysis: mining twitter for social good. In: Paper presented at the 39th International Conference on Parallel Processing (ICPP) San Diego, CA
11. Franceschet, M.: Collaboration in computer science: a network science approach. J. Am. Soc. Inform. Sci. Technol. **62**(10), 1992–2012 (2011). doi:10.1002/asi.21614
12. Freeman, L.C.: Set of measures of centrality based on betweenness. Sociometry **40**(1), 35–41 (1977). doi:10.2307/3033543
13. Freeman, L.C.: Centrality in social networks conceptual clarification. Soc. Netw. **1**(3), 215–239 (1979). doi:10.1016/0378-8733(78)90021-7
14. Granovetter, M.S.: The strength of weak ties. Am. J. Sociol. **78**(6), 1360–1380 (1973). doi:10.1086/225469
15. Jeong, H., Mason, S.P., Barabási, A.L., Oltvai, Z.N.: Lethality and centrality in protein networks. Nature **411**(6833), 41–42 (2001). doi:10.1038/35075138
16. Knuth, D.E.: The Stanford GraphBase: A Platform for Combinatorial Computing. Addison-Wesley Reading (1993)
17. Liu, P., Xia, H.: Structure and evolution of co-authorship network in an interdisciplinary research field. Scientometrics **103**(1), 101–134 (2015). doi:10.1007/s11192-014-1525-y
18. McLaughlin, A., Bader, D.A.: Scalable and high performance betweenness centrality on the GPU. In: Paper presented at the International Conference for High Performance Computing, Networking, Storage and Analysis, New Orleans, Louisana
19. Morstatter, F., Pfeffer, J., Liu, H., Carley, K.M.: Is the sample good enough? comparing data from Twitter's streaming API with Twitter's firehose. In: Paper presented at the International AAAI Conference on Web and Social Media, Cambridge, Massachusetts
20. Newman, M.E.: Scientific collaboration networks—I. Network construction and fundamental results. Phys. Rev. E **64**, 016131 (2001)
21. Newman, M.E.: Scientific collaboration networks—II. Shortest paths, weighted networks, and centrality. Phys. Rev. E **64**, 016132 (2001)
22. Newman, M.E.J.: The structure of scientific collaboration networks. Proc. Nat. Acad. Sci. USA **98**(2), 404–409 (2001). doi:10.1073/pnas.021544898
23. Newman, M.E.J.: Coauthorship networks and patterns of scientific collaboration. Proc. Nat. Acad. Sci. **101**(suppl 1), 5200–5205 (2004). doi:10.1073/pnas.0307545100
24. Shi, X., Bonner, M., Adamic, L.A., Gilbert, A.C.: The very small world of the well-connected. In: Paper presented at the Nineteenth ACM Conference on Hypertext and Hypermedia, Pittsburgh, PA, USA
25. Spearman, C.: The proof and measurement of association between 2 Things (Reprinted from Amer. J. Psychol. vol. 15, pp. 72–101, 1904). Am. J. Psychol **100**(3–4), 441–471 (1987)
26. Tomassini, M., Luthi, L.: Empirical analysis of the evolution of a scientific collaboration network. Physica A **385**(2), 750–764 (2007)
27. Vidgen, R., Henneberg, S., Naude, P.: What sort of community is the European Conference on information systems? a social network analysis 1993–2005. Eur. J. Inf. Syst. **16**(1), 5–19 (2007)

28. Watts, D.J., Strogatz, S.H.: Collective dynamics of 'small-world' networks. Nature **393**(6684), 440–442 (1998). doi:10.1038/30918
29. Yan, E., Ding, Y.: Applying centrality measures to impact analysis: a coauthorship network analysis. J. Am. Soc. Inform. Sci. Technol. **60**(10), 2107–2118 (2009). doi:10.1002/asi.21128
30. Yang, J., Chen, Y.: Fast computing betweenness centrality with virtual nodes on large sparse networks. PLoS ONE **6**(7), e22557 (2011). doi:10.1371/journal.pone.0022557
31. Yang, Y., Dong, Y., Chawla, N.V.: Predicting node degree centrality with the node prominence profile. Sci. Rep. **4**(7236) (2014). doi:10.1038/srep07236

An Empirical Study of the Diversity of Athletes' Followers on Twitter

Ricardo Silveira, Giulio Iacobelli and Daniel Figueiredo

Abstract The study of user diversity in online social networks is an important and ongoing research effort to better understand human behavior. This work takes a step in this direction by providing an empirical study of around 8,000 athletes divided into 13 categories and followed by 197 million users in Twitter. We propose a metric for follower diversity at the category level that factors the vast popularity difference between categories (e.g., soccer versus golf). Using this metric, we propose a measure for athlete heterogeneity based on the diversity of his/her followers. Our findings reveal that follower diversity is spread across two scales with the vast majority of users having very small diversity. We also find that athlete heterogeneity is inversely proportional to its number of followers. This indicates that very popular athletes are followed by users that (on average) do not follow other sports.

1 Introduction

Online social networks became highly popular in the last decade, and currently a massive amount of data is generated daily by the millions of users of such systems. This data has been leveraged to study various aspects of human behavior at unprecedented scale, such as information cascades, news bias, opinion formation and others [1–3]. Among the most widely used and studied platforms is Twitter, a system blending social media and social networks with over half a billion users and the advantage that in principle all user generated data is publicly available [4–6].

R. Silveira · G. Iacobelli (✉) · D. Figueiredo
Department of Computer and System Engineering (PESC),
Federal University of Rio de Janeiro (UFRJ), Rio de Janeiro, Brazil
e-mail: giulio@land.ufrj.br

R. Silveira
e-mail: ricardosilveira@land.ufrj.br

D. Figueiredo
e-mail: daniel@land.ufrj.br

H. Cherifi et al. (eds.), *Complex Networks VII*, Studies in Computational Intelligence 644, DOI 10.1007/978-3-319-30569-1_25

An interesting object of study is user interest diversity [2, 7, 8]. In systems where so much is available on so many topics will users' interest become more or less diverse? Will strong polarization around topics emerge out of local interactions? Can we accurately quantify user diversity in the presence of strong biases imposed by popularity and context? Prior works have focused primarily on user generated content to assess diversity and polarization. We focus on a much stronger signal which does not depend on content and topic categorization is external to the system. In particular, we focus on a target group where topics and interests are well-defined: sports, athletes and followers. In particular, we present an empirical study of diversity considering around 8 thousand athletes divided into 13 categories and followed by 197 million people on Twitter. Our main contributions and findings are:

- Characterize athlete/follower relationship in the context of different categories.
- Propose a metric to quantify user diversity and athlete heterogeneity that accounts for popularity bias of categories.
- Quantify the proposed metrics and identify various relationships between categories, popularity and diversity.
- Among other findings, we highlight that popular categories are followed by less diverse users, and that athletes with many followers have less diverse followers.

The remainder of this paper is organized as follows. Section 2 describes the data collection and presents a preliminary analysis. Sections 3 and 4 present the proposed metrics for user diversity and athlete heterogeneity and their empirical evaluation, respectively. Section 5 concludes the paper with a brief discussion.

2 Data Collection and Analysis

The website http://www.tweeting-athletes.com is a semi-public platform keeping track of athletes that have a Twitter account. Besides associating an athlete to its twitter account, the website has categorized all athletes according to their sport or league in which they play, such as soccer and NBA. Since athlete information is manually verified by the website, the available data can be taken as reliable, although not complete since athletes (or their managers) must register with the website.

We developed a web crawler to collect all athletes' profiles on the website which on February 2015 was around 8,000 athletes. We then developed a program to use the public Twitter API to collect information of each of these athletes, including the identity of all their followers (this procedure lasted several days, also on February 2015). The data collected is summarised in Table 1. The first column corresponds to the thirteen main categories listed in the website and each athlete is in exactly one of such categories with their respective sizes given in the third row. The fourth row denotes the number of followers (users) of all athletes in the corresponding category, while the second row shows the number of follows, which is the sum of the number of followers of each athlete in the category. Note that the number of follows is larger than the number of followers, since a follower may follow different athletes in the

Table 1 Summary of collected data across different categories

Category	# Follows	# Athletes	# Followers	Popularity
Soccer	649,689,003	1417	107,598,814	458,496.1
NBA	180,315,704	560	50,800,914	321,992.3
NFL	131,436,876	2187	29,500,112	60,099.2
Other sports	114,895,505	651	37,282,225	176,490.8
Motorsports	40,263,566	122	16,808,352	330,029.2
MLB	39,629,147	658	11,505,122	60,226.7
MMA	35,901,352	334	12,056,037	107,489.1
Olympic games	31,767,754	834	15,537,349	38,090.8
NHL	27,228,831	435	5,451,714	62,595.0
Golf	23,408,708	233	8,753,212	100,466.6
Cycling	13,536,674	151	6,407,248	89,646.8
Winter Olympics	7,582,522	222	5,125,114	34,155.5
Tennis	3,131,908	22	2,268,151	142,359.5

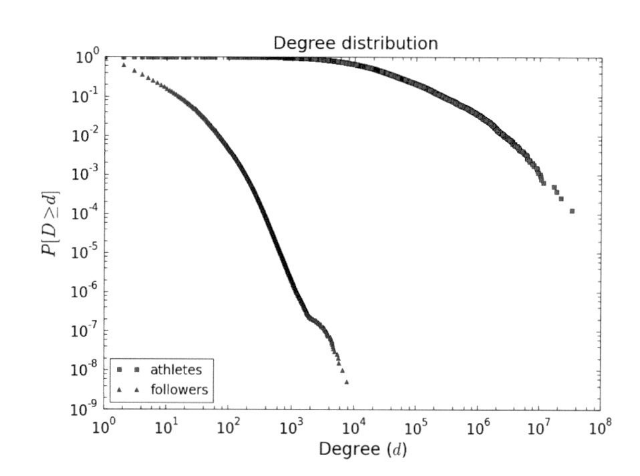

Fig. 1 Complementary Cumulative Distribution Function (CCDF) of the degree of athletes (*red curve*) and followers (*blue curve*). *Athletes' degree*: number of followers

same category. The last column presents the *popularity* of each category, computed as the number of follows per athlete in the category.

Note that categories are of very different sizes, either when considering number of follows, number of athletes or number of followers. This reflects the enormous bias induced by popularity in such systems. To better accommodate for this bias, we consider that the popularity of each category is measured as # Follows/# Athletes, rather than simply using the number of follows or followers. In other words, the popularity of a category corresponds to the average number of follows per athlete in the category.

Figure 1 depicts the Complementary Cumulative Distribution Function (CCDF) for the degree of athletes and followers in log-log scale. For an athlete, the degree corresponds to the number of followers it has, whereas for a follower the degree

Table 2 Top five followed athletes and their number of followers

Athlete	C. Ronaldo	Kaka	L. James (NBA)	Neymar Jr.	Ronaldinho
# Followers	33,657,773	22,092,978	18,978,751	17,008,330	11,608,006

Four are soccer players

is the number of athletes followed. Note that 70 % of all athletes have more than 8 thousand followers, with extremely popular athletes having over 10 million followers. Considering that the average number of followers per athlete is 20,552, the degree of very popular athletes ($\approx$10 million) is an extremely large value, more than 400 times larger than the average value, an observation also reflected in the large standard deviation of 798,766. The top five athletes in number of followers are listed in Table 2, along with the corresponding number of followers.

The degree distribution (CCDF) of followers is quite different, showing that more than 80 % of users follow less than 10 athletes and very few users follow more than 800 athletes (10 % of the total number).[1] This much shorter tail is reflected in the average (15.7) and standard deviation (6.6) of the empirical distribution.

3 Follower's Diversity

We consider that users are interested in topics by associating topics to categories and interest by a following relationship. Thus, a user is interested in a category if he/she follows at least one athlete in that category. Note that we will not consider the number of athletes a user follows within a given category. By focusing on categories rather than on athletes we can avoid the general bias of following multiple athletes in the same category. Another important aspect when measuring diversity is the popularity of each category. As shown in Table 1, some categories are more popular than others, and to properly quantify diversity we must take this into account. Our diversity measure is build on the following observations:

- diversity should increase with the number of categories followed.
- diversity should decrease with the popularity of a followed category.

We introduce the following notation to provide a clear definition of diversity. Let the data be encoded as a directed graph $G = (V, E)$, in which a vertex $k \in V$ corresponds to a Twitter user account, and there is a directed edge $(k, j) \in E$ from k to j if user k follows user j. Let $A \subset V$ denote the set of athletes, and let S denote the set of categories. For $s \in S$, we denote by A_s the set of athletes in category s. Recall that each athlete belongs to exactly one category. A vertex $k \in V$ is a follower if there exists a $j \in A$ such that $(k, j) \in E$; we denote by $F \subseteq V$ the set of followers.

[1]Note that there is a user following all athletes—most likely an account not associated with a real person.

Given a follower $k \in F$, and a category $s \in S$, we denote by $A_{s,k}$ the set of athletes in A_s which are followed by k, i.e., $A_{s,k} = \{j \in A_s \mid (k, j) \in E\}$. Moreover, for k a follower, we denote by $S_k = \{s \in S \mid A_{s,k} \neq \emptyset\}$, i.e., the set of categories followed by k. Given an athlete $j \in A$, we denote by $K_j = \{k \in F \mid (k, j) \in E\}$ the set of followers of j, and we denote by d_j^{in} its cardinality, corresponding to its degree, i.e., $d_j^{\text{in}} \triangleq |\{k \in F \mid (k, j) \in E\}|$. With a slight abuse of notation, given a category $s \in S$, we denote by d_s^{in} the total number of follows (of incoming links) for category s, i.e., $d_s^{\text{in}} = \sum_{j \in A_s} d_j^{\text{in}}$.

As mentioned above, to measure diversity the popularity of each category must be taken into account. Thus, we assign a weight ω_s to each category s which is inversely proportional to its popularity, i.e., $\omega_s \triangleq \frac{|A_s|}{d_s^{\text{in}}}$. The *diversity* of a follower k, denoted by α_k, is then defined as:

$$\alpha_k \triangleq \frac{\sum_{s \in S} \omega_s \, \mathbb{I}(A_{s,k} \neq \emptyset)}{\sum_{s \in S} \omega_s} = \sum_{s \in S_k} \rho_s \,, \tag{1}$$

where, $\rho_s = \omega_s / \sum_s \omega_s$ is the normalised weight for category s, while $\mathbb{I}(\cdot)$ is the indicator function, that is $\mathbb{I}(A_{s,k} \neq \emptyset) = 1$, if $A_{s,k} \neq \emptyset$, and 0 otherwise.

The weight and normalized weight values are listed in Table 3. Note that the diversity induced by the different categories is quite different; following Tennis contributes to user diversity much more than following NBA, while Winter Olympics contributes the most to user diversity.

Figure 2a depicts the CCDF of follower diversity as measured by α_k. Note that approximately 30 % of the followers have the smallest possible diversity (0.014) which correspond to users who only follow Soccer category, the most popular category. Figure 2a also shows that more than 60 % of the followers have diversity smaller than 0.05 and they follow athletes from one or two categories. Finally, we observe a small fraction of users (less than 10^{-5}) that follow all categories, thus having highest possible diversity of 1.

We now consider the diversity of followers of a given category with results shown in Table 4. Note that diversity is not spread uniformly across categories. In particular, followers of Winter Olympics category have the largest average diversity among all categories, with Soccer the lowest. We point out that, although following a popular category s results in a small weight ω_s, a follower may in principle follow other categories inducing a higher diversity. However, it seems that followers of popular categories do not tend to follow other categories and therefore have a small diversity. This is confirmed by considering the total and fraction of users that follow just that category (columns 4 and 5 of Table 4). Note that more than 66 % of those who follow soccer, do not follow any other category, while for NFL this number is 30 %.

Table 3 Weights (ω) and normalised weights (ρ) for each category

Category	Soccer	Motorsports	NBA	Other sports	Tennis	MMA	Golf	Cycling	NHL	MLB	NFL	Olympic games	Winter Olympics
$\omega\,(\times 10^{-6})$	2.2	3.0	3.1	5.7	7.0	9.3	10.0	11.2	16.0	16.6	16.6	26.3	29.3
$\rho\,(\times 10^{-2})$	1.4	1.9	2.0	3.6	4.5	6.0	6.4	7.1	10.2	10.6	10.6	16.8	18.7

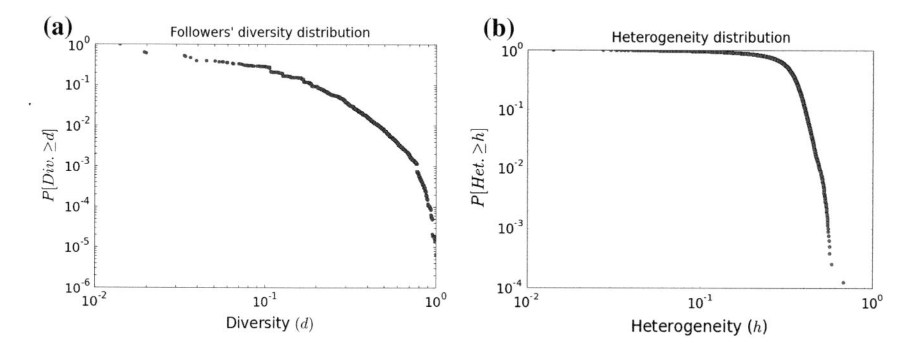

Fig. 2 Fraction of followers/athletes with diversity/heterogeneity greater than give value (CCDF). **a** CCDF of followers diversity, α_k (avg=0.097, std=0.073). **b** CCDF of athlete heterogeneity, β_j (avg=0.32, std=0.09)

Table 4 Average and standard deviation of follower diversity in each category

Category	Average	Std. deviation	# Excl. followers	Fraction excl. followers
Winter olympics	0.358307	0.181506	1,428,837	0.278791
Olympic games	0.277852	0.136397	3,499,990	0.225263
NHL	0.251917	0.175995	1,726,975	0.316777
MLB	0.248452	0.167417	3,194,130	0.277627
Golf	0.245631	0.185144	1,479,384	0.169010
Cycling	0.229478	0.181855	1,366,709	0.213307
NFL	0.200693	0.134341	8,985,095	0.304578
Tennis	0.189565	0.163223	274,485	0.121017
MMA	0.153562	0.146019	2,709,689	0.224758
Other sports	0.126630	0.141822	15,953,684	0.427917
Motorsports	0.122397	0.144113	4,288,922	0.255166
NBA	0.114913	0.136002	18,088,226	0.356061
Soccer	0.055275	0.097337	71,148,842	0.661242

Exclusive followers (# Excl. followers) corresponds to users who only follow that category, while their fraction is with respect to all followers of that category

4 Athlete's Heterogeneity

We now focus on the diversity of the followers of given athletes, a concept we refer to as athlete's *heterogeneity*. In particular, we are interested in studying the relationship between the athlete heterogeneity and other characteristics, such as number of followers or popularity. Building on the concept of follower's diversity, the heterogeneity of an athlete j is defined as:

$$\beta_j \triangleq \frac{1}{d_j^{\text{in}}} \sum_{k \in K_j} \alpha_k \ , \tag{2}$$

Table 5 Top five athletes according to heterogeneity

Athlete	Peter Bakare (Oly. G.)	Kim St-Pierre	Aja Evans	Molly Schaus	Kacey Bellamy
# Heterog.	0.566	0.564	0.560	0.557	0.555
# Followers	30,313	6,181	7,717	5,636	4,705

The most heterogeneous athlete is from Olympic Games (Volleyball) while the other four are from Winter Olympics category

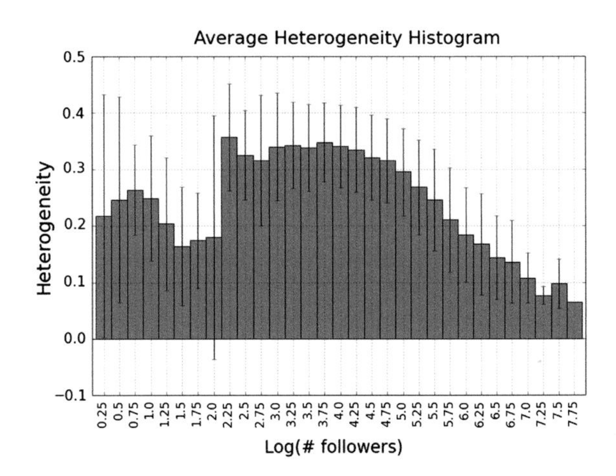

Fig. 3 Histogram of the athletes' heterogeneity as a function of log of the number of followers

where, K_j denotes the set of followers of athlete j and d_j^{in} its cardinality (degree). Note that β_j is the average follower diversity among the users that follow athlete j.

Figure 2b depicts the CCDF of the athlete heterogeneity. The vast majority of athletes (more than 70 % of the athletes) have a heterogeneity between 0.2 and 0.4, with an average value of approximately 0.32 and standard deviation 0.09. Different from follower diversity, athlete heterogeneity is much more center around its mean with few athletes being very different. Thus, followers tend to be more diverse than an athlete's followers. Table 5 shows the top five athletes according to heterogeneity. Note that they are different from the top five according to number of followers, shown in Table 2. This already suggests that more heterogeneous athletes do not have many followers, which we next investigate. Figure 3 shows a histogram of the (average) athletes' heterogeneity as a function of the logarithm of number of followers. The error bars represent the standard deviation in each bin. The figure shows a clear trend indicating that as the number of followers increases the average heterogeneity decreases. Note that, as shown in Fig. 1 (red curve) there are very few athletes in the range [0, 3] (as in the extreme tail) and thus the data is more noisy in this range and the trend is not so clear. Overall, athlete popularity (i.e., number of followers) seems to be inversely proportional to the diversity of its followers. It is worth noting that not every athlete belonging to a very popular category has a small heterogeneity value. For example, soccer players Charlie Davies (more than 100 thousand followers) and

Landon Donovan (more than 1 million followers) have heterogeneity 0.39 and 0.31, respectively. Considering that soccer weight is 0.014, this indicates that most of their followers also follow other categories.

5 Discussion and Conclusion

The study of diversity in online social networks is an ongoing research effort that can significantly contribute to the understanding of human behavior. This work takes a step in this direction by considering the context of athletes their categories and their followers in Twitter. In particular, we introduce a simple metric for follower diversity that allows a more fair comparison between categories with extremely different popularities, such as Soccer and Winter Olympics. Our analysis of around 8,000 athletes and their 197 million followers reveals very interesting findings, such as that most followers have very small diversity (60 % follow less than two categories) and that popular athletes are followed by less diverse followers. Although this work has focused on diversity at the level of categories, our dataset includes subcategories for athletes, such as the team they currently play for. For future work, we intend to measure diversity considering the subcategories followed by users. Lastly, the measure hereby introduced in the context of athletes/sports can be used to study user interest diversity across different Twitter categories, such as politicians/political parties, actors/films and news/topics.

Acknowledgments This work has been partially funded through research grants from the following Brazilian agencies: CNPq, CAPES and FAPERJ.

References

1. May, A., Chaintreau, A., Korula, N., Lattanzi, S.: Filter & follow: How social media foster content curation. In: International Conference on Measurement and Modeling Computer Systems, pp. 43–55 (2014)
2. Weng, L., Menczer, F.: Topicality and impact in social media: diverse messages, focused messengers. PloS One **10**(2), e0118,410 (2015)
3. Yang, S.H., Long, B., Smola, A., Sadagopan, N., Zheng, Z., Zha, H.: Like like alike: joint friendship and interest propagation in social networks. In: WWW, pp. 537–546. ACM (2011)
4. Cha, M., Haddadi, H., Benevenuto, F., Gummadi, P.K.: Measuring user influence in twitter: the million follower fallacy. ICWSM **10**(10–17), 30 (2010)
5. Kwak, H., Lee, C., Park, H., Moon, S.: What is twitter, a social network or a news media? In: International Conference on World Wide Web, pp. 591–600. ACM (2010)
6. Marwick, A.E., et al.: I tweet honestly, i tweet passionately: Twitter users, context collapse, and the imagined audience. New Media Soc. **13**(1), 114–133 (2011)
7. An, J., Cha, M., Gummadi, P.K., Crowcroft, J.: Media landscape in twitter: a world of new conventions and political diversity. In: ICWSM (2011)
8. Yardi, S., Boyd, D.: Dynamic debates: an analysis of group polarization over time on twitter. Bull. Sci. Technol. Soc. **30**(5), 316–327 (2010)

Forecasting a Global Air Passenger Demand Network Using Weighted Similarity-Based Algorithms

Ivan Terekhov, Antony Evans and Volker Gollnick

Abstract The aim of this study is to define an appropriate approach to forecast the appearance and disappearance of air passenger demand between cities worldwide. For the air passenger demand link forecasting, a weighted similarity-based algorithm is used, with an analysis of nine indices. The weighted resource allocation index demonstrates the best metrics. The accuracy of this method has been determined through a comparison of modeled and known data from three separate years. The known data was used to establish boundaries when applying the similarity-based algorithm. As a result, it is found that a weighted resource allocation index, with defined boundaries, should be utilized for link prediction in the air passenger demand network. Furthermore, it is shown that grouping cities within the air passenger demand network, based on socio-economic indicators, increases the accuracy of the forecast.

1 Introduction

The modular environment AIRCAST [1, 2], aims to forecast future development of the air transportation system (ATS) based on socio-economic scenarios. AIRCAST allows to simulate a range of possible outcomes for the future ATS and assess, for example, the impact of new technology on the number of demand passengers or the size and number of aircraft on particular routes. An air passenger demand (APD) forecast model of 'origin to destination air travel passenger demand between city pairs' on a global level called D-CAST [1] is the first layer in a chain of models within AIRCAST [2]. In D-CAST, the APD model forecasts the number of passengers as well as changes in the number of connected cities within the forecast period. This

I. Terekhov (✉) · V. Gollnick
German Aerospace Center (DLR) Air Transportation Systems,
Hamburg, Germany
e-mail: ivan.terekhov@dlr.de

A. Evans
University of California Santa Cruz, University Affiliated Research Center,
Moffett Field, CA, USA

H. Cherifi et al. (eds.), *Complex Networks VII*, Studies in Computational Intelligence 644, DOI 10.1007/978-3-319-30569-1_26

paper aims to define an appropriate approach to forecasting the appearance of APD between cities worldwide.

The APD network is a dynamically evolving network. This network contains a number of cities (nodes) with links between them. In this study, the APD network is considered as an undirected network [1]. The APD network is a weighted network i.e. each link is characterized by a parameter or a set of parameters. As shown, the APD has interdependences with economic and social indicators [3, 4]. Thus, the weight of a link could be considered as a combination of socio-economic indicators between cities in pairs. During the forecasting period, the socio-economic indicators of cities vary. Therefore, the weighting of links is also changing. This variation in weightings over time has an impact on the APD network and, accordingly, the topology of the network is likely to change. For example, where the socio-economic indicators of cities (e.g. GDP, population and oil price) show a rapid increase, it is probable that a number of connected cities with a significant APD will appear where no APD connections previously existed.

There are three main groups of link prediction methods [5] for forecasting connections in the network: similarity-based algorithms, maximum likelihood (ML) and probabilistic models (PM). Similarity-based algorithms are divided into local, global and quasi-local indices [5]. Similarity-based algorithms are the mainstream class of link prediction algorithms. ML methods and PM are complex and very time consuming. ML is able to handle networks with up to a few thousand nodes in a reasonable time [5]. Furthermore, ML methods do not demonstrate the best accuracy [5]. Mostly, studies consider link prediction in non-weighted networks. Studies on link prediction in weighted networks are mainly conducted using weighted local similarity indices [6, 7]. In addition, the APD network is a high clustered network as shown by Ghosh and Terekhov [2]. For highly clustered networks, the common-neighbor-based indices demonstrate relatively good prediction with low complexity [5]. Thus, in this study, only weighted local similarity indices are considered.

The underlying principle of weighted and non-weighted indices of similarity-based algorithms is the same. These algorithms assign a score to each non-existing link in a given network. Then the links are ranked in descending order according to their score. Links with the highest score should appear in the network. Here, two significant problems arise. In the network, one index can perform well while another fails [5]. Thus, the first problem is to define which weighted local similarity index shows the best performance in the APD network. The second problem is to define a criterion for adding new connections to the network with the highest score from the top of the ranking list. In other words, a boundary condition in the ranking list of non-existing links has to be defined: links from the ranking list between the first link and a boundary link will be added to the network.

In addition, as shown by Zheleva et al. [8], the combination of network structure, node attributes, and node community features improve link prediction performance. In the APD network, the network structure and node attributes are known. For node communities, cities are allocated to groups according to the proximity of their socio-economic indicators. For example, cities with large GDPs and populations are classified under the big-rich group and cities with large populations and small GDPs are classified under the big-poor group. Since cities generally possess different socio-economic indicators in these groups (clusters) [1], the process of link appearance and disappearance in each cluster pair of the APD network could be different. Thus, a similarity-based algorithm which shows the best performance in one cluster is probably different in another cluster. For instance, different weighted similarity algorithms could perform better between big-rich cities and small-poor cities, than between megacities and middle-rich cities. It is also likely that every cluster pair has its own boundary. In this paper, the performance of similarity-based algorithms for each cluster pair is analyzed. The boundary for each cluster pair is defined utilizing the algorithm with the best performance.

Two standard metrics are used to identify the appropriate index for each cluster pair: the area under the receiver operating curve (AUC) [9] and precision [10]. In this study, these metrics are applied to the APD topology for 2009. For accuracies and boundary identification, a set of forecasts of the APD network is made: from 2002 to 2012, from 2007 to 2012 and from 2011 to 2012.

For 2002, 2007, 2011 and 2012 origin to destination city pairs worldwide (topology) have been obtained from Sabre Airport Data Intelligence (ADI) [11] database. The ADI database contains two types of annual APD data: from 2002 to 2013—preliminary data, from 2009 to 2012—final data. The preliminary data contains a number of mistakes, while the final data has been corrected. For the metrics calculations, the final ADI data for 2009 are utilized. For the accuracies and boundary calculations, the preliminary APD data of 2002 and 2007 are used and the final data for 2011. For the link weighting calculations, GDP [12, 13], population [14, 15] and geographical coordinates [16, 17] of the cities have been obtained from various databases [1]. However, city populations for 2002, 2007 and 2011 are not available. Thus, this data has been obtained by extrapolation of the city population from 2012 to 2002, 2007 and 2011 based on the historical population growths for the countries. For the average air fare between cities, a simple air fare model [2] is adopted.

2 Definition of the Weighted Similarity-Based Algorithm for the APD Network

The initial set of 4,435 cities obtained from the ADI data base is divided into 9 clusters, based on their socio-economic indicators [1] in 2012: GDP, city population and GDP per capita. All economic indicators within the study are adjusted to 2005 US dollars. The detailed description of the clustering process is shown in [18]. For the purposes of the study, short hand cluster names, derived from cluster means (population, GDP and per capita GDP), are adopted (i.e. very small rich cities, small poor cities, etc.).

In the APD network, every cluster is defined as a set of cities and weighted connections. These connections link cities in one cluster with cities in other clusters and cities within a cluster. Weights in this study are considered as a combination of average air fare [2], distance between cities and main socio-economic indicators such as city GDP and city population. The weight on the connection between cities x and y is presented as follows:

$$w_{xy} = (g_x * g_y)^{\alpha} * \left(p_x * p_y\right)^{\beta} * \left(l_{xy}\right)^{\gamma} * \left(t_{xy}\right)^{\delta} * \varepsilon + \theta \tag{1}$$

where $g_{x,y}$ are the gross domestic products of city x and y; $p_{x,y}$ are the populations of city x and y; $l_{x,y}$ is the distance between city x and y; $t_{x,y}$ is the average air fare between city x and y; $\alpha, \beta, \gamma, \delta$ are elasticities of GDP, population distance and average air fare respectively; ε is a dummy variable; is a free parameter. In this study it is assumed that $\alpha = 1$, $\beta = 1$, $\gamma = 1$, $\delta = -1$, $\varepsilon = 1$ and $\theta = 0$. Thus, the Eq. (1) turns to a variation of Newton's gravity model and the weight could be interpreted as an abstract attractive force between cities. Furthermore, the gravity model has been used in a number of studies [4, 19] to predict the APD between city pairs.

Based on the simple air fare model [2] and the assumptions in Eq. (1) the weight w between cities x and y could be presented as:

$$w_{xy} = \frac{g_x * g_y * p_x * p_y}{\left(r * 2 * 10^{-4} + 0.0653\right) * l_{xy}^2 + 140 * l_{xy}} \tag{2}$$

where r represents an average oil price in a given year.

Within this study, nine indices of similarity-based algorithms are analyzed. Based on Lü and Zhou's study [5], the weighted common neighbors (WCN), the weighted Adamic-Adar index (WAA) and the weighted resource allocation index (WRA) are applied to the APD network. In addition, similarity indices for unweighted networks are adapted for weighted networks using the proposed simple method by Murata and Moriyasu [6]. These indexes are the weighted Salton index (WSA), the weighted Sorensen index (WSO), the weighted hub promoted index (WHPI), the weighted hub depressed index (WHDI), the weighted Leicht-Holme-Newman index (WLHN) and the weighed preferential attachment index (WPA). These similarity indexes are presented in Table 1.

Two standard metrics AUC [5, 9] and precision [5, 10] are used to determine the accuracy of each index. Initially, for an undirected weighted network, all existing and non-existing links are known. From this set of existing links, a group of links—the probe set—is excluded. The remaining existing links are the testing set. The score of each index in the network formed by the testing set is calculated for all non-existing links and the probe set. The AUC shows the probability that a randomly chosen link from the probe set has a higher score than a randomly chosen link from the set of non-existing links. For the precision metric, the set of probe links and non-existing links is ordered in descending order according to their scores. From this list, the top-L links are selected as the predicted once. Among these links, Lr links are correct (links from the probe set). The precision is a ratio of Lr to L. Thus, higher precision means higher prediction accuracy 5. Both metrics are numbers between 0 and 1. The closer the metric is to 1, the better the performance of the index in a given network.

Table 1 Weighted similarity-based algorithm indexes

Index name	Index formula
Weighted Common Neighbors index (WCN)	$s_{xy}^{WCN} = \sum_{z \in \Gamma(x) \cap \Gamma(y)} w(x, z) + w(z, y)$ (3)
Weighted Adamic-Adar index (WAA)	$s_{xy}^{WAA} = \sum_{z \in \Gamma(x) \cap \Gamma(y)} \frac{w(x,z) + w(z,x)}{\log(1 + s(z))}$ (4)
Weighted Recourse Allocation index (WRA)	$s_{xy}^{WRA} = \sum_{z \in \Gamma(x) \cap Gamma(y)} \frac{w(x,z) + w(z,x)}{s(z)}$ (5)
Weighted Salton index (WSA)	$s_{xy}^{WSA} = \sum_{z \in \Gamma(x) \cap \Gamma(y)} \frac{w(x,z) + w(z,x)}{\sqrt{s(x) * s(y)}}$ (6)
Weighted Sorensen index (WSO)	$s_{xy}^{WSO} = \sum_{z \in \Gamma(x) \cap \Gamma(y)} \frac{2(w(x,z) + w(z,x))}{s(x) + s(y)}$ (7)
Weighted Hub Promoted Index (WHPI)	$s_{xy}^{WHPI} = \sum_{z \in \Gamma(x) \cap \Gamma(y)} \frac{w(x,z) + w(z,x)}{\min\{s(x), s(y)\}}$ (8)
Weighted Hub Depressed index (WHDP)	$s_{xy}^{WHDI} = \sum_{z \in \Gamma(x) \cap \Gamma(y)} \frac{w(x,z) + w(z,x)}{\max\{s(x), s(y)\}}$ (9)
Weighted Leicht-Holme-Newman index (WLHN)	$s_{xy}^{\mathrm{WLHN}} = \sum_{z \in \Gamma(x) \cap \Gamma(y)} \frac{w(x,z) + w(z,x)}{s(x) * s(y)}$ (10)
Weighed Preferential Attachment index (WPA)	$s_{xy}^{WPA} = s(x) * s(y)$ (11)

In this study for AUC and precision calculations, the 2009 APD network is utilized. For that year, 3,919 cities are obtained. These cities are allocated to 9 clusters according to their socio-economic indicators, based on the 2012 cluster means. It is assumed that the cluster means remain fixed as in 2012 (base year) and do not change. In other words, city clustering in 2009 is made from the perspective of clustering in 2012. Based on city clusters, 471,824 real connections in 2009 are distributed between 45 cluster pairs. Non-existing links are obtained for each cluster pair. The total number of non-existing links in the APD network of 2009 is 7,205,497. For the calculation of the two metrics, sets of existing and non-existing links are used.

Based on existing studies [5, 7] the network has been divided into two sets: testing and probing in proportions 90 % and 10 %, respectively. Each AUC and precision value is obtained by averaging 10 realizations with independent random separations of random and probe sets. Metrics for the whole network and each cluster pair for different indexes are calculated as well as their standard deviations. AUC and precision are used to determine the accuracy of each index for the whole network and for clusters. The index with the best metrics values is then chosen for the topology forecast in the APD network. The result demonstrates that only one index—Weighted Hub Promoted Index (WHPI) has a higher precision value in the whole network than the cluster average. However, this value is low compared to other indices. All other indices show higher AUC and precision numbers in clusters than in the whole network. This proves the necessity of separating cities into groups according to their socio-economic indicators, so as to improve the link forecasting performance. The best AUC number for the whole network is WSO. But this number is smaller than AUC for the WRA in clusters. The WRA index shows the best AUC and precision results in clusters pairs. This is expected, since WRA gives a higher score to a non-existing connection between two nodes if these nodes have many common neighbors with large weights. It is important to note, that the WRA index has the best performance of AUC and precision in each cluster pair. This disproves the assumption that cluster pairs in the APD network have different similarity indices demonstrating the best performance.

Based on the aforementioned analysis, the Weighted Resource Allocation (WRA) index is chosen for the topology forecast in the APD network. The score for each non-existing link in each cluster pair will be calculated utilizing the WRA index. Next, it is necessary to validate the method based on historical data.

3 Model Validation

For the validation, the APD topologies of four years from 2002 to 2012 are utilized. Data for these years from the ADI database (2002, 2007, 2011 and 2012 APD networks) are retrieved. Socio-economic data and geographical coordinates for cities from the same databases as for 2012 are obtained. The conditions required for the appearance of new cities in the APD network are not clear and hard to predict [5]. Thus, for the analysis, sets of cities from four networks are reviewed. Cities which are presented in a given year and 2012 are allocated to the set of common cities. Thereby, there are three sets of common cities for all 3 networks. For 2002, 2010, 2011 there are 3,699, 3,896 and 3,667 common cities with 426,150, 500,020 and 521,171 connections between them respectively.

Three analyses based on modified networks with common cities to define accuracies of the APD topology forecast have been made. Within the analyses new connections are calculated utilizing the WRA index and compared with the real data. These connections are calculated for the following topologies: 2012 from 2002, 2012 from 2007 and 2012 from 2011.

For all three analyses, predicted topologies are compared with real topologies. For example, new calculated connections in 2012 from 2002 are compared with the real topology of 2012. The analysis procedure is as follows: the 2012 socio-economic indicators and cluster affiliations are assigned to cities in 2002. Thus, the 2002 APD network turns to an incomplete network of the 2012 APD network. The score for all non-existing connections in every cluster pair of the 2002 network is calculated using the WRA index. Connections are sorted in descending order by their score. The calculated data is then compared to real data. The number of real added connections for every cluster is already known. Thus, from the calculated connections in the sorted list, the same amount of connections is added to the modeled APD network. In other words, from the real data in cluster A from 2002 to 2012, for example, x connections are added. This means, that from the calculated sorted list

Table 2 The average accuracies for new predicted connections, eliminated connections and the final accuracies of the forecasted 2012 ADP networks from 2002, 2007 and 2011

	2002	2007	2011
Addition	0.344461159	0.283425859	0.261119746
Elimination	0.127025427	0.140473134	0.111642672
Total	0.718827952	0.695375928	0.748008114

in cluster A, these x connections are added from the top. All of these new calculated connections from every cluster are added to the 2002 APD network, thus forming an extended 2002 APD network. In addition, some of the connections had to be removed from the network because the socio-economic indicators changed and, as a consequence, some APD connections have disappeared. The elimination process follows a similar procedure to the connection addition process. The score is calculated for every connection from the extended 2002 APD network and the connections are sorted in descending order by their score in this second list. The number of the real added connections is already known. From the second sorted list, the same number of connections as in the number of the real added connections is added to the final network. The remaining connections are eliminated from the APD network. Thus, the APD forecasting method has two sequential steps: the connection addition process and the connection elimination process. It is possible to define the accuracy of these processes. The accuracy for the connection addition is defined as a ratio between the amount of real new added connections in 2012 and the number of real new connections in the sorted list. The accuracy for the connection elimination is defined as a ratio between the amount of real connections in 2012 and the number of real new connections in the ordered list. This number is between 0 and 1. The addition process has a higher accuracy the closer the ratio is to 1. The elimination process has a higher accuracy the closer the ratio is to 0. Accuracies for the addition and elimination processes for every cluster pair for 2012 from years 2002, 2007 and 2011 are calculated.

In this study, all average accuracies for added connections from the years 2002, 2007 and 2011 are below 0.5, meaning that the prediction contains errors amounting to more than 50 %. For connection elimination average accuracies are above 0.5, meaning that there is less than 50 % error in the elimination prediction. However, the final average accuracy for all clusters of the APD network forecasts is above 0.6 and it is higher than, for example, in T. Murata and S. Moriyasu's study4 of link prediction in a weighted network of Question-Answering Bulletin Boards. The average accuracies for new predicted connections, eliminated connections and the final accuracies of the forecasted 2012 ADP networks from 2002, 2007 and 2011 are presented in Table 2. The final accuracy of every cluster pair for 2012 from 2002, 2007 and 2011 is shown in Fig. 3.

After the addition and elimination processes, the forecasted APD networks of 2012 and the real APD network of 2012 are compared. This comparison shows that the forecasted network has accuracies higher than 0.6 (Table 2). The accuracy is high even when the average addition and elimination accuracies show poor results. The reason for the over 0.6 accuracy is that more than 50 % of connections in 2012 remain in the network from 2002, 2007 and 2011 APD networks and high accuracies in most clusters (Fig. 1).

The accuracies for 2002 in most cluster pairs are almost always higher than for 2007 and 2011. It is, in particular, significantly higher in megacities—megacities, big poor cities—megacities, etc. cluster pairs. This is probably related to the 2008 economic crisis, and predictions from 2007 and 2011 to 2012 are more likely to show the higher impact of this crisis, when the world economy was not fully recovered, than from 2002. Furthermore, there are different accuracy figures for cluster pairs. Relationships between cities with "strong" socio-economic indicators in some cluster pairs (for example small-rich—megacities) are better described using Eq. (2) than for "weak" cluster pairs (for example big-poor–small-poor). Therefore, the accuracy is higher for cities with relatively high socio-economic indicators. In addition, it should be noted that cluster pairs are not equal in terms of number of passengers. For the forecast model, it is important to have a high accuracy for connections with a high APD. Table 3 presents the final accuracies for the accumulative number of passengers by cluster pairs in Fig. 1 of the real 2012 APD networks from forecasted 2002, 2007 and 2011. Numbers in brackets indicate accumulated numbers of cluster pairs corresponding to a given accumulated percentage of passengers. The model validation on historical data shows higher accuracy compared to existing studies. At this stage of the study, the topology forecast model validation using historical data shows acceptable results and the accuracy obtained seems to be sufficient. However, the accuracy could probably be enhanced by defining appropriate coefficients in Eq. (1). Next, it is necessary to analyze the WRA index boundary criteria in the sorted lists of non-existing connections for each cluster pair.

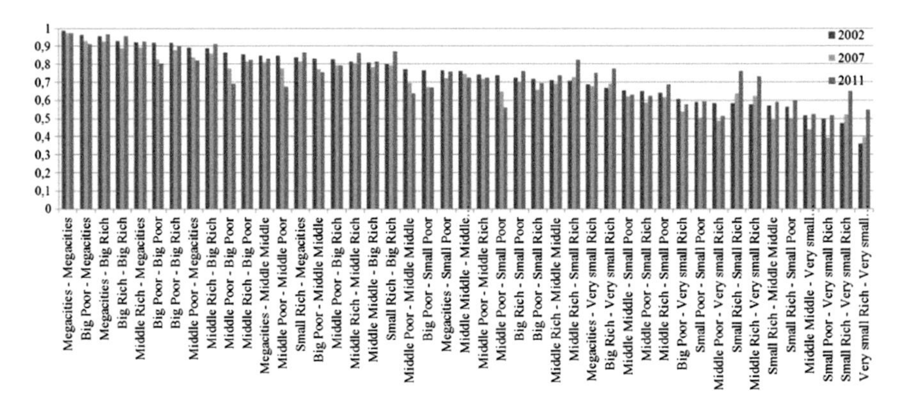

Fig. 1 Final accuracies for connection in every cluster pair in 2012 from 2002, 2007 and 2011

Table 3 Average accuracies for 2002, 2007 and 2011 for a given percentage of passengers

	50 % passengers	90 % passengers	75 % passengers	100 % passengers
2002	0.950410 (5)	0.893178 (12)	0.830114 (19)	0.741496 (45)
2007	0.911038 (5)	0.856984 (12)	0.803311 (19)	0.703929 (45)
2011	0.916903 (5)	0.884192 (12)	0.841171 (19)	0.741308 (45)

Figures in brackets indicate how many cluster pairs generate a given percentage of passengers

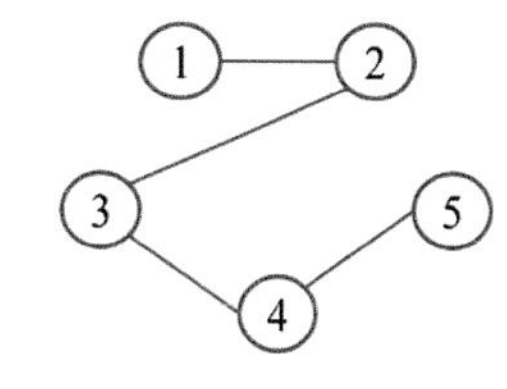

Fig. 2 The APD network topology of a cluster pair in year y

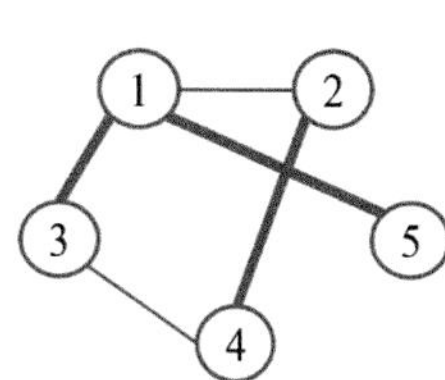

Fig. 3 The APD network topology of the cluster pair in year y + 1. *Thick lines* depict new real added connections to the network of year y

4 Boundaries

Based on the aforementioned analysis, there are two ways to define boundaries in the sorted list of connections: either using the number of new added connections and the number of real connections in each cluster pair or the boundary scores for each cluster pair. In other words, for the first method, a fixed number of connections are added to the network from the sorted list of non-exiting connections, which is in descending order according to their score. No additional manipulations are required for connection elimination. In the second method, for the addition connection process into the APD network, all connections where the score exceeds the boundary score (boundary for adding connections) in the sorted list with all possible connections is added to the APD network. For the elimination process, all connections where the score does not exceed the boundary score (boundary for eliminating connections) in the sorted list is added to the APD network.

For example, the APD network topologies of a cluster pair in year y (Fig. 2) and the next year y + 1 (Fig. 3) are known. Socio-economic indicators of cities from year y + 1 are assigned to the same cities as in year y. Utilizing the WRA index, scores for all non-existing connections are calculated (Fig. 5). Connections are sorted in descending order by their score (Fig. 6). The accuracy of the method can be defined using new real added connections to network in year y + 1. The number of forecasted links from the top of the list is equal to the number of new real added connections. The accuracy is defined as the ratio of relevant connections in the list of non-existing connections to number of new real added connections. There are two types of criteria

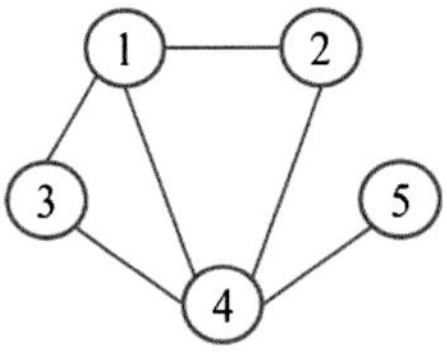

Fig. 4 The forecasted APD network of year y + 1

Fig. 5 Existing connection and all non-existing links in the APD network in year y. The score for each non-existing link is calculated. New real added connections in the APD network in year y + 1 are marked in *bold*

Existing connections	Non-existing connections	Score of non-existing connections
1-2	1-3	S13
2-3	1-4	S14
3-4	1-5	S15
4-5	2-4	S24
	2-5	S25
	3-5	S35

for adding connections to the ADP network. The first criterion is a fixed amount of connections. This amount added every year is equal to the number of new real added connections from year y + 1. The second criterion is the boundary score. Each connection with a score higher than the boundary score in year y + 1 is added to the network. The next step in forecasting is the elimination connections from the network. Forecasted links from the top of the list (from Fig. 6) are added to the existing links in year y (from Fig. 5) and sorted by their score in descending order. The number of forecasted links from the top of this new list is equal to the number of real connections in year y + 1. Connections with the score lower than the boundary score for the elimination process are eliminated from the network (Fig. 7). Thus, after the addition and elimination processes, the forecasted APD network for year y + 1 is obtained (Fig. 4).

In this study, the possibility of link addition and elimination are considered. It is assumed that APD connections are able to appear and disappear in the APD network. However, there are two connection adding and elimination approaches from the network: a fixed amount of connections or boundary scores. In both approaches, a situation could arise where all cities within a cluster pair are connected to each other. This is more likely to occur when applying the first method, i.e. using the number of new added connections in each cluster pair. For example, cluster pair middle-rich–small-rich in 2012 has 207 and 565 cities respectively. This cluster pair in 2012 has 27,628 connections including 3,538 new connections added from 2011 (shown in Fig. 3). The number of non-existing connections is 89,327. If it is assumed that the number of added connections will remain fixed, all cities in this cluster pair will be connected to each other within ~25 years. For the second method, applying boundary scores, the year when all of the cities are interconnected in the cluster is hard to predict. This will depend on various factors such as network configuration,

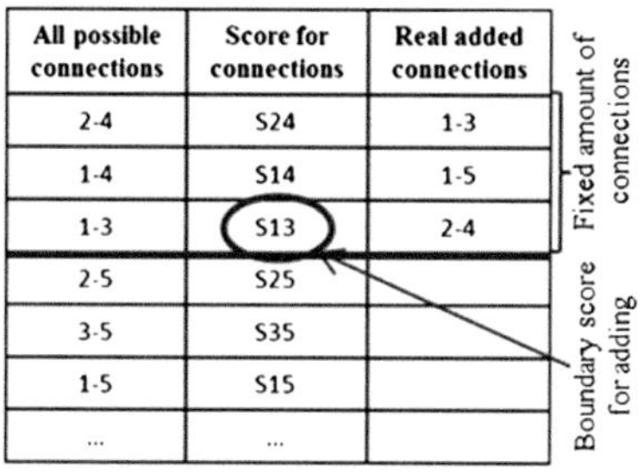

All possible connections	Score for connections	Real added connections
2-4	S24	1-3
1-4	S14	1-5
1-3	S13	2-4
2-5	S25	
3-5	S35	
1-5	S15	
...	...	

Fixed amount of connections

Boundary score for adding

Fig. 6 Non-existing connections are sorted in descending order based on their score. Two types of boundaries based on the number of new real added connections are presented: fixed amount of connections and the boundary score. The forecast predicts two actual connections out of three. Thus, the accuracy in this case is 0.6666

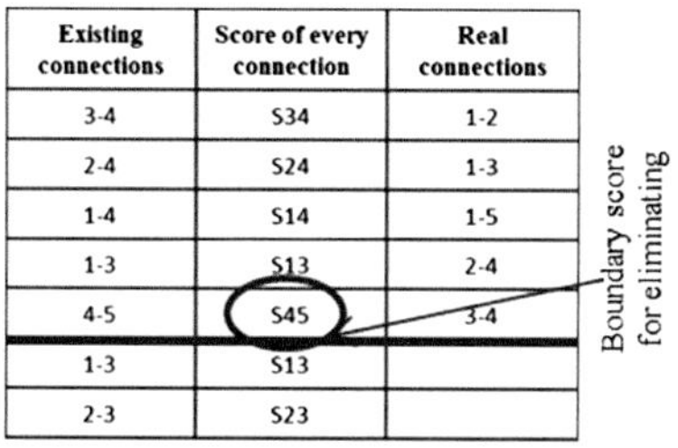

Existing connections	Score of every connection	Real connections
3-4	S34	1-2
2-4	S24	1-3
1-4	S14	1-5
1-3	S13	2-4
4-5	S45	3-4
1-3	S13	
2-3	S23	

Boundary score for eliminating

Fig. 7 The new three connections for year y + 1 from Fig. 6 are added to the existing connections of year y from Fig. 7 and sorted in descending order by their score. Connections with a score less than S45 (boundary for the elimination process) are eliminated from the network

city clustering, socio-economic scenario, etc. Nevertheless, at this stage of the study for the APD network forecasting, it seems reasonable to use the second method of boundary definition–boundary scores.

It is important to note that each cluster pair has different boundaries either for the fixed number of connections method or the boundary score method. This proves the assumption that each cluster pair has its own boundaries.

5 Conclusion

This paper presents the study of topology forecast in APD networks utilizing a socio-economic scenario. The study shows that the Weighted Resource Allocation (WRA) index demonstrates the best performance. AUC and precision metrics are higher for cluster pairs than for the whole APD network. This proves the necessity of separating cities into groups by their socio-economic indicators to improve the link forecasting performance. Thus, the WRA index is used to calculate scores for all non-existing links in each cluster pair. This disproves the assumption that cluster pairs in the APD network have different similarity indices demonstrating the best performance. For existing years, the modeling is applied and results are compared with real data. The accuracy of the similarity-based algorithm for the APD network is higher than in related studies. The study shows two methods of adding new connections from the ordered score list of non-existing connections. The first method is to add a fixed number of connections based on the historical analyses. The other method is to use a score number from the ordered list as the boundary. Both methods prove the assumption that each cluster pair has its own boundary. It seems reasonable to use the second approach with the boundary score. However, this adding process requires further investigation. Despite the low average accuracy for predicting new connections in the APD network, the validation of the topology forecast model demonstrates a high accuracy for elimination connections and the final accuracy for the forecasted APD networks (Fig. 6). Nevertheless, this accuracy is strongly related to the fact that more than 50 % of connections in 2012 stay in the network from 2002, 2007 and 2011. The APD topology forecast approach was tested for 1, 5 and 10 years. Yet, due to the lack of data, current validation is not able to estimate possible error propagation within the forecast period. For future study, the validation must be extended in order to understand the prediction error growth. One of the possible validation methods is forecasting the APD network from 2002 to 2007 and from 2007 to 2012. Comparing validation results could allow for error growth estimation. It is furthermore believed that the accuracy of the similarity-based algorithm could be enhanced. Topology forecasting can be improved by defining appropriate coefficients in the Eq. (1). It is likely that every cluster pair could have its own coefficients. In addition, the main network metrics should be analyzed (e.g. average weighted degree, average path length, modularity, etc.) and compared with the metrics obtained from historical data described in [17]. This may help to understand latent processes for the APD connections generation.

References

1. Terekhov, I., Ghosh, R., Gollnick, V.: A concept of forecasting origin-destination air passenger demand between global city pairs using future socio-economic development scenarios. In: 53rd AIAA Aerospace Sciences Meeting, Kissimmee, Florida, USA (2015)
2. Ghosh, R., Terekhov, I.: Future Passenger Air Traffic Modelling: Trend Analysis of the Global Passenger Air Travel Demand Network. In: 53rd AIAA Aerospace Science Meeting, Kissimmee, Florida (2015)
3. Boeing, Current Market Outlook 2013–2032, USA (2013). http://www.boeing.com/assets/pdf/commercial/cmo/pdf/Boeing_Current_Market_Outlook_2013.pdf [cited 19.11.2014]
4. Dray, L., Evans, A.D., Reynolds, T., Schäfer, A.: Mitigation of aviation emissions of carbon dioxide: analysis for Europe. Transp. Res. Record **2177**, 17–26 (2010)
5. Lü, L., Zhou, T.: Link prediction in complex networks: a survey. Physica A **390**, 1150–1170 (2011)
6. Murata, T., Moriyasu, S.: Link prediction of social networks based on weighted proximity measures. In: IEEE/WIC/ACM International conference on Web Intelligence, Fremont, California (2007)

7. Lü, L., Zhou, T.: Link prediction in weighted networks: the role of weak ties. EPL Lett. J. Explor. Front. Phys. **89**(2012), 18001 (2010)
8. Zheleva, E., Golbeck, J., Kuter, U.: Using Friendship Ties and Family Circles for Link Prediction. Advances in Social Network Mining and Analysis. Lecture Notes in Computer Science, vol. 5498, pp. 97–113 (2012)
9. Hanely, J.A., McNeil, B.J.: The meaning and use of the area under a receiver operating characteristic (ROC) curve. Radiology **143**, 29–39 (1982)
10. Herlocker, J.L., Konstann, J.A., Terveen, L.G., Riedl, J.T.: Evaluating collaborative filtering recommender systems. ACM Trans. Inf. Syst. **22**(1), 5–53 (2004)
11. Sabre Airline Solutions, Aviation Data Intelligence (ADI). http://www.sabreairlinesolutions.com/home/software_solutions/airports/ [cited 19.11.2014]
12. UN, National Accounts Main Aggregates Database. http://unstats.un.org/unsd/snaama/dnllist.asp [cited 19.11.2014]
13. The World Bank, World Bank Open Data. http://data.worldbank.org/indicator/NY.GDP.MKTP.CD. [cited 19.11.2014]
14. UN, World population Prospects: The 2012 Revision. http://esa.un.org/unpd/wpp/Excel-Data/population.htm [cited 19.11.2014]
15. MaxMind, Free World Cities Database. https://www.maxmind.com/en/worldcities. [cited 19.11.2014]
16. Our Airports. http://ourairports.com/data/. [cited 19.11.2014]
17. OpenFlights, Airport database. http://openflights.org/data.html. [cited 19.11.2014]
18. Terekhov, I., Gollnick, V.: Clustering of airport cities and cluster dynamic for the air passenger demand forecasting model based on a socio-economic scenario. In: CEAS conference, Delft, Netherlands (2015)
19. Grosche, T., Rothlauf, F., Heinzl, A.: Gravity models for airline passenger volume estimation. J. Air Transp. Manag. **13**, 175–183 (2007)

Influence and Sentiment Homophily on Twitter Social Circles

Hugo Lopes, H. Sofia Pinto and Alexandre P. Francisco

Abstract Web-based social relations mirror several known phenomena identified by Social Sciences, such as Homophily. Social circles are inferable from those relations and there are already solutions to find the underlying sentiment of social interactions. We present an empirical study that combines existing Graph Clustering and Sentiment Analysis techniques for reasoning about Sentiment dynamics at cluster level and analyzing the role of social influence on sentiment contagion, based on a large dataset extracted from Twitter during the 2014 FIFA World Cup. Exploiting WebGraph and LAW frameworks to extract clusters, and SentiStrength to analyze sentiment, we propose a strategy for finding moments of Sentiment Homophily in clusters. We found that clusters tend to be neutral for long ranges of time, but denote volatile bursts of sentiment polarity locally over time. In those moments of polarized sentiment homogeneity there is evidence of an increased, but not strong, chance of one sharing the same overall sentiment that prevails in the cluster to which he belongs.

1 Introduction

Twitter is a highly dynamic social environment where 316 million monthly active users generate a stream of 500 million tweets per day. It not only allows millions of users to interact among each other, but it is also a window for those interactions. Since it is an accessible and prolific source of social data, Twitter and other web-based social networks are widely used in the literature for different Social-related Analysis [8], such as Network Dynamics [15], Community Detection [16], Event

H. Lopes (✉) · H.S. Pinto · A.P. Francisco
INESC-ID/Instituto Superior Técnico, Universidade de Lisboa, Lisbon, Portugal
e-mail: hugomalopes@tecnico.pt

H.S. Pinto
e-mail: sofia@inesc-id.pt

A.P. Francisco
e-mail: aplf@tecnico.pt

H. Cherifi et al. (eds.), *Complex Networks VII*, Studies in Computational Intelligence 644, DOI 10.1007/978-3-319-30569-1_27

Detection and Prediction [7, 18], Information Flow [2], Influence and Homophily Analysis [1, 21], Sentiment Analysis [12]. Some of these study the interdependencies and possible correlations among the different topics, however we found that there is not an extensive study about sentiment prevalence on clusters and whether this sentiment can be spread by influence into a state of sentiment homophily inside those clusters. Understanding how sentiment behaves at a cluster level can be useful for mining the overall mood of communities, and it may also be useful for improving sentiment classification techniques using enriched information about surrounding emotions.

Easley and Kleinberg [8] define homophily as the principle that people tend to be similar to their friends, which may be caused by selection or social influence. We found that sometimes homophily is defined in the literature as selection itself [19], i.e., people select friends with similar characteristics. Following the first definition, we search for moments of sentiment homophily in social circles and we try to understand if they are caused by social influence. The hypotheses that motivate our work are:

- *H1*: The sentiment expressiveness inside clusters is highly dynamic over time.
- *H2*: Clusters show moments of sentiment prevalence.
- *H3*: During moments of sentiment homogeneity in a cluster, there is an increased chance that a user is influenced by the surrounding emotion and shows a similar sentiment to the one prevailing at that moment.

Regarding some specific terms related with Twitter, a tweet is a message with a maximum size of 140 characters that can include photos and videos. By retweeting a tweet, a user is forwarding that tweet to his own followers. A mention is an explicit reference to a user using the tag "@" followed by his unique username. For instance, typing "@maria" is a mention to the user "maria". A reply is a particular case of a mention in which the mention is located at the bottom of the tweet. Replies are used to comment or answer something that the mentioned user has tweeted.

Using existing clustering and sentiment classification techniques, we propose to measure the overall sentiment of clusters based on the frequency of tweets for each possible sentiment value, regarding their sentiment classification. We found that the neutral value is the most frequent classification during the clusters' lifetime, however different sentiment values appear, usually in spikes and with different polarities over time, confirming the highly dynamic nature of clusters' sentiment (*H1*). We also observed moments of sentiment homophily (*H2*), for instance in chains of retweets or topic-related discussions and we describe a systematic strategy for finding those moments. Finally, we used dubious sentiment classifications for testing the role of influence in the origin of those moments of sentiment homophily by comparing the extrapolation of the clusters' overall sentiment with human-coders' evaluations. With this strategy we found a tendency for ambiguous classifications being correctly relabeled with the prevalent sentiment of respective clusters (*H3*).

2 Related Work

Fowler and Christakis [10] conducted a study about the spread of happiness within social networks, using data from the Framingham Heart Study,[1] collected between 1983 and 2003. From a network of 4,739 individuals, each person was weekly asked how often they experienced certain feelings during the previous week: "I felt hopeful about the future", "I was happy", "I enjoyed life", "I felt that I was just as good as other people". They used this information to measure the state of happiness of individuals throughout a period of time. According to their results there is happiness homophily up to three degrees of separation between nodes. This study not only found evidence of sentiment contagion through influence, it also suggests that it may cause sentiment homophily at a cluster level.

Thelwall [22] searched for homophily in social network sites using data extracted from MySpace, concluding that there was a highly significant evidence of homophily for several characteristics such as ethnicity, age, religion, marital status. Then, he conducted another study on emotion homophily [23], based on the same type of data. Using an initial version of SentiStrength [24] for sentiment classification, two different methods were tested to seek emotion homphily between pairs of friends: a direct method and an indirect method. The direct method compares only the sentiment of the conversational comments between each pair of friends. The indirect method compares the average emotion classification of comments directed to each node, independently, in each pair of friends. Weak but statistically significant levels of homophily were found with both methods. However, the direct method can only give insight of the average homophily at a maximum distance of 1, while the indirect method covers a maximum distance of 3.

Gruzd et al. [11] followed the study of Fowler and Christakis with web-based social network data, focusing on the potential propagation factors for sentiment contagious instead of searching for evidence of sentiment homophily. They performed a topic-oriented data extraction from Twitter in order to minimize possible bias caused by the occurrence of multiple events that generate multiple unrelated discussions, and they found on the 2010 Winter Olympics a well covered and very popular event on Twitter, from which they got strong emotional content. Using SentiStrength for tweets' sentiment classification, they found that a tweet is more likely to be retweeted through a network of follow relations if its tone and content are both positive. Fan et al. [9] decomposed sentiment into four emotions: angry, joyful, sad and disgusting. They used a bayesian classifier to infer these emotions based on emoticon occurrence in interactions extracted from Weibo. Considering pairs of direct friends in a follow-relation network, they only found evidence of emotion homophily regarding anger and joy, observing that anger was the most influential emotion and the chance of contagion was higher in stronger ties. Using a follow-relation network extracted from Twitter, Bollen et al. [5] also found sentiment homophily but regarding sentiment polarity, which they called subjective well-being assortativity. They observed that pairs of friends connected by strong ties are more assortative, however they did not

[1] Medical study about cardiovascular disease—https://www.framinghamheartstudy.org/.

identify whether this phenomenon was caused by selection or social influence. None of these studies analyzed sentiment dynamics over time nor looked into an overall sentiment at community level.

Following these findings, we propose to look for signs of sentiment homophily at a cluster level and understand whether prevalent sentiment in social circles can be used for estimating individuals' sentiment.

3 Dataset Overview

To find social circles and analyze their behavior over time, a large amount of data needs to be extracted during a period of several weeks. We extracted the dataset using Twitter Public Streaming API,[2] filtering the data according to a list of keywords related to 2014 FIFA World Cup. Extraction started on March 13th of 2014 and it ended on July 15th of 2014, covering the entire event that took place from June 12th to July 13th of 2014. It resulted in 166 GB of compressed data containing a collection of 339,702,345 tweets, having missed an estimated amount of at most 30 million tweets about the topic according to the limit messages received from the API. Due to the large amount of countries participating in the World Cup, we only considered a subset of the entire dataset for our analysis. This subset covers the knock-out stage of the event, from June 27th until July 15th, with 97,403,564 tweets that represent 28.7 % of the entire data. We did this to minimize the sparsity of the information, since only 16, from the initial 32 participating countries, were still in competition. English is the most spoken language in the subset, representing 45.8 % of the tweets, followed by Spanish with 24.2 %, and Portuguese with 10.2 %. Regarding the distribution of each type of tweets in the subset we found that 38.2 % are simple tweets, 55.3 % are retweets, and 6.5 % are replies. We also noticed that 64.7 % of all tweets have at least one mention, which makes it the most frequent type of strong relations in the dataset, followed by retweets and then replies. However, the set of mentions contains the set of replies and also intersects the set of retweets.

4 Approach

Our approach is divided into four stages: User Clustering; Tweet Clustering; Sentiment Analysis; and Influence and Homophily Analysis in time series, as it is outlined in Fig. 1. The first three stages integrate existing solutions for clustering and sentiment analysis with several scripts for data transformation. They were used to process the extracted dataset into time-series of sentiment information about social circles. With the preprocessed data obtained from these three stages, we propose a set of metrics to evaluate the extent of sentiment homophily. Then, we propose a strategy

[2]https://stream.twitter.com/1.1/statuses/filter.json.

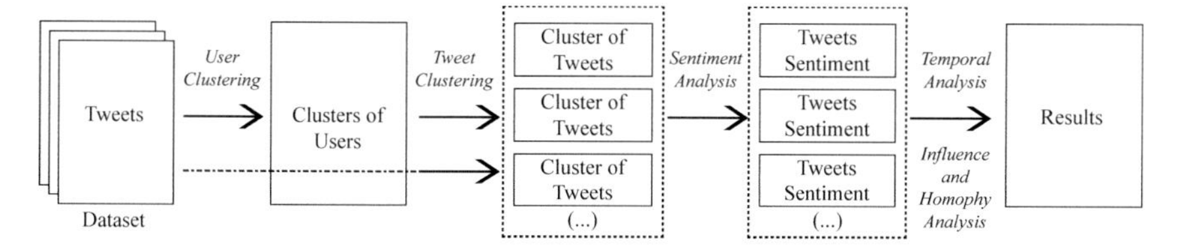

Fig. 1 High-level view of the designed workflow

to ascertain a possible relation between influence and sentiment, which can eventually improve the sentiment classification of tweets in clusters that denote sentiment homophily.

4.1 User Clustering

Before finding the social circles, we needed to find the social network that comprises them. We decided to build the network's graph considering only strong ties, which the literature states to be found in retweets and mentions [6, 13, 20]. However, we chose to use only replies because retweets and replies are mutually exclusive and replies represent direct conversations, which may not be necessarily true with mentions. We started by filtering all retweets and replies from the dataset, converting them from JSON to the condensed format: *"type tweetID userID receiverID timestamp"*. To analyze the clusters in different periods of time, the set of retweets and replies were individually filtered and sorted by their timestamp values, for independent analysis.

Once we were dealing with networks with millions of nodes and edges, we chose to use Webgraph[3] [4] to build and analyze underlying graphs, and used Layered Label Propagation (LLP) algorithm in LAW software library[4] for clustering them. Besides compressing the ASCIIGraph to the WebGraph's format BVGraph, we had to symmetrize it to an undirected and loop-less graph to be used by LLP algorithm, to do user clustering. The symmetric graph was also used to calculate the connected components of the network. LLP [3] is an iterative strategy that reorders the graph such that nodes with the same label are close to one another. This node reordering is useful for graph compression, however, for our purposes we only require the node labeling assignment produced by the label propagation algorithm that returns a clustering configuration of the graph. The clustering result is mappable with a sorted list of user IDs, and all these steps are outlined in Fig. 2.

[3]http://webgraph.di.unimi.it/.

[4]http://law.di.unimi.it/software.php.

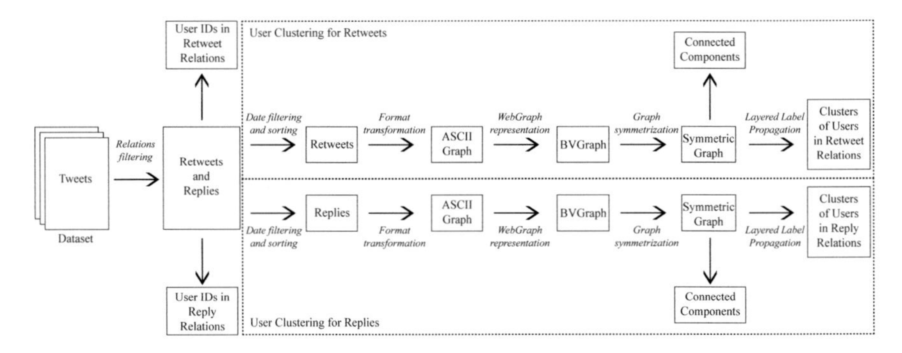

Fig. 2 User clustering process

4.2 Tweet Clustering

Our strategy to classify the sentiment of a cluster was getting the tweets that the users in that cluster tweeted during the lifetime of the cluster, and then classifying each one independently to sum up an overall result. For that we clustered the tweets according to the previously obtained clusters of users, i.e., we extracted from the dataset all the tweets of each user in the cluster, created in the same period of time used to cluster the users. Then we converted these tweets to the shorter format: *"userID tweetID language epochTimestamp hashtagCounter URLCounter mentionCounter tweetText"*. All the clusters with only one or two tweets were removed. Each cluster of tweets was filtered and divided by its prevalent language, in order to perform the sentiment classification without mixed languages (Fig. 3).

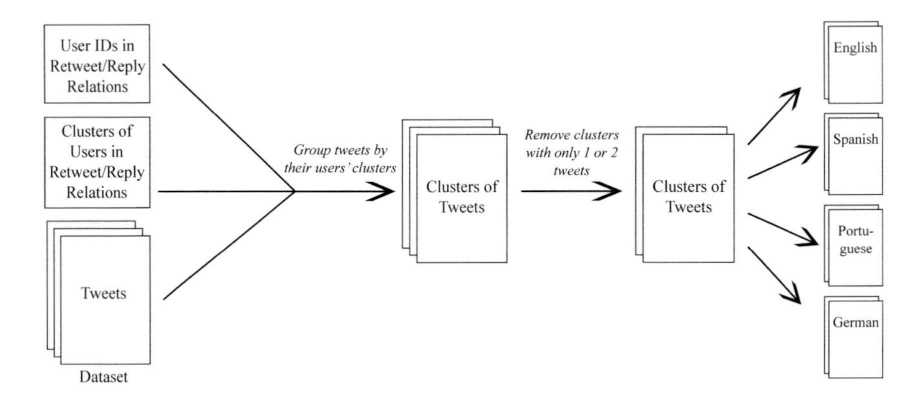

Fig. 3 Tweet clustering process

4.3 Sentiment Analysis

We chose the lexicon-based SentiStrength tool [24] to perform automatic sentiment classification of the tweets, because (1) it does not require training data when working in unsupervised mode; (2) it has good performance and it is able to process more than 16,000 tweets/s in standard machines; (3) and has good results on Twitter datasets [11, 24]. Giving a text file as input, SentiStrength outputs another file with each line of text of the input file annotated with two sentiment values: a positive integer $s_+ \in \{1, \ldots, 5\}$ and a negative integer $s_- \in \{-5, \ldots, -1\}$. The higher the absolute value, the higher the polarity strength of that value.

To classify the tweets in each cluster of tweets we filtered only the tweet text. To avoid words out of context that could be matched by SentiStrength, we removed all the mentions, retweet indicatives and URLs occurrences in the text. After running SentiStrength over the clusters of tweets we got, for each cluster, a matching file with the classified sentiment annotated for each tweet.

4.4 Influence and Sentiment Homophily Analysis over Time

The user clustering, tweet clustering and sentiment analysis stages were scripted to extract the information about the clusters in the network and their sentiment, during desired time intervals. For our analysis we performed a round-based clustering for each round of the knock-out stage subset, which includes the round of 16, quarter-finals, semi-finals and final stage of the World Cup.

Since we were seeking an overall sentiment, we chose to condensate the two sentiment values in one unique value, calculating the Absolute Sentiment value,

$$|s| = s_+ + s_-, \in \{-4, \ldots, 0, \ldots, 4\} \quad (1)$$

This way, a tweet is positive with a strength between 1 and 4, neutral when 0, or negative with a strength between -1 and -4. This approach promotes clearly polarized sentiment results and penalizes balanced strength results. Thus, the results $(5, -5)$, $(4, -4)$, $(3, -3)$, $(2, -2)$, which we consider ambiguous results, have the same absolute sentiment of 0 as the SentiStrength neutral result $(1, -1)$.

We focused on polarity changes over time and we calculated the distribution of the absolute sentiment values per hour, in each cluster, by counting the number of tweets for each absolute sentiment result. By analyzing these distributions over time we were able to observe sentiment dynamics and detect sentiment homophily, when existing.

To systematically find periods of polarity homophily, assuming that sentiment homophily is found locally in time, we defined a time window t, a minimum number of tweets m needed to consider a sentiment prevalence in t, and minimum rate of polarity prevalence p in t, as metric for sentiment homogeneity. Let $\Delta t(x_1, x_2)$ be the time interval between two tweets, and $pol(x_1, \ldots, x_n)$ be the rate of the prevalent polarity in a sequence of tweets, there is sentiment homophily for a sequence of tweets $x_1, x_2, \ldots, x_n$ when,

$$n \geq m \wedge pol(x_1, \ldots, x_n) \geq p \wedge \forall \{x_i, x_{i+1}, \ldots, x_{i+m}\} \in \{x_1, x_2, \ldots, x_n\}, \Delta t(x_i, x_{i+m}) \leq t. \quad (2)$$

However, finding time intervals that satisfy this metric does not show if there is an increased chance of any user in that cluster of sharing the same befitting sentiment with the overall sentiment that surrounds him, i.e., being influenced by his peers' mood. Our approach to evaluate whether moments of sentiment homophily are caused by influence is to look for ambiguous tweets in moments of prevalent polarized sentiment in the cluster, to which we assign that same prevalent polarization, and then we compare this updated sentiment classification with human coders classifications.

Lets assume the pairs $(1, -1)$, $(2, -2)$, $(3, -3)$, $(4, -4)$, $(5, -5)$ as ambiguous results in polarized clusters. The reason for this assumption regarding $(2, -2)$, $(3, -3)$, $(4, -4)$, $(5, -5)$ is that they reveal sentiment strength but not a decided polarization, even in a polarized environment. We also include $(1, -1)$ because SentiStrength outputs this value both for neutral sentences and for sentences that do not match any word in the lexicon, which gives a dubious meaning to this value. This way, we trust more in polarized classifications.

After identifying ambiguous results, we search for an ambiguity a that has a number of surrounding tweets equal or greater than m, with a prevalence of a certain polarity equal or greater than p during a period of time t that includes a. For each ambiguity a found in a context with these characteristics, we set its polarity to be the same as the prevalent polarity of the tweets surrounding it. We propose two algorithms to do this sentiment extrapolation, that only differ in the position that the ambiguity occupies in the context configuration. The first algorithm searches for ambiguities that have a central position in the polarized context, being fixed at the center of the time window. For a set of ambiguities A found in a sequence of tweets $T = \{x_1, \ldots, x_n\}$, when $x_a \in A \wedge x_a \in T$, and

$$\begin{aligned} \exists x_b, x_e \in T, (b \leq a < e \vee b < a \leq e) \wedge \Delta t(x_b, x_a) \leq t/2 \wedge \\ \Delta t(x_a, x_e) \leq t/2 \wedge e - b \geq m \wedge pol(x_b, x_e) \geq p, \end{aligned} \quad (3)$$

then the sentiment polarity of x_a is relabeled with the prevalent sentiment polarity in $x_b, \ldots, x_e$.

The second algorithm considers any ambiguity that belongs to a sliding time window t that fulfills those restrictions, independently of its position towards the context. For a set of ambiguities A found in a sequence of tweets $T = \{x_1, \ldots, x_n\}$, when $x_a \in A \wedge x_a \in T$, and

$$\exists x_b, x_e \in T, (b \leq a < e \vee b < a \leq e) \wedge \Delta t(x_b, x_e) \leq t \wedge e - b \geq m \wedge pol(x_b, x_e) \geq p, \quad (4)$$

then the sentiment polarity of x_a is relabeled with the prevalent sentiment polarity in $x_b, \ldots, x_e$.

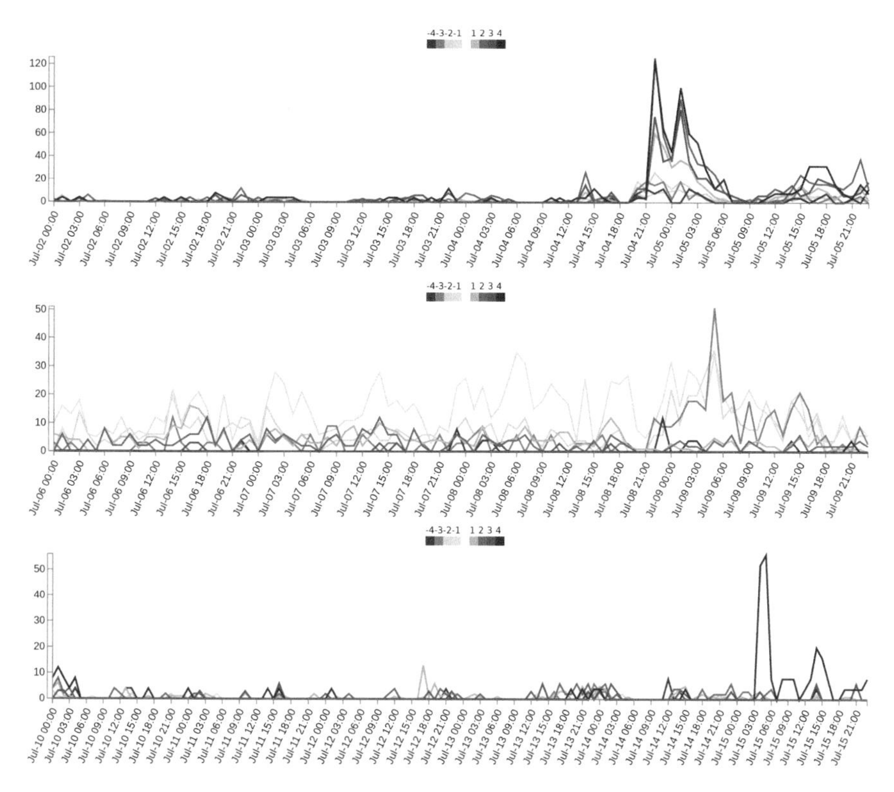

Fig. 4 Time-line of tweets' frequency of absolute sentiment for each accumulation of 3 h. Cluster "413547" from the Spanish-speaking set of reply-based clusters over the quarter-finals stage, cluster "1000883" from the English-speaking set of reply-based clusters over the semi-finals, and cluster "2049176" from the Spanish-speaking set of retweet-based clusters over the final stage

5 Results and Discussion

We used the modularity coefficient Q, that measures the division of the nodes in a graph into different clusters and the strength of their connections [17], to evaluate the quality of the clusters obtained with LLP algorithm. For clusters obtained from retweet-relation graphs we got an average of $Q = 0.620$, while for reply-relation graphs this value increased for $Q = 0.800$. This denotes that reply-relations are more restrict than retweets and generate smaller but denser clusters. The size distribution of all sets of clusters followed a power-law, regardless the round, language, or type of relation of the graphs. Considering hypothesis *H1* and *H2* we can observe in Fig. 4 that sentiment is highly dynamic, especially for reply-based clusters. With periods of sentiment neutrality interleaved with periods of sentiment polarity, there are moments in which a certain polarity prevails, where we can find periods of local sentiment homophily. Even though these moments appear to be quite ephemeral. The majority of clusters have sentiment peaks during their lifetime, which represents

85.5 % of the clusters obtained from retweet-relation graphs and 95.2 % of the clusters obtained from reply-relation graphs, confirming *H1*. Clusters that showed total neutral sentiment were mainly small size clusters. We detected that chains of polarized retweets generate moments of sentiment homogeneity (*H2*), as well as some topic-related conversations, which are respectively more frequent (but not exclusively) on retweet-based clusters, and reply-based clusters. If we assume that, when some user retweets a certain status, there is a chance of that user being also sharing the inherent sentiment of that status' message, then we may say that there is sentiment influence on cascades of retweets.

Regarding *H3*, we gathered 24 human-coders, in which 23 of them are Portuguese native-speakers and the remaining one is a Spanish native-speaker. All of them are able to read and interpret English, and 18 are also able to read and interpret Spanish. We shuffled them into 8 groups of 3, and each group evaluated two sets of 100 ambiguous tweets. This way, each ambiguity was classified by three different human-coders. The testing samples were randomly collected from the set of ambiguous tweets found with the sliding window algorithm, using the fixed parameters $t = 6\,\text{h}$, $m = 10$, and $p = 0.7$. These samples sum a total of 1,600 ambiguous tweets, divided into 800 for English, 600 for Spanish, and 200 for Portuguese. Half of the sets of each language was extracted from retweet-based clusters, and the other half from reply-based clusters. Each person was asked to classify the sentiment expressed in the tweet message, as *positive*, *neutral*, or *negative*. We chose to only ask for the polarity and not the sentiment strength to simplify the classification process. We included the *neutral* option assuming that there are indeed some tweets that do not express any kind of polarization. The results in Table 1 suggest a tendency for the real sentiment of ambiguous tweets to match the overall sentiment of their clusters, over having a neutral or mismatching sentiment polarity, and this value is clearly higher than it would be assigned by chance. However, this matching rate is not sufficient to claim that when there is a period of sentiment homophily there is a strong chance of a user in that cluster sharing a tweet with an equivalent polarity. We evaluated the reliability of the human coder classifications in terms of agreement using the Krippendorff's alpha-coefficient [14], which varied between 0.24703 and 0.53167, i.e., they are statistically reliable but with a certain level of disagreement, unveiling the subjective nature of this task.

Table 1 Manual evaluation results regarding the approach implemented in the sliding window algorithm, and comparison with a random approach, considering also the level of agreement between human-coders

Set of ambiguities		Human-coders agreement (%)			Cluster sentiment polarity mismatch (%)			Neutral sentiment (%)			Cluster sentiment polarity match (%)		
		≥2	Unanimity	Total dis-agreement	≥2	Unanimity	Random	≥2	Unanimity	Random	≥2	Unanimity	Random
en	*RT*	92.50	36.25	7.50	30.00	28.28	29.00	24.86	20.69	37.75	45.14	**51.03**	33.25
	RE	93.75	39.00	6.25	16.80	20.51	27.00	36.53	33.97	34.75	**46.67**	45.51	38.25
	Total	93.13	37.63	6.88	23.36	24.25	28.00	30.74	27.57	36.25	45.91	**48.17**	35.75
es	*RT*	86.33	32.00	13.67	32.43	32.29	36.33	15.83	7.29	30.67	51.74	**60.42**	33.00
	RE	90.33	37.67	9.67	34.32	33.63	36.33	17.71	7.96	33.00	47.97	**58.41**	30.67
	Total	88.33	34.83	11.67	33.40	33.01	36.33	16.79	7.66	31.83	49.81	**59.33**	31.83
pt	*RT*	87.00	29.00	13.00	32.18	27.59	33.00	29.89	24.14	32.00	37.93	**48.28**	35.00
	RE	95.00	42.00	5.00	36.84	42.86	35.00	20.00	9.52	37.00	43.16	**47.62**	28.00
	Total	91.00	35.50	9.00	34.62	36.62	34.00	24.73	15.49	34.50	40.66	**47.89**	31.50
Global	*RT*	89.50	33.75	10.50	31.15	29.63	32.25	22.21	16.30	34.38	46.65	**54.07**	33.38
	RE	92.63	38.88	7.38	25.78	28.30	31.50	27.53	21.22	34.38	46.69	**50.48**	34.13
	Total	91.06	36.31	8.94	28.41	28.92	31.88	24.91	18.93	34.38	46.67	**52.15**	33.75

6 Conclusion and Future Work

With this work we observed that sentiment reveals a highly dynamic behavior at cluster level, having ephemeral spikes of polarity usually lasting for a few hours. We were able to locally find moments of sentiment homogeneity during those spikes by setting a time window t, a minimum number of tweets m needed to consider a sentiment prevalence in t, and minimum rate of polarity prevalence p in t. Using a sample with 97,403,564 tweets where 14,075,547 different users participate in retweet/reply relations, we found similar results for clusters of different languages (English, Spanish, Portuguese) during different periods of time. For understanding if an existing overall sentiment in a cluster may influence the sentiment of its individuals, we relabeled the sentiment of ambiguous classifications surrounded by a context of sentiment homophily with the prevalent sentiment of that cluster during t and we evaluated this extrapolation with human coders. The matching rate between the human-coders classification and the clusters' sentiment polarity extrapolation always shows higher and more stable expressiveness over mismatching and neutral rates. However, with the best matching result around 60 %, we can only say we found a weak but significant tendency of a user sharing a befitting sentiment in a cluster during a period of sentiment homogeneity.

Given the level of disagreement between human coders it would be desirable to use a higher odd number of coders for each evaluation set. In the future it would be interesting to separate neutral sentiment classifications from undecidable sentiment classifications, which have the same value $(1, -1)$ when classified by SentiStrength, and see what would happen to the rate of neutral classifications among the human coder classifications. It could also be interesting testing the repeatability of the results when exploring different techniques for building and clustering the network's graph, such as using ego-networks and local clustering methods, respectively.

Acknowledgments This work was partly supported by national funds through Fundação para a Ciência e Tecnologia (FCT), under projects EXCL/EEI-ESS/0257/2012 and UID/CEC/50021/2013.

References

1. Bakshy, E., Hofman, J.M., Mason, W.A., Watts, D.J.: Everyone's an influencer: quantifying influence on twitter. In: Proceedings of the Fourth ACM International Conference on Web Search and Data Mining. WSDM '11, pp. 65–74. ACM, NY (2011)
2. Bakshy, E., Rosenn, I., Marlow, C., Adamic, L.: The role of social networks in information diffusion. In: Proceedings of the 21st International Conference on World Wide Web. WWW '12, pp. 519–528. ACM, NY (2012)
3. Boldi, P., Rosa, M., Santini, M., Vigna, S.: Layered label propagation: a multiresolution coordinate-free ordering for compressing social networks. In: Proceedings of the 20th International Conference on World Wide Web. WWW '11, pp. 587–596. ACM, NY (2011)
4. Boldi, P., Vigna, S.: The webgraph framework I: compression techniques. In: Proceedings of the 13th International Conference on World Wide Web. WWW '04, pp. 595–602. ACM, NY (2004)

5. Bollen, J., Gonçalves, B., Ruan, G., Mao, H.: Happiness is assortative in online social networks. Artif. Life **17**(3), 237–251 (2011)
6. Cha, M., Haddadi, H., Benevenuto, F., Gummadi, K.P.: Measuring user influence in twitter: the million follower fallacy. In: ICWSM '10: Proceedings of international AAAI Conference on Weblogs and Social (2010)
7. Conover, M., Gonçalves, B., Ratkiewicz, J., Flammini, A., Menczer, F.: Predicting the political alignment of twitter users. In: SocialCom/PASSAT, pp. 192–199. IEEE (2011)
8. Easley, D., Kleinberg, J.: Networks, Crowds, and Markets: Reasoning About a Highly Connected World. Cambridge University Press, NY (2010)
9. Fan, R., Zhao, J., Chen, Y., Xu, K.: Anger is more influential than joy: sentiment correlation in weibo. PLoS ONE **9**(10), e110184 (2014)
10. Fowler, J., Christakis, N.: Dynamic spread of happiness in a large social network: longitudinal analysis over 20 years in the framingham heart study. Br. Med. J. **337**, a2338 (2008)
11. Gruzd, A., Doiron, S., Mai, P.: Is happiness contagious online? A case of twitter and the 2010 winter olympics. In: Proceedings of the 2011 44th Hawaii International Conference on System Sciences. HICSS '11, pp. 1–9. IEEE Computer Society, Washington (2011)
12. Hodeghatta, U.R.: Sentiment analysis of hollywood movies on twitter. In: Proceedings of the 2013 IEEE/ACM International Conference on Advances in Social Networks Analysis and Mining. ASONAM '13, pp. 1401–1404. ACM, NY (2013)
13. Huberman, B., Romero, D., Wu, F.: Social networks that matter: twitter under the microscope. First Monday **14**(1) (2008)
14. Krippendorff, K.: Computing krippendorff's alpha reliability. Technical report, University of Pennsylvania, Annenberg School for Communication (2011)
15. Leskovec, J., Huttenlocher, D., Kleinberg, J.: Signed networks in social media. In: Proceedings of the SIGCHI Conference on Human Factors in Computing Systems. CHI '10, pp. 1361–1370. ACM, NY (2010)
16. Mcauley, J., Leskovec, J.: Discovering social circles in ego networks. ACM Trans. Knowl. Discov. Data **8**(1), 4:1–4:28 (2014)
17. Newman, M.: Networks: An Introduction. Oxford University Press Inc, NY (2010)
18. Sakaki, T., Okazaki, M., Matsuo, Y.: Earthquake shakes twitter users: real-time event detection by social sensors. In: Proceedings of the 19th International Conference on World Wide Web. WWW '10, pp. 851–860. ACM, NY (2010)
19. Shalizi, C.R., Thomas, A.C.: Homophily and contagion are generically confounded in observational social network studies (2010)
20. Tang, J., Chang, Y., Liu, H.: Mining social media with social theories: a survey. SIGKDD Explor. Newsl. **15**(2), 20–29 (2014)
21. Tang, J., Gao, H., Hu, X., Liu, H.: Exploiting homophily effect for trust prediction. In: Proceedings of the Sixth ACM International Conference on Web Search and Data Mining. WSDM '13, pp. 53–62. ACM, NY (2013)
22. Thelwall, M.: Homophily in myspace. J. Am. Soc. Inf. Sci. Technol. **60**(2), 219–231 (2009)
23. Thelwall, M.: Emotion homophily in social network site messages. First Monday **15**(4) (2010)
24. Thelwall, M., Buckley, K., Paltoglou, G.: Sentiment strength detection for the social web. J. Am. Soc. Inf. Sci. Technol. **63**(1), 163–173 (2012)

Comparative Network Analysis Using KronFit

Gupta Sukrit, Puzis Rami and Kilimnik Konstantin

Abstract Comparative network analysis is an emerging line of research that provides insights into the structure and dynamics of networks by finding similarities and discrepancies in their topologies. Unfortunately, comparing networks directly is not feasible on large scales. Existing works resort to representing networks with vectors of features extracted from their topologies and employ various distance metrics to compare between these feature vectors. In this paper, instead of relying on feature vectors to represent the studied networks, we suggest fitting a network model (such as Kronecker Graph) to encode the network structure. We present the directed *fitting-distance* measure, where the distance from a network A to another network B is captured by the quality of B's fit to the model derived from A. Evaluation on five classes of real networks shows that KronFit based distances perform surprisingly well.

Keywords Complex networks · Comparative analysis · Generative models · Distance metrics

1 Introduction

Comparative network analysis and network classification on the basis of structural similarity are at a nascent stage holding great potential. The topology of a network often encompasses important information on the functionality and dynamics of the system it represents. As case in point, structural similarity of road networks and fungal networks are the result of low cost and robustness being the main driving

G. Sukrit (✉) · P. Rami · K. Konstantin
Department of Information Systems Engineering,
Ben Gurion University of the Negev, 8410501 Beersheba, Israel
e-mail: gupta@post.bgu.ac.il

P. Rami
e-mail: puzis@bgu.ac.il

K. Konstantin
e-mail: kilimnik@post.bgu.ac.il

H. Cherifi et al. (eds.), *Complex Networks VII*, Studies in Computational Intelligence 644, DOI 10.1007/978-3-319-30569-1_28

forces in the network development [1]. Evaluation of network similarity is important in diverse research fields, particularly in computational biology, where it reveals previously unknown interactions and biological function of protein complexes.

So far network similarity does not have a concrete, rigorous definition. Therefore, researchers employ vectors of features extracted from the networks at various scales to compare networks. Existing features range from microscopic properties that describe interactions between individual nodes to macroscopic features that describe the network as a whole, e.g. average path length, degree distribution exponent, etc. Lately, mesoscopic features such as network motifs [2], graphlets [3], and backbones [4] have been utilized.

An important limitation of comparing networks based on feature vectors is that no single set of features can be claimed universal. Some network properties are good for comparing between protein interaction networks, other work well for social networks. In addition, feature extraction is computationally expensive in large networks [5]. This makes use of approaches that do not involve feature vectors a lucrative research problem. In this paper, we take a direction orthogonal to the conventional one and represent a network by its model rather than a set of features.

There are several statistical models that can capture the topology of a given network by fitting a small set of parameters [6–9]. The result of such fitting can be regarded as a compressed (imperfect) representation of the original network i.e. features of the network topology are recapitulated by a small number of metrics. *Similar networks should have similar models.* We explore this claim using series of distance metrics which are based on Kronecker Graphs model fitting algorithm (KronFit) [9]. We also take into account distance metrics derived from network features and compare them with KronFit based distance metrics in performing unsupervised clustering. We analyze the quality of clusters produced using each of these distance metrics, by evaluating against a number of cluster quality metrics. We show that *log-likelihood* (LL) of one network being generated by a Kronecker Graph model fitted to another network performs surprisingly well as a measure of similarity between networks.

The rest of this paper is structured as follows: We recapitulate the Kronecker Graphs generative model in Sect. 2. We proceed and develop series of network similarity estimators based on KronFit algorithm in Sect. 3, followed by description of baseline network distances in Sect. 4. In Sect. 5, we show that augmented network similarity estimators derived from KronFit perform surprisingly well based on the results of experiments performed with 5 classes of real networks. Sect. 6 discusses related works. Our conclusions on generative models as a tool in comparative analysis of networks are summarized in Sect. 7.

2 Background on KronFit

Let $G = (V, E)$ denote a network where $V = \{1, \ldots, n\}$ is a set of n vertices and $E \subseteq V^2$ is a set of m directed unweighted edges. We represent all undirected networks using the directed edges semantic ($(u, v) \in E \Leftrightarrow (v, u) \in E$). Although, we assume unweighted networks, the proposed approach can easily be extended to weighted networks as well. Let A represent the $n \times n$ adjacency matrix of G such that for any $u, v \in V$, $A_{uv} = 1$ if $(u, v) \in E$ and $A_{uv} = 0$, otherwise.

Let $M^{n \times n}$ be a matrix. Let $M^{[k]}$ be the kth Kronecker product of the matrix, then its Kronecker square [10] is a matrix $M^{[2]}$ of dimensions $n^2 \times n^2$. The items of $M^{[2]}$ are

$$M_{ij}^{[2]} = M_{div(i,n),div(j,n)} \cdot M_{mod(i,n),mod(j,n)}$$

where $div(i, n)$ is the integer quotient of i divided by n and $mod(i, n)$ is the remainder. Leskovec et al. [9] have suggested using $I^{2\times2}$ initiator matrices, where $I_{ij} \in (0, 1)$, raised to a Kronecker power of k as the basic building block in generation of large scale probabilistic adjacency matrices $P = I^{[k]}$. The dimension of all generated probabilistic adjacency matrices P is thus $2^k \times 2^k$ where k is some integer. Every item P_{ij} is the probability of having an edge between the vertices i and j in the generated Kronecker networks. The log-likelihood of a given adjacency matrix A to be generated by drawing each edge (i, j) with the probability P_{ij} is:

$$LL = \sum_{i \in V} \sum_{j \in V} \log \left(P_{ij}^{(A_{ij})} \left(1 - P_{ij}\right)^{(1-A_{ij})} \right) \tag{1}$$

The KronFit algorithm, suggested by the authors, finds the optimal initiator matrix such that LL is maximized. The authors observed that the 2×2 initiator matrices are sufficient for a good match and choosing larger initiator matrices does not improve the results significantly.

Leskovec et al. also suggest a method to reduce the complexity of LL calculation to $O(m)$ by first computing the probability of a network with 2^k vertices and no edges being generated from P. However, the number of vertices in real networks is not a power of two. Therefore, the network G is padded with $2^k - n$ disconnected vertices. In the following discussions, we will assume a padded adjacency matrix $\hat{A}$ whose dimensions are $2^k \times 2^k$ where $k = \lceil \log_2 n \rceil$. $\hat{A}$ is constructed from A by padding $2^k - n$ rows and $2^k - n$ columns with zeros. $\hat{A}$ can be interpreted as an adjacency matrix of a network that is similar to G but has additional $2^k - n$ isolated vertices, which we refer to as synthetic vertices in following discussions.

3 KronFit Network Distances

Model distance (MD). An initiator matrix $I^{2\times2} = \begin{bmatrix} a & b \\ c & d \end{bmatrix}$ contains four values that lie in the range of [0, 1] and are considered equally important for fitting a correct model. Lescovec et al. [9] proposed to compare networks by comparing their initiator matrices. We use *euclidean distance (ED)* to perform such kind of comparison. An *ED* heatmap on 500 networks of five different types[1] is presented in Fig. 1a.

[1]Details on evaluated data set are presented in Sect. 5.1.

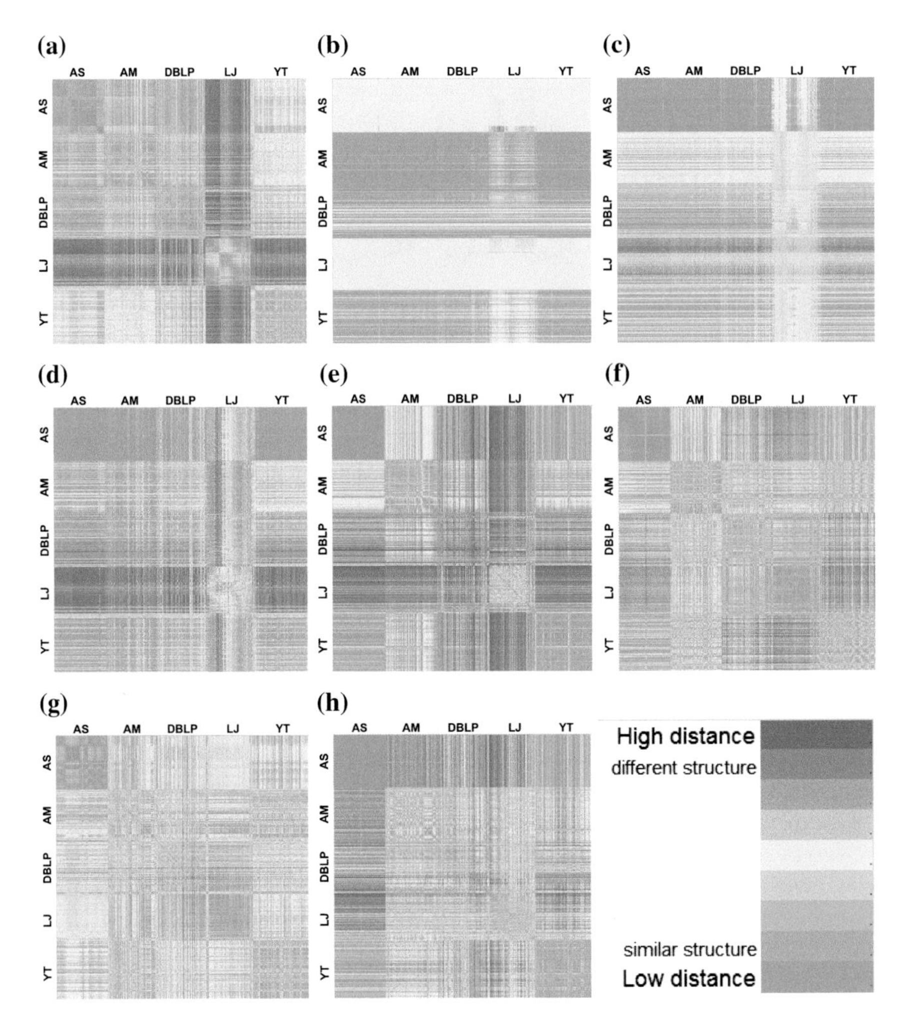

Fig. 1 Network similarity metrics. **a** *MD*. **b** *FD*. **c** *SFD*. **d** *NFD*. **e** Density. **f** Clust.Coef. **g** Diameter. **h** Degree distr.

Fitting Distance (FD). Given a network G, an initiator matrix I found by the KronFit algorithm maximizes the log-likelihood of G being generated using I. The likelihood of similar networks being generated by the same model is maximized as well. Structurally dissimilar networks should, therefore, have lower likelihood of being generated by the model. In the rest of this paper we denote the initiator matrix fit to a network G as I_G. We will refer to the network G as *prototype*. We denote by LL_{GS} the log-likelihood of a *subject network* S being generated from I_G.

These LL_{GS} values can be used as inverse distance measure between the networks. LL_{GS} values are negative numbers ranging from $-2.6E + 08$ to -376 in current study. The value $LL_{GS} = -2.6E + 08$ was obtained for a prototype being a community in the Live Journal social network tested against one of the Autonomous Systems topologies. In the rest of this paper, we denote the negation of LL_{GS} as the *fit-distance* (FD) measure between networks.

LL computation time is linear at the number of edges. Thus the time complexity of computing the FD matrix is $O(k^2\hat{m})$ where k is the number of networks in the data set and $\hat{m}$ is the average number of edges. In contrast, MD is easier to compute because for each pair of networks it requires calculating the euclidean distance between the respective initiator matrices, an operation which can be regarded as constant time.

See Fig. 1b for a heatmap of FD. The columns in this figure correspond to prototype networks and rows correspond to subject networks. We can clearly notice the large horizontal strips in this heatmap. In contrast to MD which is a symmetric measure, LL_{GS} can differ significantly from LL_{SG}. There are several factors that affect this asymmetry. For example, KronFit does not work equally well for all kinds of networks. We notice that types of prototype networks with high average FD (the yellow horizontal stripes in Fig. 1b) are better differentiated from other networks using this measure. Network size is another significant factor that affects the LL calculation. The more nodes a network S has the lower is its LL_{GS} value w.r.t. any prototype G.

Scaled Fitting Distance (SFD). The log likelihood metric for a network crudely measures how well the synthetic networks generated from the initiator matrix will match the original network. The larger the networks we strive to generate, the more variations can be there and thus, the likelihood of generating a particular subject network of the same size drops. Figure 2 shows that LL scales as the number of elements in the padded adjacency matrix $\hat{A}^{2^k \times 2^k}$.

Next, we adjust the $-LL_{GS}$ values by the factor of $1/2^{2k}$. The resulting distance measure is presented in Fig. 1c. It is clear that the adjusted LL measure is more informative but the asymmetry is still there. The fitting distance obtained from SLL is referred to as Scaled Fitting Distance (SFD), in the rest of the paper.

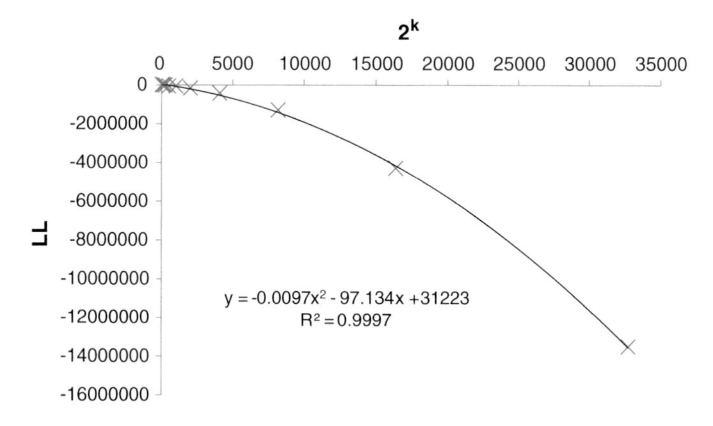

Fig. 2 Log-likelihood (averaged over all prototype networks and subject networks having the same $k = \lceil \log_2 n \rceil$) as a function of the dimension of the padded adjacency matrix (2^k)

Normalized Fitting Distance (NFD). Equation 1 inadvertently incorporates an extra factor that affects the magnitude of log likelihood. While calculating log likelihood, they also consider the probability of edges formed by vertices that weren't part of the initial network. The largest deviations are found in networks whose size is just above a power of two ($n = 2^k + i$ where i is some small integer). In such networks, the number of synthetically added vertices (zero padding discussed in Sect. 2) grows to the scale of original network size and notably affects the log likelihood value. Kim and Leskovec [11] propose a corrected log-likelihood LL_{N_Z} that disregards padded nodes. In this paper we focus on KronFit and the original LL rather than LL_{N_Z} and KronEM proposed in [11].

While comparing a subject network to the existing prototypes one would like to find the closest prototypes. In this case, maintaining consistency among subject networks is not required. We define a Normalized Fitting Distance (NFD) such that the most similar network (self) is at distance zero and the farthest network is at distance unity.

$$NLL_{GS} = \frac{LL_{GS} - \max_{G'} LL_{G'S}}{\min_{G'} LL_{G'S} - \max_{G'} LL_{G'S}}$$

Since all values are negative, NLL_{GS} is a positive real number in range [0, 1]. The distance metric obtained from NLL is termed as Normalized Fit-Distance (NFD).

4 Baseline Network Distances

We compare the KronFit-based network distances to four common baseline metrics: Density, Clustering Coefficient, Diameter, and Degree distribution [12].

Density of a network is the fraction of existing edges out of possible edges in the network. For undirected networks $Density = \frac{2m}{n(n-1)}$ where m is the number of edges and n is the number of nodes. The $Density$ network distance is simply the absolute difference between the densities of two networks.

Clustering Coefficient (CC) of a node is the fraction of existing edges between node's neighbors out of all possible edges between them. CC of a node can be regarded as the density of its ego network. Average CC is a common macroscopic network feature. Similar to $Density$, the CC network distance is the absolute difference between the networks' average CCs.

Diameter is the distance between the two farthest nodes in the network. We consider the absolute difference between network diameters as the $Diameter$ network distance.

Degree Distribution is one of the the most common microscopic properties that are used to describe networks. In this paper, we consider an euclidean distance between normalized degree distributions [3] as the eighth network distance measure.

Table 1 Details of the evaluated networks (all values are averaged over 100 networks of each type)

Network type	n	m	Avg. degree	Avg. CC	Diameter	LL_{GG}
AS	6,060	23,891	7.78	0.362	10	−188,666
Amazon	96	326	7.12	0.560	8.7	−1,422
DBLP	662	2,224	10.1	0.832	10	−13,406
LiveJournal	270	14,320	116.2	0.857	3.5	−38,279
Youtube	288	862	5.64	0.366	6.7	−5,098

5 Evaluation

5.1 Data and Procedure

We conducted experiments on 500 networks containing five different types of networks. 100 snapshots of the Autonomous Systems relationships between the years 1997 and 2000 were obtained from [13]. The rest of the networks were obtained from the SNAP network collection [14] where we selected 100 largest network files from each one of the following data sets: Amazon (AM), DBLP, LiveJournal (LJ), and YouTube (YT). The evaluated data is briefly summarized in Table 1.

First we have computed the degree distribution, diameter, average clustering coefficient, and density for each network. We normalized the degrees of vertices to fit the range (0, 1]. Then for every pair of consecutive deciles of the normalized degrees (d_i, d_{i+1}), we calculated the fraction of vertices whose normalized degrees are between d_i and d_{i+1}. Along with diameter, average clustering coefficient, and density this results in 13 features that describe each network.

We proceed by calculating the initiator matrices for all 500 networks in the data set. We set the initiator matrix, $I^{[1]} = \begin{bmatrix} 0.9 & 0.5 \\ 0.5 & 1 \end{bmatrix}$ (these are the default values used in Leskovec et al. [9] and these represent the general trend in fitted values for most networks) and configure the KronFit algorithm with the following parameters: 50 iterations for gradient descent; learning rate equal to 10^5; minimal and maximal gradient steps equal to 0.005 and 0.05 respectively; 100,000 samples per gradient estimation; and 10,000 warm-up samples. The initiator matrices add four features to the description of each network. Based on this data set we compute the distances between each pair of network using the eight metrics defined in Sects. 3 and 4, namely: Model Distance *MD*, Fitting Distance (*FD*), Scaled Fitting Distance (*SFD*), Normalized Fitting Distance (*NFD*), *Density*, Clustering Coefficient (*CC*), *Diameter*, and Degree Distribution (*Deg*). In the next subsections we analyze the quality of the distance metrics.

Table 2 Comparison of quality of clusters obtained using different distance metrics using purity, prediction strength (PS), adjusted Rand index (Rand), and Fowlkes-Mallows index (FM)

Distance Metric	Purity	PS	Rand	FM
MD	0.608	0.340	0.398	0.555
Density	0.502	0.268	0.192	0.426
Degree	0.592	0.360	0.348	0.486
CC	0.634	0.432	0.381	0.518
Diameter	0.516	0.288	0.215	0.412
FD	**0.765**	**0.503**	**0.637**	**0.724**
SFD	0.435	0.258	0.162	0.437
NFD	0.393	0.237	0.104	0.518

5.2 Cluster Analysis

The standard way of evaluating the quality of a distance metric is through application of unsupervised clustering or supervised classification algorithms which require distances between entities to be evaluated. Examples of such algorithms are k-means [15] or Ward's algorithm [16] for unsupervised or k-nearest neighbors [17] for supervised methods. Here we focus on the clustering algorithms.

In order to measure the quality of a clustering algorithm with respect to a given distance metric, a variety of measures can be used. Average inter-cluster distance (ICD_O), average intra-cluster distance (ICD_I), the Dunn index [18], Calinski and Harabasz index [19, 20] are only a few examples of such measures. Some algorithms directly optimize one of the cluster quality measures. For example, k-means minimizes the sum of square distances between elements and centers of their clusters. FD is clearly superior according to all these metrics as depicted in Table 2.

Given a dissimilarity matrix the objective of a clustering algorithm is partitioning the elements into a set of clusters such that every element is similar to other elements within its cluster and not similar to elements of other clusters. Next, we evaluated the distance metrics with all hierarchical clustering algorithms available in standard distribution of the R programming language. The best overall results were obtained with the Ward's clustering algorithm which takes a square of the input dissimilarities (distances). Thus, in the rest of this subsection we depict the results using this algorithm.

Figure 3 compares the cluster hierarchies based on the Model Distance (MD) and the proposed Fit Distance (FD). The FD hierarchy is strict with a clear partition into four clusters while the MD hierarchy is more detailed and results clusters of uneven size when cut at any level. We cut all hierarchies to produce five clusters due to the five network types in the data set. By color-coding the leaf nodes based on the types of the respective networks we can visualize the purity of the produced clusters. The networks in the MD hierarchy are mixed up while in the FD hierarchy

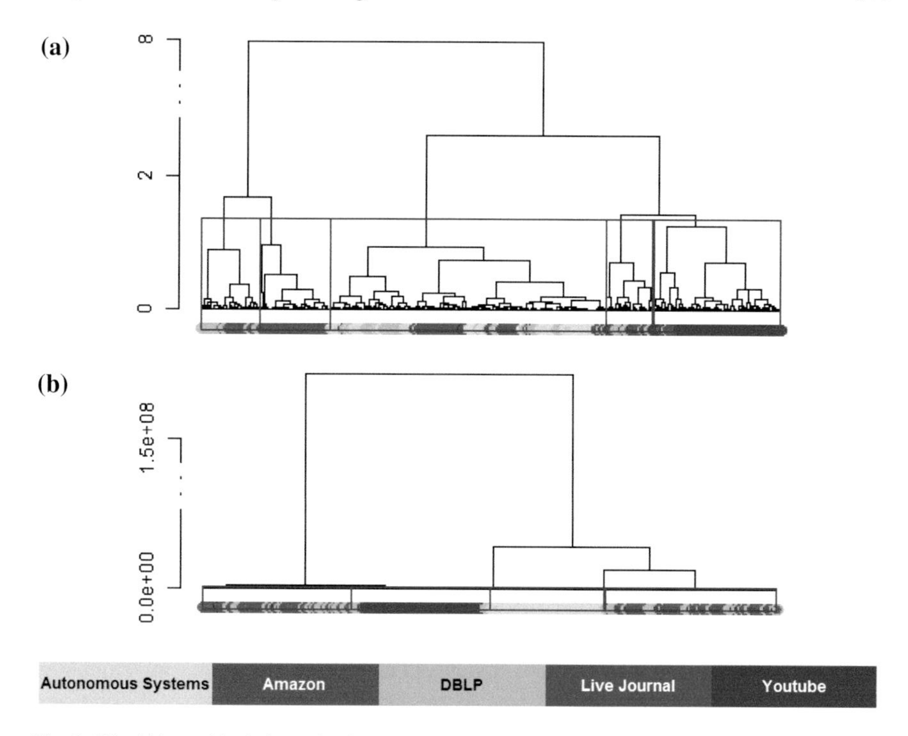

Fig. 3 Ward hierarchical clustering based on **a** the Model Distance and **b** the Fit Distance

the clusters are much more pure with YouTube cluster being the easiest to identify in both hierarchies.

In order to quantitatively evaluate the accuracy of the clustering, we calculate cluster purity, prediction strength [21], adjusted Rand index [22], and Fowlkes-Mallows index [23]. Purity evaluates the extent to which clusters are homogeneous. Purity of a clustering C_1, ..., C_5 of the 500 networks is

$$Purity = \sum_j \max_i \frac{n_{ij}}{500}$$

where n_{ij} is the number of networks of type i in a cluster j. Other measures also evaluate the produced clustering versus a gold standard, which is derived from the types of the networks as listed in Table 1.

From Table 2, we can clearly see that FD is superior to MD and all other distance metrics. Purity of 0.76 means that only 24 % of the networks were included as a minority group in their cluster. Note that SFD and NFD are both extremely inefficient performing worse than the baseline in most cases. Based on these results we can conclude that FD is better suited for unsupervised clustering than other metrics and that normalizing the log-likelihood to produce a good fit distance should be performed with a great care.

Table 3 Average distances between networks and the inter-cluster/intra-cluster distance ratio for then gold standard and the random reference

Distance Metric	Gold			Random		
	ICD_O	ICD_I	ICD_O/ICD_I	ICD_O	ICD_I	ICD_O/ICD_I
MD	0.374	0.169	2.210	0.334	0.334	1.000
Density	0.339	0.154	2.200	0.303	0.303	1.000
Degree	0.822	0.408	2.010	0.740	0.741	0.999
CC	0.324	0.119	2.710	0.283	0.283	1.000
Diameter	5.594	3.737	1.500	5.230	5.230	1.000
FD	1,165,784	53,503	**21.800**	943,813	950,231	0.993
SFD	0.207	0.186	2.320	0.328	0.325	1.009
NFD	0.181	0.061	2.930	0.842	0.842	1.000

5.3 Separation of Network Types

In previous subsection we showed that fit-distance is a good distance (a.k.a. dissimilarity) measure for unsupervised analysis of network collections. Unfortunately, many clustering algorithms are sensitive to the distributions and scale of the distance values. We, therefore, strive to evaluate the quality of the distance metrics directly, without proxies such as classification or clustering algorithms.

In the following analysis we use the gold standard, data set partitioned by the network types, for evaluation of the distance metric quality using standard cluster quality metrics (ICD_I, ICD_O, Dunn Index, etc.). We use random partition as a reference point. An appropriate cluster quality metric should produce the same value for the random partition regardless the scale of the distance metric used. This requirement is especially important in current study because the log-likelihood values differ by several orders of magnitude from other distance metrics as presented in Table 3. Therefore we employ the intra-cluster/inter-cluster distance ratio (ICD_O/ICD_I), as it the most stable according to the random reference (see rightmost column in Table 3). This measure can be regarded as a normalized separation index since it is agnostic to the scale of the evaluated distance metric.

High ICD_O/ICD_I value indicates good separation between the clusters. In our case, ICD_O/ICD_I is the upper bound on the cluster quality that can be produced by any clustering algorithm. For example, if the degree distribution distance is accurate, then for this particular set of 500 networks the best partition can be at most twice as good as the random partition.

For this purpose, inter-cluster distance and intra-cluster distance of clusters formed by different distance metrics were calculated. A low value for ICD_I implies that the networks in the same cluster were very similar to each other, which is desirable. Similarly, a higher ICD_O implies that different clusters were far away from each other and easily distinguishable. Note that different scales are introduced with clustering using different distance metrics and we divided ICD_O and ICD_I, to normalize the scales. From Table 3, it can be easily seen that FD performs better than all the other distance metrics by a large factor. Comparing with the model distance, FD produces clusters that are considerably more distinguishable.

No clustering algorithm can perform better than the gold standard. Thus, cluster quality measure applied to the gold standard with respect to a distance metric results in the highest quality that can be achieved.

6 Related Work

Pržulj [3] uses 73 constraints in the form of graphlets to compare PPI networks with their synthetically generated counterparts. The approach is especially appropriate for partially known networks where global topology characteristics are biased but some parts of the networks are well studied and contain reliable local information. This approach also requires munificent computing resources.

In [24], the focus is on comparing mesoscopic properties of networks. Networks are decomposed into communities of different sizes, starting with a single node and a maximum of n nodes. They compare networks based on different parameters calculated for different community sizes. In [25], networks were compared based on a measure called n-tangle density, which is basically the edge density in sub-graphs of n nodes from the network. The n-tangle density is calculated for different values of n and these densities are compared to evaluate network similarity. This method cannot find node to node correspondence between similar networks, therefore, making it difficult to pinpoint the source of anomaly (if any) in a network.

Aliakbary et al. [26] used several network features like average shortest path, degree distribution, average density, average clustering, etc. to compare networks for unsupervised machine learning task. They also compare the effectiveness of their approach to Euclidean distance metric based on KronFit initiator matrix and demonstrate its inferiority. It may be noted here that there are many different initiator matrices that can encode the original network with the same likelihood. For example, two initiator matrices $\begin{bmatrix} 1 & 1 \\ 1 & 0 \end{bmatrix}$ and $\begin{bmatrix} 0 & 1 \\ 1 & 1 \end{bmatrix}$ have non zero euclidean distance of $\sqrt{2}$, while both encode isomorphic graphs yielded by Kronecker product of any degree. As demonstrated in Sect. 5, the correct way of utilizing the full power of KronFit is by calculating the likelihood of a network being generated from an initiator matrix and not by comparing the initiator matrices directly.

7 Discussion and Future Work

Network comparison is an emerging research area with wide applications in social and biological networks analysis. In this paper, we propose a fit-distance (FD) distance metric between a subject network and the model derived from a prototype network. We demonstrated *FD* with the Kronecker Graphs model and showed that it is superior to direct comparison between the models using euclidean distance and to the baseline network distance measures.

One of the interesting features of *FD* is that prototype networks that can easily be distinguished from other types of networks using this measure, receive high *FD* values in general (see the yellow stripes in Fig. 1b). Although, the differences between subject networks are not comprehensible to the eye in this sub-figure, they are quite significant as can bee seen from the normalized values in Fig. 1d. We attribute the success of *FD* in clustering and its extremely high ICD_O/ICD_I value to this natural weighting of "easy" and "hard" prototype networks.

The primary objective of this paper was to draw the attention of scholars to the network comparison opportunities opened by some generative and descriptive network models. In the nearest future, more accurate fitting distance measures should be developed based on new network models such as MAGFit [7] or KronEM [8]. With the advance of model-based network comparison methods, we expect to see machine learning models that classify networks directly, without the need for feature extraction.

References

1. Bebber, D.P., Hynes, J., Darrah, P.R., Boddy, L., Fricker, M.D.: Biological solutions to transport network design. Proceedings of the Royal Society of London B: Biological Sciences **274**(1623), 2307–2315 (2007)
2. Milo, R., et al.: Network motifs: simple building blocks of complex networks. Science **298**, 824827 (2002)
3. Pržulj, Natasa: Biological network comparison using graphlet degree distribution. Bioinformatics **23**(2), e177–e183 (2007)
4. Serrano, M.Ă., Bogună, M., Vespignani, A.: Extracting the multiscale backbone of complex weighted networks. Proc. Nat. Acad. Sci. **106**(16), 6483–6488 (2009)
5. Baskerville, Kim: Paczuski, Maya: Subgraph ensembles and motif discovery using an alternative heuristic for graph isomorphism. Phys. Rev. E **74**(5), 051903 (2006)
6. Airoldi, E.M., Blei, D.M., Fienberg, S.E., Xing, E.P.: Mixed membership stochastic blockmodels. In: Advances in Neural Information Processing Systems, pp. 33–40 (2009)
7. Myunghwan, K., Leskovec, J.: Multiplicative attribute graph model of real-world networks. Internet Math. **8**(1–2), 113–160 (2012)
8. Davis, M., Liu, W., Miller, P., Hunter, R.F., Kee, F.: AGWAN: A Generative Model for Labelled, Weighted Graphs. In: New Frontiers in Mining Complex Patterns, pp. 181–200. Springer International Publishing (2014)
9. Leskovec, J., Chakrabarti, D., Kleinberg, J., Faloutsos, C., Ghahramani, Z.: Kronecker graphs: an approach to modeling networks. J. Mach. Learn. Res. **11**, 985–1042 (2010)
10. Neudecker, H.: A note on Kronecker matrix products and matrix equation systems. SIAM J. Appl. Math. **17**(3), 603–606 (1969)
11. Kim, M., Leskovec, J.: The network completion problem: inferring missing nodes and edges in networks. In: SDM, pp. 47–58 (2011)
12. Newman, M.: Networks: An Introduction. Oxford University Press, Oxford (2009)
13. U. of Oregon Route Views Project. Online data and reports: http://www.routeviews.org. The CAIDA UCSD, AS Relationships Dataset (years 1997–2000). http://www.caida.org/data/active/as-relationships/

14. Leskovec, J., Krevl, A.: Stanford Large Network Dataset Collection, June 2014. http://snap.stanford.edu/data
15. MacQueen, J.: Some methods of classification and analysis of multivariate observations. In: LeCam, L.M., Neyman, J., (eds.), Proceedings of 5th Berkeley Symposium on Mathematical Statistics and Probability, p. 281. University of California Press, Berkeley, CA (1967)
16. Ward, Jr., J.H.: Hierarchical grouping to optimize an objective function. J. Am. Stat. Assoc. **58**(301), 236–244 (1963)
17. Cover, T.M., Hart, P.E.: Nearest neighbor pattern classification. IEEE Trans. Inf. Theory **13**(1), 21–27 (1967)
18. Halkidi, M., Batistakis, Y., Vazirgiannis, M.: On clustering validation techniques. J. Intell. Inf. Syst. **17**, 107–145 (2001)
19. Calinski, T., Harabasz, J.: A Dendrite method for cluster analysis. Commun. Stat. **3**, 1–27 (1974)
20. Hennig, C., Liao, T.: How to find an appropriate clustering for mixed-type variables with application to socio-economic stratification. J. Roy. Stat. Soc. Ser. C. Appl. Stat. **62**, 309–369 (2013)
21. Tibshirani, R., Walter, G.: Cluster validation by prediction strength. J. Comput. Graph. Stat. **14**(3), 511528 (2005)
22. Gordon, A.D.: Classification, 2nd edn. Chapman & Hall/CRC, Boca Raton, FL (1999)
23. Fowlkes, E.B., Mallows, C.L.: A method for comparing two hierarchical clusterings. J. Am. Stat. Assoc. **78**, 553569 (1983)
24. Onnela, J.-P., et al.: Taxonomies of networks from community structure. Phys. Rev. E **86**(3), 036104 (2012)
25. Gallos, L.K., Fefferman, N.H.: Revealing effective classifiers through network comparison. EPL (Europhys. Lett.) 108(3), 38001 (2014)
26. Aliakbary, S., Motallebi, S., Rashidian, S., Habibi, J., Movaghar, A.: Distance metric learning for complex networks: towards size-independent comparison of network structures. Chaos: An Interdisciplinary. J. Nonlinear Sci. **25**(2), 023111 (2015)

Author Index

H. Cherifi et al. (eds.), *Complex Networks VII*, Studies in Computational Intelligence 644, DOI 10.1007/978-3-319-30569-1